Ordinary Differential Equations and Boundary Value Problems

Volume II: Boundary Value Problems

TRENDS IN ABSTRACT AND APPLIED ANALYSIS

ISSN: 2424-8746

Series Editor: John R. Graef
The University of Tennessee at Chattanooga, USA

This series will provide state of the art results and applications on current topics in the broad area of Mathematical Analysis. Of a more focused nature than what is usually found in standard textbooks, these volumes will provide researchers and graduate students a path to the research frontiers in an easily accessible manner. In addition to being useful for individual study, they will also be appropriate for use in graduate and advanced undergraduate courses and research seminars. The volumes in this series will not only be of interest to mathematicians but also to scientists in other areas. For more information, please go to http://www.worldscientific.com/series/taaa

Published

Vol. 8 *Ordinary Differential Equations and Boundary Value Problems Volume II: Boundary Value Problems*
by John R. Graef, Johnny Henderson, Lingju Kong & Xueyan Sherry Liu

Vol. 7 *Ordinary Differential Equations and Boundary Value Problems Volume I: Advanced Ordinary Differential Equations*
by John R. Graef, Johnny Henderson, Lingju Kong & Xueyan Sherry Liu

Vol. 6 *The Strong Nonlinear Limit-Point/Limit-Circle Problem*
by Miroslav Bartušek & John R. Graef

Vol. 5 *Higher Order Boundary Value Problems on Unbounded Domains: Types of Solutions, Functional Problems and Applications*
by Feliz Manuel Minhós & Hugo Carrasco

Vol. 4 *Quantum Calculus: New Concepts, Impulsive IVPs and BVPs, Inequalities*
by Bashir Ahmad, Sotiris Ntouyas & Jessada Tariboon

Vol. 3 *Solutions of Nonlinear Differential Equations: Existence Results via the Variational Approach*
by Lin Li & Shu-Zhi Song

More information on this series can be found at https://www.worldscientific.com/series/taaa

Trends in Abstract and Applied Analysis
Volume 8

Ordinary Differential Equations and Boundary Value Problems

Volume II: Boundary Value Problems

John R Graef
University of Tennessee at Chattanooga, USA

Johnny Henderson
Baylor University, USA

Lingju Kong
University of Tennessee at Chattanooga, USA

Xueyan Sherry Liu
St Jude Children's Research Hospital, USA
University of New Orleans, USA

World Scientific

NEW JERSEY • LONDON • SINGAPORE • BEIJING • SHANGHAI • HONG KONG • TAIPEI • CHENNAI • TOKYO

Published by

World Scientific Publishing Co. Pte. Ltd.
5 Toh Tuck Link, Singapore 596224
USA office: 27 Warren Street, Suite 401-402, Hackensack, NJ 07601
UK office: 57 Shelton Street, Covent Garden, London WC2H 9HE

British Library Cataloguing-in-Publication Data
A catalogue record for this book is available from the British Library.

First published 2019 (Hardcover)
Reprinted 2020 (in paperback edition)
ISBN 978-981-122-126-2 (pbk)

Trends in Abstract and Applied Analysis — Vol. 8
ORDINARY DIFFERENTIAL EQUATIONS AND BOUNDARY VALUE PROBLEMS
Volume II: Boundary Value Problems

ISBN 978-981-3274-02-0

For any available supplementary material, please visit
https://www.worldscientific.com/worldscibooks/10.1142/11091#t=suppl

Desk Editors: V. Vishnu Mohan/Lai Fun

Typeset by Stallion Press
Email: enquiries@stallionpress.com

Dedication

John Graef's dedication is to his wife Frances and to his long time friend and collaborator Johnny Henderson.

Johnny Henderson's dedication is to his long time friends, Paul W. Eloe and David L. Skoug.

Lingju Kong's dedication is to his Ph.D. advisor Qingkai Kong.

Xueyan Liu's dedication is to her Ph.D. advisor Johnny Henderson, MS advisor Binggen Zhang, postdoctoral supervisor Hui Zhang, colleague Deo Kumar Srivastava, former colleagues John Graef, Lingju Kong, Min Wang, Culian Gao, Jin Wang, Andrew Ledoan, Xuhua Liu, Qin Sheng, Lance Littlejohn, academic friends Richard Avery, Douglas Anderson, Yu Tian, Liancheng Wang, and academic brothers Jeffrey Thomas Neugebauer, Jeffrey Wayne Lyons, and Shawn Michael Sutherland.

Preface

Although this is a companion to *Ordinary Differential Equations and Boundary Value Problems Volume 1: Advanced Ordinary Differential Equations*, the contents here do not directly depend on Volume 1. In fact, each chapter here has its own list of references, and with some care, can be read somewhat independently of the others. This gives great flexibility to the independent reader as well as for use in a classroom setting. The relaxed writing style is intended to make progression through the book relatively easy. Especially in the early part of the book, there are exercises to help readers solidify their understanding of the topics.

This work begins with an introduction to boundary value problems (BVPs) including a discussion of disconjugacy and basic Sturm–Liouville theory. The second chapter introduces Green's functions for linear BVPs and includes detailed methods for their construction. These play an important role throughout the study of boundary value problems for both linear and nonlinear problems.

Chapters 3 and 4 are concerned with the existence of solutions to second and higher order BVPs. Important mathematical tools to prove existence of solutions such as the Contraction Mapping Principle and the Schauder–Tychonoff Fixed Point Theorem are introduced and then applied, as is the method of lower and upper solutions. Uniqueness implies existence results are also discussed.

The important and useful technique of solution matching is introduced in Chapter 5. It is first applied to third-order three-point problems to obtain uniqueness and existence of solutions. This is followed by an application to higher order problems.

Comparison of eigenvalues is studied via u_0-positive operators in Chapter 6. This is first considered for two-point conjugate problems, then for nonlocal BVPs, and finally for second-order multi-point problems. It is here where the reader is introduced to the well-known Krein–Rutman Theorem.

Chapter 7 is devoted to BVPs for functional differential equations (FDEs). As models of phenomena depending on past history of the system, such equations play an important role in many applications. Periodic solutions are often important in such discussions and so after describing the basic setting, in the next two sections we turn our attention to periodic solutions of first-order equations. The elements of degree theory and the Leray–Schauder Nonlinear Alternative are introduced for the purpose of this development.

Existence of positive solutions are of special interest in real-world applications so it is appropriate that we devote space to this important topic in Chapter 8. The well-known Guo–Krasnoselskii cone expansion–compression fixed point theorem, the Leggett and Williams triple fixed point theorem, and the Gatica, Oliker, and Waltman fixed point theorem for decreasing operators defined on a cone are important tools in this regard. Each of these approaches are described and used to obtain the existence of positive solutions for various problems. Also shown is how these methods can be applied to difference equations. The existence of multiple positive solutions has been a very active area of research in recent years so it seems appropriate to show how these fixed point theorems apply here as well. Special attention is paid to singular problems and to problems involving the p-Laplacian. Schauder's fixed point theorem and fixed point index theory play important roles in these types of problems.

In Chapter 9, the differentiability of solutions of boundary value problems with respect to their boundary conditions is discussed. This forms a companion to the discussion in Chapter 3 of Volume 1 on the continuous dependence of solutions on initial conditions.

The final chapter in this work is devoted to the study of nodal solutions. The techniques of proof used include a bifurcation theorem due to Rabinowitz, the shooting method, and an energy function approach. Problems with separated boundary conditions and nonlocal problems are considered.

John R. Graef
Johnny Henderson
Lingju Kong
Xueyan "Sherry" Liu

Contents

Chapter 1

Introduction to Boundary Value Problems

1.1 Introduction

A boundary value problem (BVP) for an ordinary differential equation (ODE) will consist of an ODE together with conditions specified at more than one point. In particular, we will be concerned with solving scalar differential equations, $y^{(n)} = f(x, y, y', \ldots, y^{(n-1)})$, $n \geq 2$, where f is real-valued and boundary conditions (BC's) on solutions of the equation are specified at k, (with $k \geq 2$), points belonging to some interval of the reals. Let us first consider some difficulties which might occur.

Example 1.1. Linear equations (Initial Value Problems (IVP's) have unique solutions which extend to maximal intervals of existence).

Consider $y'' + y = 0$, $y(0) = 0$, $y(\pi) = 1$. The general solution is $y = c_1 \cos x + c_2 \sin x$ and it exists on $\mathbb{R}$. Also, from the boundary conditions, we have $0 = y(0) = c_1$ but $y(\pi) = 0 \neq 1$. So, the BVP has no solution.

Exercise 1. If equations are not linear, solutions may not extend to an interval.

Consider $y'' = 1 + (y')^2$. Solve this equation. Show that if $x_2 - x_1 \geq \pi$, then no solution starting at x_1, (i.e., $y(x_1) = y_1$), ever reaches x_2. (i.e., we cannot have a BVP with $y(x_1) = y_1$ and $y(x_2) = y_2$, whenever $x_2 - x_1 \geq \pi$).

Example 1.2. Another question involves uniqueness of solutions of BVP's when solutions exist.

Consider $y'' = 0$, $y'(0) = 0$, $y'(1) = 0$. Then $y(x) \equiv k$, where k is a constant, constitutes infinitely many solutions.

Thus, questions we will be concerned with center around the existence and uniqueness of solutions of BVP's for ODE's. We will first consider such questions for BVP's of linear differential equations. In this case, let $L : C^{(n)}(I) \to C(I)$ be the linear differential operator given by

$$Ly := y^{(n)} + \sum_{i=0}^{n-1} a_i(x)y^{(i)},$$

where the a_i, $i = 0, 1, \ldots, n-1$, are continuous on the interval I. Also consider the differential equation:

$$Ly = y^{(n)} + \sum_{i=0}^{n-1} a_i(x)y^{(i)} = f(x), \tag{1.1}$$

where $f(x)$ is continuous on the interval I. IVP's for (1.1) are uniquely solvable on I and all solutions of (1.1) extend to I.

Let us first examine the setting where $f(x) \equiv 0$:

$$y^{(n)} + \sum_{i=0}^{n-1} a_i(x)y^{(i)} = 0, \tag{1.2}$$

and let us deal primarily with conjugate type boundary conditions.

1.2 Disconjugacy

Consider the third order differential equation $y''' = 0$ and the BC's, $y(a) = y(b) = y(c) = 0$, where $a,\ b,\ c$ are distinct. It follows that $y \equiv 0$ on all of $\mathbb{R}$.

Exercise 2. Show that each of the BVP's

$$\begin{cases} y''' = 0, \\ y(a) = y'(a) = y(b) = 0, \end{cases} \qquad \begin{cases} y''' = 0, \\ y(a) = y(b) = y'(b) = 0, \end{cases}$$

has only the trivial solution, where $a \neq b$.

It follows then that the only solution of $y''' = 0$ having 3 zeros counting multiplicities is the trivial solution. We say that the equation $y''' = 0$ is *disconjugate* on any subinterval of $\mathbb{R}$. Our attention will be focused on corresponding problems for Eq. (1.2).

Definition 1.1. Equation (1.2) is said to be *disconjugate* on an interval I if every nontrivial solution has less than n zeros on I, counting multiplicities of zeros, (i.e., the only solution of (1.2) having n zeros on I counting multiplicities is $y \equiv 0$).

Exercise **3.** Prove that $y^{(n)} = 0$ is disconjugate on any interval.

Is it possible that Eq. (1.2) is disconjugate on some subintervals of $\mathbb{R}$ and not on others? This is a question which has aroused much interest and from which many papers have been produced.

Example 1.3. The differential equation $y'' + y = 0$ is not disconjugate on $[0, \pi]$, for the solution $y(x) = \sin x$ has zeros at $x = 0$ and at $x = \pi$. Yet, if I is any proper subinterval of $[0, \pi]$, $y'' + y = 0$ is disconjugate on I. To see this, observe $y(x) = c_1 \cos x + c_2 \sin x$ is the general solution of the differential equation. Assume there exist $a, b \in \mathbb{R}$ with $0 < |b - a| < \pi$ so that $y(a) = y(b) = 0$. Then

$$c_1 \cos a + c_2 \sin a = 0, \quad \text{and} \quad c_1 \cos b + c_2 \sin b = 0.$$

Hence, $\tan a = \tan b$, which is contrary to $0 < |b - a| < \pi$.

Exercise **4.** Let I be a subinterval of $\mathbb{R}$. Show $y'' + y = 0$ is disconjugate on I if either (i) $I = [a, b]$ and $b - a < \pi$, or (ii) $I = [a, b)$, (a, b), or $(a, b]$, and $b - a \leq \pi$. Show $y'' + y = 0$ is not disconjugate on I if neither (i) nor (ii) hold.

What is special about the zeros of (1.2)? What does disconjugacy mean in terms of solving BVP's? First, it says something about the number of solutions certain BVP's for (1.2) can have.

Theorem 1.1. *Suppose a_i, $0 \leq i \leq n-1$, are continuous on I. Then* (1.2) *is disconjugate on I if and only if for any $k \leq n$ distinct points $x_1, \ldots, x_k \in I$, for all positive integers $m_1, \ldots, m_k$ such that $m_1 + \cdots + m_k = n$ and n arbitrary reals $y_{11}, y_{21}, \ldots, y_{m_1 1}, y_{12}, y_{22}, \ldots, y_{m_2 2}, \ldots, y_{1k}, y_{2k}, \ldots, y_{m_k k}$, there exists at most one solution $y(x)$ of* (1.2) *such that*

$$y^{(i-1)}(x_j) = y_{ij}, \quad 1 \leq i \leq m_j, \quad 1 \leq j \leq k. \tag{1.3}$$

These are called ***conjugate type*** *BC's in* (1.3).

Proof. Assume first that (1.2) is disconjugate on I, yet assume the conclusion in this situation is false. Then there exists points $x_1, \ldots, x_k \in I$, $(k \leq n)$, positive integers $m_1, \ldots, m_k$ such that $\sum_{i=1}^{k} m_i = n$, n real values $y_{11}, \ldots, y_{m_k k}$ and distinct solutions y and z of (1.2) such that y and z both satisfy BC's (1.3). If we let $w \equiv y - z$, then from the linearity of (1.2), w is a solution of (1.2) and satisfies

$$w^{(i)}(x_j) = 0, \quad 0 \leq i \leq m_j - 1, \quad 1 \leq j \leq k;$$

in particular w has "n" zeros at $x_1, \dots, x_k$, counting multiplicities; a contradiction. Hence, the conclusion for this direction is justified. □

Exercise **5.** Prove the converse of Theorem 1.1.

What we have proven is that disconjugacy of (1.2) is equivalent to the uniqueness of solutions of all BVP's (1.2), (1.3) when solutions exist. More importantly, solutions will exist for all BVP's (1.2), (1.3).

Theorem 1.2. *Suppose a_i, $0 \le i \le n-1$, are continuous on I. Then* (1.2) *is disconjugate on I if and only if for any $k \le n$ distinct points $x_1, \dots, x_k \in I$, for all positive integers $m_1, \dots, m_k$ such that $m_1 + \dots + m_k = n$ and for n reals $y_{11}, \dots, y_{m_k k}$, there is a solution $y(x)$ of* (1.2) *such that* (1.3) *holds.*

Proof. The solutions of (1.2) constitute an n-dimensional vector space, hence let $y_1(x), \dots, y_n(x)$ be linearly independent solutions of (1.2). Then all solutions of (1.2) assume the form

$$y(x) = c_1 y_1(x) + \dots + c_n y_n(x).$$

Hence, in this theorem, we want to choose a vector

$$\alpha = \begin{bmatrix} c_1 \\ c_2 \\ \vdots \\ c_n \end{bmatrix}$$

so that $A\alpha = \beta$, where

$$A = \begin{bmatrix} y_1(x_1) & y_2(x_1) & \cdots & y_n(x_1) \\ y_1'(x_1) & y_2'(x_1) & \cdots & y_n'(x_1) \\ \ddots & \ddots & \ddots & \ddots \\ y_1^{(m_1-1)}(x_1) & y_2^{(m_1-1)}(x_1) & \cdots & y_n^{(m_1-1)}(x_1) \\ y_1(x_2) & y_2(x_2) & \cdots & y_n(x_2) \\ \ddots & \ddots & \ddots & \ddots \\ y_1^{(m_2-1)}(x_2) & y_2^{(m_2-1)}(x_2) & \cdots & y_n^{(m_2-1)}(x_2) \\ \ddots & \ddots & \ddots & \ddots \\ y_1(x_k) & y_2(x_k) & \cdots & y_n(x_k) \\ \ddots & \ddots & \ddots & \ddots \\ y_1^{(m_k-1)}(x_k) & y_2^{(m_k-1)}(x_k) & \cdots & y_n^{(m_k-1)}(x_k) \end{bmatrix}, \quad \beta = \begin{bmatrix} y_{11} \\ y_{21} \\ \vdots \\ y_{m_1 1} \\ \vdots \\ y_{1k} \\ \vdots \\ y_{m_k k} \end{bmatrix}.$$

But from linear algebra, for each β, there exists α such that $A\alpha = \beta$ if and only if the homogeneous equation $A\alpha = 0$ has only the trivial solution (i.e., if and only if (1.2) is disconjugate). □

Because of their simplicity we shall discuss in some detail the disconjugacy of second order linear equations; more precisely, we will be concerned with the disconjugacy of the second order equation known as the Sturm-Liouville equation

$$Ly := (p(x)y')' + q(x)y = 0, \tag{1.4}$$

where p, q are continuous on an interval I, $p(x) > 0$ and $q(x)$ is real on I. Note that p may not be differentiable.

Notice that (1.4) is disconjugate if every nontrivial solution has *at most* one zero.

Now if p is differentiable, (1.4) can be written in the form (1.2), since

$$(p(x)y')' + q(x)y = p(x)y'' + p'(x)y' + q(x)y = 0$$

is equivalent to

$$y'' + \frac{p'(x)}{p(x)}y' + \frac{q(x)}{p(x)}y = 0. \text{ (Recall } p(x) > 0.)$$

Moreover, any second order equation $y'' + a_1(x)y' + a_0(x)y = 0$ can be written in the form (1.4), with

$$p(x) = e^{\int_{x_0}^{x} a_1(s)\,ds}, \quad q(x) = a_0(x)p(x).$$

Exercise 6. Verify this last statement.

The following lemma has proven to be quite useful in the study of disconjugacy of (1.2). We shall exploit it with respect to Eq. (1.4).

Lemma 1.1. *Let $u(x)$ and $v(x)$ be differentiable on the interval $[a, b]$ such that $u(a) = u(b) = 0$, and $v(x) \neq 0$ for all $a \leq x \leq b$. Then there exists a constant α such that $u + \alpha v$ has a double zero in (a, b).*

Proof. If we let $w \equiv vu' - uv'$, then it follows that $\left(\frac{u}{v}\right)' = \frac{vu'-uv'}{v^2} = \frac{w}{v^2}$. Since $\frac{u}{v}(a) = \frac{u}{v}(b) = 0$, it follows from Rolle's Theorem that there exists $c \in (a, b)$ so that $\frac{w}{v^2}(c) = 0$. Hence

$$v(c)u'(c) - u(c)v'(c) = 0;$$

in particular,

$$\begin{vmatrix} v(c) & u(c) \\ v'(c) & u'(c) \end{vmatrix} = 0,$$

and so there exists $\lambda, \mu \in \mathbb{R}$, not both zero, such that

$$\lambda v(c) + \mu u(c) = 0,$$
$$\lambda v'(c) + \mu u'(c) = 0.$$

Since $v(c) \neq 0$, it follows that $\mu \neq 0$; hence taking $\alpha = \frac{\lambda}{\mu}$, we have that $u + \alpha v$ has a double zero at c. □

The famous Sturm Separation Theorem is an immediate consequence of Lemma 1.1.

Theorem 1.3 (Sturm). *If $y_1(x)$, $y_2(x)$ are linearly independent solutions of (1.4), then their zeros are interlaced; i.e., between two consecutive zeros of one solution is **exactly one** zero of the other solution.*

Proof. First, it must be the case that $y_1(x)$ and $y_2(x)$ have no common zero. To see this, suppose there exists an a such that $y_1(a) = y_2(a) = 0$. Of course, by uniqueness of solutions of IVP's, $y_1'(a), y_2'(a) \neq 0$. Consider the solution of (1.4), $y(x) = y_2'(a)y_1(x) - y_1'(a)y_2(x)$. Then $y(a) = y'(a) = 0$ and hence $y(x) \equiv 0$ for all x. This in turn implies $y_1(x) \equiv \frac{y_1'(a)}{y_2'(a)}y_2(x)$; a contradiction to the linear independence of y_1 and y_2. Thus, y_1 and y_2 have no common zero.

Exercise 7. Verify that no nontrivial linear combination of y_1 and y_2 has a double zero.

Hence, if a and b are consecutive zeros of $y_1(x)$, by Lemma 1.1, $y_2(x)$ has *at least one zero* between a and b, say at $x = c$. By interchanging the roles of y_1 and y_2 in Lemma 1.1, if y_2 has another zero at $d \neq c$, where $a < d < b$, then y_1 has at least one zero between c and d, contradicting the consecutiveness of a and b. Thus, y_2 has *exactly* one zero between a and b. □

Exercise 8. Prove the following.

Corollary 1.1. *Suppose $I = [a, b)$. If one nontrivial solution of (1.4) has infinitely many zeros on I, then every solution has infinitely many zeros on I. If no nontrivial solution has infinitely many zeros on I, then (1.4) is disconjugate on some subinterval $[c, b)$.*

If solutions of (1.4) behave as in the first part of the corollary, you might say that the solutions oscillate.

Definition 1.2. Equation (1.4) is said to be *oscillatory* on I, if every solution has infinitely many zeros on I.

Thus, the corollary states that if (1.4) is nonoscillatory on $[a, b)$, then we might say it is *eventually disconjugate.*

Theorem 1.4. *Equation* (1.4) *is disconjugate on an interval* I *if it has a solution without zeros on* I. *If* $I = [a, b]$ *or* $I = (a, b)$, *this condition is also a necessary condition, i.e., if* $I = [a, b]$ *or* (a, b), *then* (1.4) *is disconjugate on* I *if and only if* (1.4) *has a solution without zeros on* I.

Proof. For the first part, let I be an interval and suppose (1.4) has a solution y such that $y(x) \neq 0$, for all $x \in I$. Clearly, from Lemma 1.1, if there is a nontrivial solution z such that $z(a) = z(b) = 0$, $a \neq b \in I$, then for some constant α, $z + \alpha y$ has a double zero between a and b. Thus, $z + \alpha y \equiv 0$ on I, or $z = -\alpha y$; a contradiction unless $z \equiv 0$. Therefore, (1.4) is disconjugate on I.

For the converse considerations, let's first take care of the situation $I = [a, b]$. Assume now that (1.4) is disconjugate on I, and let $y_1(x)$, $y_2(x)$ be solutions of (1.4) satisfying $y_1(a) = 0$, $y_1'(a) > 0$, $y_2(b) = 0$, $y_2'(b) < 0$. By the disconjugacy of (1.4), $y_1(x) > 0$ on $(a, b]$ and $y_2(x) > 0$ on $[a, b)$. Thus, the solution $y(x) \equiv y_1(x) + y_2(x) > 0$ on $[a, b]$.

For the other case assume $I = (a, b)$ and that (1.4) is disconjugate on I. Let $c \in (a, b)$ be arbitrary but fixed, and let $\{a_n\}$, $\{b_n\}$ be sequences such that, for each n, $a_n < c < b_n$ and $a_n \downarrow a$, $b_n \uparrow b$. Then from the case just disposed of, for each n, there exists a solution $y_n(x)$ of (1.4) such that $y_n(x) > 0$ on $[a_n, b_n]$. By multiplying $y_n(x)$ by a suitable positive constant and relabeling, we can suppose that $y_n^2(c) + (y_n'(c))^2 = 1$, for all n.

Exercise 9. Show how such constants can be found.

Then the sequences $\{y_n(c)\}$, $\{y_n'(c)\}$ are bounded, and hence there exists a subsequence $\{n_k\} \subseteq \{n\}$ such that $y_{n_k}(c) \to \alpha$, $y_{n_k}'(c) \to \beta$, for some $\alpha, \beta \in \mathbb{R}$. Now let $y(x)$ be the solution of (1.4) satisfying the initial conditions,

$$y(c) = \alpha, \quad y'(c) = \beta.$$

Then, $y_{n_k}(x) \to y(x)$ and $y_{n_k}'(x) \to y'(x)$, for all $x \in I$. (Note: The convergence is uniform on compact subintervals by the Kamke Convergence

Theorem.) Consequently, it is clear that $y(x) \geq 0$, for all $x \in I$. Now if there exists $\tau \in I$ such that $y(\tau) = 0$, then $y'(\tau) = 0$ and so by uniqueness of solutions of IVP's, $y \equiv 0$. This is a contradiction, since $y^2(c) + (y'(c))^2 = 1$. Hence, $y(x) > 0$, for all $x \in I = (a, b)$. □

Exercise **10.** Give an example showing that if $I = [a, b)$ or $(a, b]$, then the condition *is not* a necessary condition.

However, we do have the following.

Theorem 1.5. *Equation* (1.4) *is disconjugate on the half-open interval* $I = [a, b)$, *if it is disconjugate on its interior* (a, b).

Proof. It suffices to show that if $y(x)$ is the solution of (1.4) satisfying the initial conditions

$$y(a) = 0, \quad y'(a) = 1,$$

then $y(x) \neq 0$, for all $x \in (a, b)$. Suppose on the contrary that for some $a < c < b$, $y(c) = 0$. Since $y'(c) \neq 0$, $y(x)$ assumes some negative values in a neighborhood of c. Now by Kamke's Theorem, solutions of (1.4) depend continuously upon initial conditions, hence, if $\varepsilon > 0$ is sufficiently small, the solution $z(x)$ of (1.4) satisfying,

$$z(a + \varepsilon) = 0, \quad z'(a + \varepsilon) = 1,$$

has a zero in the neighborhood of c. This contradicts the fact that (1.4) is disconjugate on (a, b). □

It is next shown that if Eq. (1.4) has a solution without zeros, then the second order linear differential operator $L(y) = (py')' + qy$ can be factored into first order linear differential operators.

Theorem 1.6. *Suppose* (1.4) *has a solution* $u(x)$ *without zeros on the interval* I. *Then there exists first order linear differential operators* L_1, L_1^* *such that, for any* $y \in C^{(2)}(I)$,

$$L(y) \equiv L_1^*[pL_1(y)].$$

Proof. Let $r(x) = \frac{u'(x)}{u(x)}$ and define L_1 and L_1^* by $L_1 y = y' - ry$ and $L_1^* y = y' + ry$. Note also that $q(x) = \frac{-(pu')'}{u}$. First, $pL_1(y) = py' - p(\frac{u'}{u})y$

and so

$$\begin{aligned}L_1^*[pL_1(y)] &= (py')' - \left(p\left(\frac{u'}{u}\right)y\right)' + \left(\frac{u'}{u}\right)\left[py' - p\left(\frac{u'}{u}\right)y\right]\\
&= (py')' - \frac{(pu')'}{u}y + p\left(\frac{u'}{u}\right)^2 y - p\left(\frac{u'}{u}\right)y'\\
&\quad + p\left(\frac{u'}{u}\right)y' - p\left(\frac{u'}{u}\right)\\
&= L(y) - qy + qy = L(y).\end{aligned}$$

This completes the proof. □

Note: L_1 and L_1^* are said to be *adjoint operators* of each other.

Exercise 11. Show that the result above can also be expressed in the equivalent form

$$L(y) \equiv \frac{1}{u}D\Big[pu^2 D\Big(\frac{y}{u}\Big)\Big],$$

where $D = \frac{d}{dx}$.

The next theorem illustrates the fact that the solutions of (1.4) having no zeros are very closely tied to solutions of the Riccati equation,

$$R[w] = w' + q(x) + \frac{w^2}{p(x)} = 0. \tag{1.5}$$

Theorem 1.7. *Equation* (1.4) *has a solution with no zeros on the interval I if and only if the Riccati equation* (1.5) *has a solution defined throughout I.*

Proof. Let us first suppose that (1.4) has a solution $u(x) \neq 0$, for all $x \in I$. If we set $w(x) = \frac{p(x)u'(x)}{u(x)}$, we have (Note $q = \frac{-(pu')'}{u}$)

$$\begin{aligned}w' + q + \frac{w^2}{p} &= \left[\frac{(pu')}{u}\right]' - \frac{(pu')'}{u} + \frac{(pu')^2}{pu^2}\\
&= \frac{p(pu')'u - p^2(u')^2 - (pu)(pu')' + (pu')^2}{pu^2}\\
&= 0.\end{aligned}$$

Since $u(x) \neq 0$, for all $x \in I$, $w(x)$ is defined on all of I.

Conversely, suppose the Riccati equation (1.5) has a solution $w(x)$ defined on all of I. Let $y(x)$ be a nontrivial solution of the first order linear equation

$$y' = \left[\frac{w(x)}{p(x)}\right] y.$$

Then $y(x)$ is a solution of (1.4) and $y(x) \neq 0$, for all $x \in I$. □

Exercise 12. In the proof above, verify that any nontrivial solution of $y' = \left[\frac{w}{p}\right] y$ is also a nonvanishing solution of (1.4).

It turns out, as the next theorem shows, that Eq. (1.4) can have a solution without zeros under weaker hypotheses than those concerning the existence of solutions of (1.5). Before we state and prove the theorem, we will state as a lemma an elementary result from *differential inequalities*.

Lemma 1.2. *Let $f(x, y)$ be continuous on an open set $D \subseteq \mathbb{R} \times \mathbb{R}$ and $(x_0, y_0) \in D$. Let $y_m(x)$ be* **the** *maximal solution of*

$$\begin{cases} y' = f(x, y), \\ y(x_0) = y_0, \end{cases} \tag{1.6}$$

(i.e., $y_m(x) \geq z(x)$, for all solutions $z(x)$ of (1.6), on any common interval of existence), on a maximal interval (α, ω). Let $v(x)$ be continuous and real-valued on $[x_0, x_0 + a]$ satisfying:

1) $(x, v(x)) \in D$, for all $x \in [x_0, x_0 + a]$,
2) $v'(x) \leq f(x, v(x))$, for all $x \in [x_0, x_0 + a)$, and
3) $v(x_0) \leq y_0 = y_m(x_0)$.

Then $v(x) \leq y_m(x)$ on $[x_0, x_0 + a) \cap (\alpha, \omega)$.

Theorem 1.8. *If the Riccati inequality,*

$$R[v] = v' + q(x) + \frac{v^2}{p(x)} \leq 0, \tag{1.7}$$

has a solution $v(x)$ defined on all of I, then Eq. (1.4) has a solution without zeros on I.

Proof. We shall show that Theorem 1.7 is satisfied by applying Lemma 1.2. Let c belong to the interior (I°) of I, and let $w(x)$ be a solution of the *Riccati equation* (1.5) such that $w(c) = v(c)$. Note here that solutions of IVP's for the Riccati equation (1.5) are unique on their maximal intervals of existence. Consequently, $w(x)$ is the maximal solution referred to in Lemma 1.2, with $x_0 = c$ and $y_0 = w(c)$. Now $R[v] \leq 0 = R[w]$.

Hence by Lemma 1.2, $v(x) \leq w(x)$ for all $x > c$ for which $w(x)$ is defined. On the other hand, if

$$u(x) = v(c) - \int_c^x q(s)\,ds,$$

then $R[u] = \frac{u^2}{p} \geq 0$, for all x in I. Thus, we have $R[u] = \frac{u^2}{p} \geq 0 = R[w]$, $u(c) = v(c) = w(c)$, and by an inequality result analogous to Lemma 1.2, it follows that $w(x) \leq u(x)$, for all $x \geq c$ for which $w(x)$ is defined. It follows then that $w(x)$ is defined and satisfies the inequalities $v(x) \leq w(x) \leq u(x)$, for all $x > c$ in I. Similarly $u(x) \leq w(x) \leq v(x)$, for all $x < c$ in I and hence the solution $w(x)$ of (1.5) exists on all of I.

Therefore, by Theorem 1.7, Eq. (1.4) has a solution without zeros on I. □

The next result indicates that if (1.4) is disconjugate on a half-open interval $[a, b)$, then there is a solution which is the smallest near b.

Theorem 1.9. *Suppose that* (1.4) *is disconjugate on the half-open interval* $I = [a, b)$. *Then there exists a solution* $y_1(x)$ *such that for any linearly independent solution* $y_2(x)$,

$$\lim_{x \to b} \frac{y_1(x)}{y_2(x)} = 0.$$

Moreover $\int^b \frac{dx}{py_1^2} = \infty$, $\int^b \frac{dx}{py_2^2} < \infty$. *If* $w_1 = \frac{py_1'}{y_1}$ *and* $w_2 = \frac{py_2'}{y_2}$, *then* $w_1(x) < w_2(x)$, *for all* x *in some left neighborhood of* b.

Proof. At the outset of the proof, we observe that for any solution z of (1.4) , it follows that

$$(pz')' = -qz. \tag{1.8}$$

Now we will construct the solution y_1 described in the theorem. Let $u(x)$ and $v(x)$ be linearly independent solutions of (1.4) and set

$$\widetilde{W} \equiv pW(x; v, u) = p\begin{vmatrix} v & u \\ v' & u' \end{vmatrix} = p(vu' - uv').$$

Observe that from (1.8),

$$\begin{aligned} \widetilde{W}' &= (pvu' - puv')' \\ &= v'(pu') + v(pu')' - u'(pv') - u(pv')' \\ &= \; -vqu + uqv \equiv 0, \end{aligned}$$

and hence, $\widetilde{W} \equiv c$, c constant. Moreover $W \neq 0$ and $p \neq 0$, so $\widetilde{W} = c \neq 0$.

Further, $\widetilde{W}(x; v, u) = pv^2 \left(\frac{u}{v}\right)'$ and the disconjugacy of (1.4) implies v vanishes at most once on $[a, b)$; it follows that $\frac{u}{v}$ is monotonic near b, and so $\lim_{x\to b^-} \frac{u(x)}{v(x)} = \gamma$ exists in the extended reals. We can assume without loss of generality that $\gamma = 0$ for the following reasons:

i) If $\gamma = \pm\infty$, by interchanging the roles of u and v, we would have $\lim_{x\to b^-} \frac{v(x)}{u(x)} = 0$.

ii) On the other hand, if $|\gamma| < \infty$, by replacing u by $u - rv$, we would have

$$\lim_{x\to b^-} \frac{(u - \gamma v)(x)}{v(x)} = \gamma - \gamma = 0.$$

Thus, by a suitable choice of u and v, we can assume

$$\lim_{x\to b^-} \frac{u(x)}{v(x)} = \gamma = 0.$$

Now, in the theorem, let $y_1(x) \equiv u(x)$ and let $y_2(x)$ be any linearly independent solution of (1.4). Then $y_2(x) = \lambda v(x) + \mu u(x)$, where $\lambda \neq 0$. It follows that

$$\frac{y_1(x)}{y_2(x)} = \frac{u(x)}{\lambda v(x) + \mu u(x)} = \frac{\frac{u(x)}{v(x)}}{\lambda + \mu\frac{u(x)}{v(x)}} \to \frac{0}{\lambda + 0} = 0, \quad \text{as } x \to b^-.$$

The first statement is proven.

Recall that $\widetilde{W} = pv^2 \left(\frac{u}{v}\right)'$, and so $\left(\frac{u}{v}\right)' = \frac{\widetilde{W}}{pv^2}$. Integrating this last equation,

$$\int_c^b \left(\frac{u}{v}\right)'(x)dx = \int_c^b \frac{\widetilde{W}}{pv^2}(x)dx, \quad \text{where } c \in [a, b)$$

and we have

$$\frac{u(b^-)}{v(b^-)} - \frac{u(c)}{v(c)} = \gamma - \frac{u(c)}{v(c)} = \widetilde{W} \int_c^b \frac{dx}{pv^2(x)},$$

whether γ is finite or not; i.e.,

$$\gamma = \frac{u(c)}{v(c)} + \widetilde{W} \int_c^b \frac{dx}{pv^2(x)}. \tag{1.9}$$

If $u = y_1$, $v = y_2$, then $\gamma = 0$ and hence

$$0 = \frac{y_1(c)}{y_2(c)} + \widetilde{W} \int_c^b \frac{dx}{py_2^2(x)}, \text{ which implies } \int^b \frac{dx}{py_2^2} < \infty.$$

On the other hand, if $u = y_2$, $v = y_1$, then $\gamma = \pm\infty$, and we have $\int^b \frac{dx}{py_1^2} = +\infty$.

For the final part of the proof, choose c near b such that $y_1(x) \neq 0$, $y_2(x) \neq 0$, for all $c \leq x < b$.

Exercise **13.** Show that w_k, $k = 1, 2$, of the theorem are unaltered if $y_k(x)$ is replaced by $-y_k(x)$, $k = 1, 2$, respectively.

Since w_k is unaltered if y_k is replaced by $-y_k$, we can suppose $y_1(x) > 0$, $y_2(x) > 0$ on $c \le x < b$. If we take $u = y_2$, $v = y_1$, then

$$w_1 - w_2 = \frac{p(y_1' y_2 - y_1 y_2')}{y_1 y_2} = \frac{-\widetilde{W}}{y_1 y_2}.$$

But $\widetilde{W} > 0$ by (1.9) since $\gamma = +\infty$ in this case. Thus $w_1(x) < w_2(x)$, for $c \le x < b$, and this completes the proof. □

Exercise **14.** Show that the "minimal solution" $y_1(x)$ (constructed above) is uniquely determined up to a constant multiple; i.e., any other solution possessing its properties is of the form $\alpha y_1(x)$, where α constant.

Definition 1.3. The solution $y_1(x)$ above is called the *principal solution* of (1.4).

Note: It is the solution which is the "smallest" near b.

The next theorem illustrates that if a sequence of differential equations of the form (1.4), which are uniformly convergent on compact subintervals of I (I an appropriate interval), are disconjugate, then the limit equation is also disconjugate on I.

Theorem 1.10. *Suppose that we are given a sequence of equations*

$$L_\nu[y] \equiv (p_\nu(x)y')' + q_\nu(x)y \equiv 0, \tag{1.4$_\nu$}$$

each of which is disconjugate on the open or half-open interval, I. Suppose that

$$p_\nu(x) \to p(x) > 0, \quad q_\nu(x) \to q(x)$$

uniformly on compact subintervals of I, as $\nu \to +\infty$. Then the limit equation

$$L[y] = (p(x)y')' + q(x)y \equiv 0, \tag{1.4}$$

is also disconjugate on I.

Proof. For definiteness, we will assume that I is open at its right endpoint. Let's suppose that, on the contrary, (1.4) has a nontrivial solution $y(x)$ with two zeros at $x = a$, $x = b$, where $a < b$. Now, let $y_\nu(x)$ be the solution of $(1.4)_\nu$ satisfying the conditions,

$$\begin{aligned} y_\nu(a) &= 0, \qquad (= y(a)) \\ y'_\nu(a) &= y'(a). \end{aligned}$$

From the Kamke Convergence Theorem and uniqueness of solutions of IVP's, it follows that $\{y_\nu^{(i)}(x)\}$ converges uniformly to $y^{(i)}(x)$ on compact subintervals of I, $i = 0, 1$. Since $y(x)$ changes sign at $x = b$, it follows that for all ν sufficiently large, $y_\nu(x)$ has a zero in a neighborhood of b. But $y_\nu(a) = 0$ also, which contradicts the disconjugacy of $(1.4)_\nu$ on I. Thus the conclusion holds. □

Exercise 15. Show that the result need not hold if I is a compact interval. For example, $y'' + \lambda^2 y = 0$ is disconjugate on the interval $[0, \pi]$, for $\lambda^2 < 1$, but what about for $\lambda^2 = 1$?

Our concluding results concerning disconjugacy for second order equations will be focused on the famous Sturm Comparison Theorem for Eq. (1.4) and the two equations

$$L_1[y] = (p_1(x)y')' + q_1(x)y = 0, \tag{1.10}$$

and

$$L_2[y] = (p_2(x)y')' + q_2(x)y = 0. \tag{1.11}$$

Prior to establishing these results, we will have need of some machinery from the *Calculus of Variations*. Disconjugacy plays an important role in the Calculus of Variations. In the Calculus of Variations, one is often concerned with minimizing or maximizing a functional of the form

$$Q[y] = \int_a^b f(x, y(x), y'(x))\, dx$$

subject to some boundary conditions on $y(x)$. The second order partial differential equation which is to be satisfied by each function giving an extremum of Q is called the *Euler-Lagrange equation* and is given by

$$\frac{\partial f(x,y,y')}{\partial y} - \frac{d}{dx}\left[\frac{\partial f(x,y(x),y'(x))}{\partial y'}\right] = 0.$$

Hence, if we let Q be the quadratic functional

$$Q[y] = \int_a^b [p(y')^2 - qy^2]\, dx,$$

for $f = p(y')^2 - qy^2$, we have $\frac{\partial f}{\partial y} = -2qy$ and $\frac{\partial f}{\partial y'} = 2py'$, and so the Euler-Lagrange equation is in this case

$$-2qy - (2py')' = 0,$$

or

$$L[y] = (py')' + qy = 0. \tag{1.4}$$

Definition 1.4. A function $y(x)$ is *admissible* on $[a, b]$ if it is piecewise continuously differentiable and $y(a) = y(b) = 0$. We will say that the quadratic functional $Q[y]$ is *positive* if $Q[y] > 0$, for all nontrivial admissible functions $y(x)$.

The following lemma is well-known from the Calculus of Variations.

Lemma 1.3. *Let $u(x)$ be twice differentiable and $v(x)$ a differentiable function. If $y(x) = u(x)v(x)$, then*

$$p(y')^2 - qy^2 = pu^2(v')^2 + (pu'vy)' - L[u]vy.$$

Proof. The right hand side of the expression equals

$$\begin{aligned} &pu^2(v')^2 + pu'(vy)' + (pu')'vy - (pu')'vy - quvy \\ &= pu^2(v')^2 + pu'(uv^2)' - qu^2. \end{aligned}$$

The same is true of the left hand side, since

$$\begin{aligned} u^2(v')^2 + u'(uv^2)' &= u^2(v')^2 + 2uu'vv' + (u')^2v^2 \\ &= (uv' + u'v)^2 = (y')^2. \end{aligned}$$

This completes the proof. □

Corollary 1.2. *For any twice differentiable function $y(x)$,*

$$Q[y] = [pyy']|_a^b - \int_a^b yL[y]\,dx.$$

(In particular, $Q[y] = 0$, if $y(a) = y(b) = 0$ and $L[y] = 0$.)

Exercise 16. Prove Corollary 1.2 to Lemma 1.3 by letting $v(x) \equiv 1$ in Lemma 1.3.

(Note: Twice differentiable is not required in Lemma 1.3 nor Corollary 1.2.)

Theorem 1.11. *Suppose there exists an admissible function $y(x)$ such that $Q[y] \leq 0$. If $u(x)$ is a nontrivial solution of (1.4) on $[a, b] = I$ and if $y(x)$ is not a constant multiple of $u(x)$, then $u(c) = 0$, for some $c \in (a, b)$.*

Proof. Our proof will be by contradiction. Suppose that $u(x) \neq 0$, for all $a < x < b$ and let $v(x) = \frac{y(x)}{u(x)}$. By Lemma 1.3, if $a < a_1 < b_1 < b$, then

$$\int_{a_1}^{b_1} [p(y')^2 - qy^2]\, dx = \int_{a_1}^{b_1} pu^2(v')^2\, dx + pu'vy\Big|_{a_1}^{b_1} - \int_{a_1}^{b_1} L[u]vy\, dx.$$

We claim that $pu'vy = \frac{pu'y^2}{u} \to 0$, as $x \to a$.

i) It is obvious if $u(a) \neq 0$, since $y(a) = 0$.

ii) If $u(a) = 0$, then $u'(a) \neq 0$ and by L'Hospital's Rule,

$$\lim_{x\to a} \frac{pu'y^2}{u} = \lim_{x\to a} \frac{2pu'yy' - quy^2}{u'} = 0.$$

Similarly $pu'vy \to 0$ as $x \to b$. Therefore letting $a_1 \to a$ and $b_1 \to b$, we have

$$Q[y] = \int_a^b pu^2(v')^2 dx.$$

Now the left side is nonpositive and the right side is nonnegative thus their common value is zero. Hence $v' = 0$ and $v =$constant; i.e., $y(x)$ is a constant multiple of $u(x)$; a contradiction. □

Theorem 1.12. *Equation* (1.4) *is disconjugate on the compact interval* $I = [a, b]$ *if and only if the quadratic functional* $Q[y]$ *is positive.*

Proof. Let's first suppose that the functional is positive, and let $u(x)$ be a solution of (1.4) such that $u(c) = u(d) = 0$, where $a \leq c < d \leq b$. (We will show $u(x) \equiv 0$.) Set

$$y(x) = \begin{cases} u(x), & c \leq x \leq d, \\ 0, & \text{otherwise.} \end{cases}$$

Then $y(x)$ is admissible and by Corollary 1.2

$$Q[y] = \int_c^d (p(u')^2 - qu^2)\, dx = 0.$$

By the positivity of Q, we must have $y(x) \equiv 0$, hence $u(x) \equiv 0$ and so (1.4) is disconjugate on $I = [a, b]$.

Conversely, suppose (1.4) is disconjugate on $I = [a, b]$ and let $u(x)$ be the solution of (1.4) which satisfies $u(a) = 0$, $u'(a) = 1$. If $Q[y] \leq 0$ for some nontrivial admissible function $y(x)$, then by Theorem 1.11, either $u(c) = 0$ for some $c \in (a, b)$, or $u(x)$ is a constant multiple of $y(x)$, in which case $u(b) = 0$. In either case we have a contradiction to the disconjugacy of (1.4). Thus Q is positive. □

We will now present one form of the Sturm Comparison Theorem as a corollary of Theorem 1.12. Recall the equations

$$L_1[y] = (p_1(x)y')' + q_1(x)y = 0, \tag{1.10}$$

and

$$L_2[y] = (p_2(x)y')' + q_2(x)y = 0. \tag{1.11}$$

Corollary 1.3. *Suppose that both Eqs.* (1.10) *and* (1.11) *are disconjugate on an interval I and suppose*

$$\begin{aligned} p(x) &= \lambda_1 p_1(x) + \lambda_2 p_2(x), \\ q(x) &= \lambda_1 q_1(x) + \lambda_2 q_2(x), \end{aligned}$$

where $\lambda_1,\ \lambda_2 > 0$. Then Eq. (1.4) *is also disconjugate on I.*

Proof. Let $[a, b]$ (compact) $\subseteq I$, let $Q_1,\ Q_2$ be quadratic functionals corresponding to Eqs. (1.10) and (1.11). Then it is not difficult to see that $Q[y] = \lambda_1 Q_1[y] + \lambda_2 Q_2[y]$. Now let $y(x)$ be a nontrivial admissible function on $[a, b]$. Then $Q[y] = \lambda_1 Q_1[y] + \lambda_2 Q_2[y] > 0$, by Theorem 1.12, since both (1.10) and (1.11) are disconjugate. Then by Theorem 1.12 again, (1.4) is disconjugate on $[a, b]$. But $[a, b]$ was an arbitrary compact subinterval of I, and hence, (1.4) is disconjugate on I. □

Corollary 1.4. (Sturm Comparison Theorem). *Suppose that Eq.* (1.11) *is disconjugate on I, that $p_1(x) \geq p_2(x) > 0$, and that $q_1(x) \leq q_2(x)$, for all $x \in I$. Then* (1.10) *is also disconjugate on I. (In particular, if $p(x)$ and $q(x)$ are as in Corollary* 1.3, *then* (1.4) *is also disconjugate on I.)*

Proof. Again let $[a, b]$ (compact) $\subseteq I$, let Q_1 and Q_2 be as in Corollary 1.3, and let $y(x)$ be a nontrivial admissible function on $[a, b]$. By Theorem 1.12, $Q_2[y] > 0$.

Now

$$\begin{aligned} Q_1[y] &= \int_a^b [p_1(y')^2 - q_1(y)^2]\, dx \\ &\geq \int_a^b [p_2(y')^2 - q_2 y^2]\, dx \\ &= Q_2[y] > 0. \end{aligned}$$

Thus, by Theorem 1.12, Eq. (1.10) is disconjugate on $[a, b]$. Therefore, (1.10) is also disconjugate on I. □

A somewhat stronger form of the Sturm Comparison Theorem can be obtained by appealing directly to Theorem 1.11.

Theorem 1.13. *Let $y_1(x)$ and $y_2(x)$ be nontrivial solutions on $I = [a, b]$ of Eqs.* (1.10) *and* (1.11) *respectively, where $p_1(x) \geq p_2(x) > 0$ and $q_1(x) \leq q_2(x)$, for all $x \in I$. If $y_1(a) = y_1(b) = 0$ and if $y_1(x)$ is not a constant multiple of $y_2(x)$, then $y_2(c) = 0$, for some $c \in (a, b)$.*

Exercise 17. Prove Theorem 1.13. Hint: First $Q_1[y_1] = 0$. Why? Then apply Theorem 1.11, where y_1 plays the role of y, y_2 plays the role of u, and Q_1 plays the role of Q.

1.3 Disconjugacy for Equations of Arbitrary Order

In this last section concerning disconjugacy we will show that the nth order linear homogeneous differential equation (1.2) is disconjugate on $I = (a, b)$ if and only if it possesses no nontrivial solution having n distinct zeros on I. Thus recall

$$L[y] \equiv y^{(n)} + a_{n-1}(x)y^{(n-1)} + \cdots + a_0(x)y = 0, \tag{1.2}$$

where a_i, $1 \leq i \leq n-1$, are continuous on an interval I.

Our first theorem of this section says that on small enough subintervals of an interval $I = [a, b]$, (1.2) is disconjugate.

Theorem 1.14. *If $I = [a, b]$, then there exists $\delta > 0$ such that* (1.2) *is disconjugate on every subinterval of I of length less than δ.*

Proof. Put

$$M = \max_{1 \leq i \leq n-1} \max_{x \in I} |a_i(x)|,$$

and then take

$$\delta = \min\left\{1, \frac{1}{nM}\right\}.$$

Now let J be a subinterval of I of length $< \delta$, and let $\hat{a} < \hat{b}$ be endpoints of J. Now if there exists a nontrivial solution $y(x)$ having n zeros, counting multiplicities, on J, then by repeated applications of Rolle's Theorem, $y^{(k)}(x)$ has at least $n - k$ zeros, counting multiplicities, on J, $1 \leq k \leq n$. Now, for all $0 \leq k \leq n$, let

$$\mu_k = \sup_{x \in J} \left|y^{(k)}(x)\right|, \qquad \text{(it exists)}.$$

Then by the Mean Value Theorem, $\mu_k \le \mu_{k+1}\delta$, and in fact if $\mu_k > 0$ and s is a zero of $y^{(k)}(x)$, by the Mean Value Theorem,

$$\left|y^{(k)}(x)\right| = \left|y^{(k+1)}(\xi_x)\right|\left|x-s\right| \le \mu_{k+1}(\hat{b}-\hat{a}).$$

Hence,

$$\mu_k \le \mu_{k+1}(\hat{b}-\hat{a}) < \mu_{k+1}\delta,$$

i.e., $\mu_k > 0$ yields $\mu_k < \mu_{k+1}\delta$, for all k. Since $\mu_0 > 0$, it follows that

$$0 < \mu_0 < \mu_1\delta < \mu_2\delta^2 < \cdots < \mu_k\delta^k < \cdots < \mu_n\delta^n,$$

or

$$0 < \mu_k < \delta^{n-k}\mu_n, \quad 0 \le k \le n-1.$$

On the other hand, $y(x)$ is a solution of (1.2), so

$$\begin{aligned}\mu_n &\le M(\mu_{n-1} + \cdots + \mu_0)\\ &< M(\delta^1 + \delta^2 + \cdots + \delta^n)\mu_n\\ &\le nM\delta\mu_n. \quad (\text{by } \delta \le 1)\end{aligned}$$

Thus $1 < nM\delta$, or $\delta > \frac{1}{nM}$; a contradiction. □

A much sharper estimate for δ than above can be obtained, but we will not concern ourselves with that.

Exercise **18.** Show that if $I = [a,b]$, then there exists a positive integer N such that every nontrivial solution of (1.2) has less than N zeros on I. Hint: partition I into sufficiently small intervals, then count the number of possible zeros on each subinterval and count the number of subintervals.

Definition 1.5. Given $2 \le h \le n$ and positive integers $m_1, \ldots, m_h$ such that $m_1 + \cdots + m_h = n$, a nontrivial solution $y(x)$ of (1.2) is said to *have an* $(m_1, \ldots, m_h)$ *distribution of zeros* on an interval I, provided there are points $t_1, \ldots, t_h \in I$, with $t_1 \le \cdots \le t_h$, such that $y(x)$ has a zero at each t_i of multiplicity at least m_i, $1 \le i \le h$.

In a manner completely analogous to the one used in Theorem 1.2, one can prove.

Theorem 1.15. *Let $2 \le h \le n$ and positive integers $m_1, \ldots, m_h$, such that $m_1 + \cdots + m_h = n$, be given. If no nontrivial solution of* (1.2) *has an $(m_1, \ldots, m_h)$ distribution of zeros on I, then given distinct points $x_1 < \cdots < x_h$ belonging to I and reals $y_{11}, \ldots, y_{m_h h}$, there is a unique solution $y(x)$ of* (1.2) *satisfying*

$$y^{(i-1)}(x_j) = y_{ij}, \quad 1 \le i \le m_j, \quad 1 \le j \le h.$$

We will also make use of the following theorems.

Theorem 1.16. *For any interval I, there exists a nontrivial solution of* (1.2) *with $n-1$ arbitrarily prescribed zeros on I.*

Proof. Let $y_1(x), \ldots, y_n(x)$ be a fundamental system of solutions of (1.2). Then any solution of (1.2) is of the form $y(x) = \alpha_1 y_1(x) + \cdots + \alpha_n y_n(x)$. Hence given $n-1$ arbitrarily prescribed zeros, we must choose coefficients $\alpha_1, \ldots, \alpha_n$ such that $y(x)$ has the prescribed zeros. But then $\alpha_1, \ldots, \alpha_n$ must satisfy a system of $n-1$ homogeneous equations (i.e. $n-1$ equations and n unknowns). The corresponding system of equations will always have a solution, $\alpha_1, \ldots, \alpha_n$, not all zero. □

Theorem 1.17. *If the interval $I = (a, b)$ and if every nontrivial solution of* (1.2) *has less than n distinct zeros, then every such nontrivial solution has less than n zeros, counting multiplicities.*

Proof. Suppose on the contrary that there exists a solution with zeros of total multiplicity at least n. Among all such solutions, let $y(x)$ be one with the greatest number of distinct zeros. Suppose $y(x)$ has h, $1 \le h < n$, distinct zeros t_k of multiplicity r_k $(1 \le r_k < n)$, where $a < t_1 < \cdots < t_h < b$ and $r_1 + \cdots + r_h \ge n$, (i.e., for each k, $y(t_k) = y'(t_k) = \cdots = y^{(r_k-1)}(t_k) = 0$). Let $a < t_0 < t_1$ and $t_h < t_{h+1} < b$ be fixed.

There are two cases:

Case (i): $h = n-1$.

Suppose first that, for some $1 \le k \le h$, r_k is even. (i.e. $y(t_k) = \cdots = y^{(r_k-1)}(t_k) = 0$ and $r_k = 2m$). We may assume $y^{(r_k)}(t_k) > 0$.

Now, by the hypotheses and either of Theorem 1.15 or Theorem 1.16, there exists a unique solution $u(x)$ such that

$$u(t_k) = 1, \quad u(t_i) = 0, \quad 1 \le i \le h+1,\ i \ne k.$$

Then, given $\varepsilon > 0$, the nontrivial solution $y_\varepsilon(x) = y(x) - \varepsilon u(x)$ has $n-2$ zeros at t_i, $1 \le i \le h$, $i \ne k$. If we choose α, β such that $t_{k-1} < \alpha < t_k < \beta < t_{k+1}$, for $\varepsilon > 0$ sufficiently small, $y_\varepsilon(\alpha) > 0$, $y_\varepsilon(\beta) > 0$, $y_\varepsilon(t_k) = -\varepsilon < 0$, and so $y_\varepsilon(x)$ has *at least* two zeros in (t_{k-1}, t_{k+1}). (See Figure 1.1). Thus, $y_\varepsilon(x)$ has at least n distinct zeros; a contradiction.

Suppose next that r_i is odd for any $1 \le i \le h$. Then for some k, $r_k > 1$. (Let's note here in this case that $y(x)$ changes sign at t_i, for all $1 \le i \le h$.) Let $\varepsilon > 0$ be given, and let $y_\varepsilon(x)$ be the solution of (1.2) satisfying the

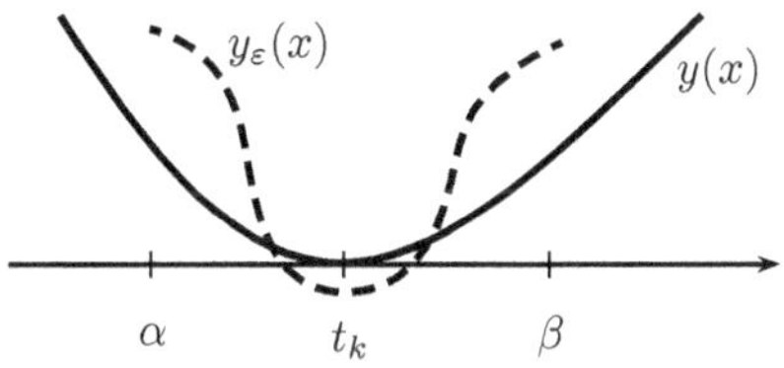

Fig. 1.1 $y_\varepsilon(x)$ has at least two zeros in (t_{k-1}, t_{k+1}).

initial conditions,

$$y_\varepsilon^{(j)}(t_k) = y^{(j)}(t_k), \;\; 0 \le j \le n-1, \;\; j \ne r_k - 1,$$
$$y_\varepsilon^{(r_k-1)}(t_k) = \varepsilon.$$

For ε sufficiently small, $y_\varepsilon(x)$ is uniformly close to $y(x)$ on $[t_0, t_{h+1}]$ and since $y(x)$ changes sign at each of its zeros, it follows that $y_\varepsilon(x)$ has $n-2$ zeros near t_i, $1 \le i \le h$, $i \ne k$, and a zero of multiplicity $r_k - 1$ at $x = t_k$. Moreover, since $r_k > 1$ and odd, and $y_\varepsilon^{(r_k-1)}(t_k) = \varepsilon > y^{(r_k-1)}(t_k)$, it follows that $y_\varepsilon(x)$ has at least one zero in a deleted neighborhood of t_k. Thus $y_\varepsilon(x)$ has at least n distinct zeros; again, a contradiction.

Case (ii): $h < n-1$.

The argument in the second part of (i) shows that if $h-1$ zeros of $y(x)$ are simple, (hence $y(x)$ would have a sign change at each of those zeros), then we can find another solution $y_\varepsilon(x)$ with more than h distinct zeros of total multiplicity at least r (and $r \ge n$). This contradicts the definition of h.

Thus, it only remains to consider the case where at least two zeros, say t_k and t_m, are not simple. Without loss of generality, we may assume $y^{(r_k)}(t_k) > 0$. Now choose integers $q_1, \ldots, q_h$ and $s \ge 1$ such that,

$$1 \le q_i \le r_i, \;\; 1 \le i \le h, \;\; i \ne k,$$
$$0 \le q_k = r_k - 2s,$$

and

$$q = q_1 + \cdots + q_h = n - 2.$$

By the maximality of h and Theorem 1.15, there exists a unique solution $u(x)$ of (1.2) such that

$$u^{(j)}(t_i) = 0, \;\; 0 \le j \le q_i - 1, \;\; 1 \le i \le h, \;\; i \ne k,$$
$$u(t_{h+1}) = 0,$$
$$u^{(j)}(t_k) = 0, \;\; 0 \le j \le q_k - 1,$$
$$u^{(q_k)}(t_k) = 1,$$

(i.e., $u(x)$ exists since by the definition of h, no nontrivial solution satisfies the corresponding homogeneous boundary conditions.) Now, given $\varepsilon > 0$, the nontrivial solution, $y_\varepsilon(x) = y(x) - \varepsilon u(x)$, has a zero at t_i of multiplicity at least q_i, $1 \le i \le h$, $i \ne k$. If $q_k \ge 1$, $y_\varepsilon(x)$ has a zero of order q_k at t_k. Moreover, since $y^{(r_k)}(t_k) > 0$, $q_k = r_k - 2s$, and $y_\varepsilon^{(q_k)}(t_k) = -\varepsilon < 0$, it follows from an argument similar to the one used in the second part of (i) that $y_\varepsilon(x)$ has at least two zeros in $(t_{k-1}, t_{k+1}) \setminus \{t_k\}$. Thus $y_\varepsilon(x)$ has at least $h+1$ distinct zeros of total multiplicity at least$n$, which again contradicts the definition of h. □

We end this section and chapter with a necessary and sufficient condition for the disconjugacy of (1.2) as a corollary to Theorem 1.17.

Corollary 1.5. (1.2) *is disconjugate on $I = (a, b)$ if and only if no nontrivial solution of* (1.2) *has n distinct zeros on I.*

Bibliography

Aliev R. G. (1964). On certain properties of solutions of ordinary differential equations of fourth order, *Sb. Aspirantsh*, Kazan Institute, pp. 15–30 (Russian).

Azbelev N. and Tsalyuk Z. (1960). On the question of the distribution of zeros of a solution of athird order linear differential equation, *Math. Sb. (N. S.)* **52**, pp. 475–486.

Barrett J. H. (1959). Disconjugacy of second-order linear differential equations with nonnegative coefficients, *Proc. Amer. Math. Soc.* **10**, pp. 552–561.

Coppel W. A. (1971). *Disconjugacy*, Lecture Notes in Mathematics, Vol. 220 (Springer-Verlag, Berlin).

Gelfand I. M. and Fomin S. V. (1963). *Calculus of Variations*, (Russian: translated by R. A. Silverman) (Prentice-Hall, Englewood Cliffs, NJ).

Hartman P. (1958). Unrestricted n-parameter families, *Rend. Circ. Mat. Palermo* **7**, No. 2, 123-142.

Hartman P. (1964). *Ordinary Differential Equations* (Wiley, New York).

Hartman P. (1969). Principal solutions of disconjugate n-th order linear differential equations, *Amer. J. Math.* **91**, pp. 306–362.

Henderson J. (1984). K-point disconjugacy and disconjugacy for linear differential equations, *J. Differential Equations* **54**, No. 1, pp. 87–96.

Kelley W. G. and Peterson A. C. (2010). *Theory of Differential Equations: Classical and Qualitative*, (Second Edition), Universitext (Springer, New York).

Kim W. J. (2971). Simple zeros of solutions of nth-order linear differential equations, *Proc. Amer. Math. Soc.* **28**, No. 2, pp. 557–561.

Leighton W. (1949). Principal quadratic functionals, *Trans. Amer. Math. Soc.* **67**, pp. 253–274.

Leighton W. and Nehari Z. (1958). On the oscillation of solutions of self-adjoint linear differential equations of the fourth order, *Trans. Amer. Math. Soc.* **89**, pp. 325–377.

Muldowney J. S. (1984). On disconjugacy and k-point disconjugacy for linear ordinary differential equations, *Proc. Amer. Math. Soc.* **92**, No. 1, pp. 27–30.

Opial Z. (1967). On a theorem of O. Arama, *J. Differential Equations* **3**, pp. 88–91.

Peterson A. C. (1969). A theorem of Aliev, *Proc. Amer. Math. Soc.* **23**, No. 2, pp. 364–366.

Peterson A. C. (1973). On a relation between a theorem of Hartman and a theorem of Sherman, *Canad. Math. Bull.* **16**, pp. 275–282.

Reid W. T. (1946). A matrix differential equation of Riccati type, *Amer. J. Math.* **68**, pp. 237–246.

Reid W. T. (1963). Riccati matrix differential equations and non-oscillation criteria for associated linear differential systems, *Pacific J. Math.* **13**, pp. 665–685.

Sherman T. L. (1965). Properties of solutions of nth order linear differential equations, *Pacific J. Math.* **15**, pp. 1045–1060.

Sherman T. L. (1969). Conjugate points and simple zeros for ordinary linear differential equations, *Trans. Amer. Math. Soc.* **146**, pp. 397–411.

Chapter 2

Linear Problems and Green's Functions

This chapter is devoted to solutions of BVP's for linear ODE's, to the construction of Green's functions, and to the role of Green's functions in solving BVP's for ODE's.

2.1 Introduction to Linear ODE's

In this section, we will be primarily concerned with the form of solutions for the linear differential equation

$$y^{(n)} + \sum_{i=0}^{n-1} a_i(x)y^{(i)} = f(x). \tag{2.1}$$

Before pursuing these results, we review some basic results concerning solutions of matrix differential equations which can be related to Eq. (2.1). In the following review, the independent variable will be denoted by t and some equations previously seen in Chapter 1 will now bear a different number designation.

Let $A(t)$ be a continuous $n \times n$ matrix and denote the $n \times n$ fundamental matrix solution (nonsingular) of the IVP,

$$\begin{cases} X' = A(t)X, \\ X(t_0) = I, \end{cases}$$

by $X(t, t_0)$, (i.e. $X(t_0, t_0) = I$). (Note here that $n \times n$ matrix solutions will be designated by capital letters and that vector solutions will "usually" be denoted by lower case letters. Usually the context will make such designations clear.)

It follows, by uniqueness of solutions of IVP's for linear systems, that the solution of

$$\begin{cases} X' = A(t)X, \\ X(t_0) = C, \end{cases}$$

is $X(t,t_0)C$. Moreover if $Y(t)$ is any solution of $X' = A(t)X$, then $Y(t) = X(t,t_0)Y(t_0)$.

So, if $X(t)$ is any fundamental matrix solution of $X' = A(t)X$, then

$$X(t,s) = X(t)X^{-1}(s), \quad \text{for all } t, s \text{ in the interval of interest,}$$

or

$$X(t,t_0) = X(t)X^{-1}(t_0).$$

We first state the *Variation of Constants Formula* for solutions of the IVP

$$\begin{cases} x' = A(t)x + f(t), & (2.2) \\ x(t_0) = 0, & (2.3) \end{cases}$$

where $x(t)$ is an n-vector.

Theorem 2.1. *Let $X(t)$ be a fundamental matrix solution of $X' = A(t)X$. Then the solution $z(t)$ of (2.2), (2.3) is given by*

$$z(t) = X(t)\int_{t_0}^{t} X^{-1}(s)f(s)\,ds,$$

or

$$z(t) = \int_{t_0}^{t} X(t,s)f(s)\,ds.$$

Finally the solution of

$$\begin{cases} x' = A(t)x + f(t), & (2.2) \\ x(t_0) = c, & (2.4) \end{cases}$$

can be written as

$$x(t) = X(t,t_0)c + \int_{t_0}^{t} X(t,s)f(s)\,ds.$$

We want to examine what these solutions say in relation to the equations:

$$x^{(n)} + a_{n-1}(t)x^{(n-1)} + a_{n-2}(t)x^{(n-2)} + \cdots + a_0(t)x = 0, \tag{2.5}$$

$$x^{(n)} + a_{n-1}(t)x^{(n-1)} + a_{n-2}(t)x^{(n-2)} + \cdots + a_0(t)x = h(t), \tag{2.6}$$

$$y' = A(t)y, \tag{2.7}$$

$$y' = A(t)y + f(t), \qquad (2.2)$$

where

$$A(t) = \begin{bmatrix} 0 & 1 & 0 & \cdots & 0 \\ 0 & 0 & 1 & \cdots & 0 \\ \vdots & \vdots & \vdots & \ddots & \vdots \\ 0 & 0 & 0 & \cdots & 1 \\ -a_0(t) & -a_1(t) & -a_2(t) & \cdots & -a_{n-1}(t) \end{bmatrix}, \quad f(t) = \begin{bmatrix} 0 \\ 0 \\ \vdots \\ h(t) \end{bmatrix}.$$

If $x_1(t), \ldots, x_m(t)$ are solutions of (2.5), then

$$y^1(t) = \begin{bmatrix} x_1(t) \\ x_1'(t) \\ \vdots \\ x_1^{(n-1)}(t) \end{bmatrix}, \ldots, y^m(t) = \begin{bmatrix} x_m(t) \\ x_m'(t) \\ \vdots \\ x_m^{(n-1)}(t) \end{bmatrix} \quad \text{are solutions of (2.7).}$$

The converse is also true.

It follows that a matrix solution $X(t)$ of (2.7) (i.e. $X' = A(t)X$) has the form

$$X(t) = \begin{bmatrix} x_1(t) & x_2(t) & \cdots & x_n(t) \\ x_1'(t) & x_2'(t) & \cdots & x_n'(t) \\ \vdots & \vdots & \ddots & \vdots \\ x_1^{(n-1)}(t) & x_2^{(n-1)}(t) & \cdots & x_n^{(n-1)}(t) \end{bmatrix},$$

where $x_1(t), \ldots, x_n(t)$ are solutions of (2.5).

Consequently $X(t)$ is a fundamental matrix solution of (2.7) if and only if $X(t)$ is nonsingular if and only if $X(t)c = 0$ implies $c = 0$ if and only if the columns of $X(t)$ are linearly independent if and only if $x_1(t), \ldots, x_n(t)$ are linearly independent solutions of (2.5).

In fact, the fundamental matrix solution $X(t)$ of (2.7) satisfying $X(t_0) = I$, (i.e., $X(t) = X(t, t_0)$), has as the jth element of its first row the solution $x_j(t)$ of (2.5) which satisfies

$$x_j^{(i)}(t_0) = 0, \;\; 0 \le i \le n-1, \;\; i \ne j-1,$$
$$x_j^{(j-1)}(t_0) = 1, \;\; \text{(this is clear since } X(t_0) = I).$$

From the Variation of Constants Formula, the unique solution of

$$\begin{cases} y' = A(t)y + f(t), \\ y(t_0) = 0, \end{cases} \tag{2.2}$$

is

$$y(t) = \int_{t_0}^{t} X(t,s) f(s)\, ds,$$

where $f(s) = [0, \cdots, 0, h(s)]^T$ and $X(t,s)$ is the solution of

$$\begin{cases} X' = A(t)X, \\ X(s) = X(s,s) = I. \end{cases}$$

Now we want to use this solution $y(t)$ to describe corresponding solutions of Eq. (2.6). Let's say that

$$y(t) = \begin{bmatrix} y_1(t) \\ y_2(t) \\ \vdots \\ y_n(t) \end{bmatrix}.$$

We will show that the first component $y_1(t)$ is the solution of (2.6) and satisfies $y_1^{(i)}(t_0) = 0$, $0 \le i \le n-1$.

To see this, first consider $y(t)$ as a solution of

$$\begin{cases} y' = A(t)y + f(t), \\ y(t_0) = 0. \end{cases} \tag{2.2}$$

Hence,

$$\begin{bmatrix} y_1(t) \\ y_2(t) \\ \vdots \\ y_n(t) \end{bmatrix}' = \begin{bmatrix} 0 & 1 & 0 & \cdots & 0 \\ 0 & 0 & 1 & \cdots & 0 \\ \vdots & \vdots & \vdots & \ddots & \vdots \\ 0 & 0 & 0 & \cdots & 1 \\ -a_0(t) & -a_1(t) & -a_2(t) & \cdots & -a_{n-1}(t) \end{bmatrix} \begin{bmatrix} y_1(t) \\ y_2(t) \\ \vdots \\ y_n(t) \end{bmatrix} + \begin{bmatrix} 0 \\ 0 \\ \vdots \\ h(t) \end{bmatrix}.$$

Thus equating last components yields

$$y_n'(t) = -a_0(t)y_1(t) - a_1(t)y_2(t) - \cdots - a_{n-1}(t)y_n(t) + h(t).$$

But also $y_i'(t) = y_{i+1}(t)$ and so by successively substituting we have

$$y_1^{(n)}(t) = -a_0(t)y_1(t) - a_1(t)y_1'(t) - \cdots - a_{n-1}(t)y_1^{(n-1)}(t) + h(t).$$

Hence, $y_1(t)$ is a solution of (2.6). Moreover, $y(t_0) = 0$ and now

$$y(t) = \begin{bmatrix} y_1(t) \\ y_2(t) \\ \vdots \\ y_n(t) \end{bmatrix} = \begin{bmatrix} y_1(t) \\ y_1'(t) \\ \vdots \\ y_1^{(n-1)}(t) \end{bmatrix},$$

so we have

$$\begin{bmatrix} y_1(t_0) \\ y_1'(t_0) \\ \vdots \\ y_1^{(n-1)}(t_0) \end{bmatrix} = \begin{bmatrix} 0 \\ 0 \\ \vdots \\ 0 \end{bmatrix}.$$

Thus the first component $y_1(t)$ of $y(t)$ is the solution of Eq. (2.6) and satisfies $y_1^{(i)}(t_0) = 0$, $0 \le i \le n-1$.

We now look more closely at the solution $y(t)$ for more detail on its first component $y_1(t)$, (i.e., we look for an explicit expression for $y_1(t)$).

Recall

$$y(t) = \int_{t_0}^{t} X(t,s) f(s)\, ds,$$

that is,

$$\begin{bmatrix} y_1(t) \\ y_1'(t) \\ \vdots \\ y_1^{(n-1)}(t) \end{bmatrix} = \int_{t_0}^{t} \begin{bmatrix} x_1(t,s) & x_2(t,s) & \cdots & x_n(t,s) \\ x_1'(t,s) & x_2'(t,s) & \cdots & x_n'(t,s) \\ \vdots & \vdots & \vdots & \vdots \\ x_1^{(n-1)}(t,s) & x_2^{(n-1)}(t,s) & \cdots & x_n^{(n-1)}(t,s) \end{bmatrix} \begin{bmatrix} 0 \\ 0 \\ \vdots \\ h(s) \end{bmatrix} ds,$$

where $x_1(t,s), \ldots, x_n(t,s)$ are linearly independent solutions of (2.5). Since $X(s,s) = I$, notice that

$$x_i^{(j-1)}(s,s) = \delta_{ij}, \quad 1 \le i, j \le n.$$

Writing the right hand side of the above equation as a vector, we now have

$$\begin{bmatrix} y_1(t) \\ y_1'(t) \\ \vdots \\ y_1^{(n-1)}(t) \end{bmatrix} = \begin{bmatrix} \int_{t_0}^{t} x_n(t,s) h(s)\, ds \\ \int_{t_0}^{t} x_n'(t,s) h(s)\, ds \\ \vdots \\ \int_{t_0}^{t} x_n^{(n-1)}(t,s) h(s)\, ds \end{bmatrix}.$$

That is, the first component $y_1(t)$ is explicitly given as

$$y_1(t) = \int_{t_0}^{t} x_n(t,s) h(s)\, ds,$$

where $x_n(t,s)$ is the solution of (2.5) satisfying $x_n^{(i)}(s,s) = 0$, $0 \le i \le n-2$, $x^{(n-1)}(s,s) = 1$.

In the above setting, for notation purposes, let $y_1(t) = x(t)$ and $x_n(t,s) = u(t,s)$. [Since $X(s,s) = I$, $u^{(i)}(s,s) = 0$, $0 \le i \le n-2$ and $u^{(n-1)}(s,s) = 1$.]

Thus summarizing, we have: The solution $x(t)$ of (2.6) satisfying $x^{(i)}(t_0) = 0$, $0 \le i \le n-1$, is given by the first component of $y(t) = \int_{t_0}^t X(t,s)f(s)\,ds$ and is

$$x(t) = \int_{t_0}^t u(t,s)h(s)\,ds,$$

where $u(t,s)$ is the solution of

$$x^{(n)} + a_{n-1}(t)x^{(n-1)} + \cdots + a_0(t)x = 0, \tag{2.5}$$

and satisfies $x^{(i)}(s) = 0$, $0 \le i \le n-2$, $x^{(n-1)}(s) = 1$.

Here, $u(t,s)$ is frequently called the *Cauchy Function* for (2.5).

Example 2.1. 1). $x^{(n)} = 0$. The Cauchy Function is

$$u(t,s) = \frac{(t-s)^{n-1}}{(n-1)!},$$

and therefore, the unique solution of $x^{(n)} = h(t)$, $x^{(i)}(t_0) = 0$, $0 \le i \le n-1$, is

$$x(t) = \int_{t_0}^t \frac{(t-s)^{n-1}}{(n-1)!} h(s)\,ds.$$

2). $x'' + x = 0$. The Cauchy Function is $u(t,s) = \sin(t-s)$. Then the unique solution of $x'' + x = h(t)$, $x(t_0) = x'(t_0) = 0$ is

$$x(t) = \int_{t_0}^t \sin(t-s)h(s)\,ds.$$

Exercise **19.** Determine the Cauchy function $u(t,s)$ for $x''' - x = 0$ and express the solution of

$$\begin{cases} x''' - x = \cos t, \\ x(1) = x'(1) = x''(1) = 0, \end{cases}$$

in terms of $u(t,s)$.

2.2 Linear BVP's

We now return to our linear equation

$$y^{(n)} + \sum_{i=0}^{n-1} a_i(x)y^{(i)} = f(x), \tag{2.1}$$

and we will be interested in solutions of (2.1) satisfying linear boundary conditions. Let us assume that the coefficient functions, $a_i(x)$, $0 \le i \le n-1$, and $f(x)$ are continuous on some interval $I \subseteq \mathbb{R}$. Then IVP's for (2.1) are uniquely solvable, exist on all of I, and a solution $y(x)$ has n continuous derivatives on I.

Now, what do we mean by linear boundary conditions? Let $[a,b]$ be a compact subinterval of I. Then a solution $y(x)$ belongs to $C^{(n-1)}[a,b]$, (actually to $C^{(n)}[a,b]$), which is a Banach space with respect to the norm

$$\|h\| = \sum_{i=0}^{n-1}\left\{\max_{a\le x\le b}\left|h^{(i)}(x)\right|\right\}, \quad \text{for} \quad h \in C^{(n-1)}[a,b].$$

Definition 2.1. A *linear functional* on a Banach space, $\mathcal{B}$, is a function $\varphi : \mathcal{B} \to \mathbb{R}$ which is linear; i.e., $\varphi(\alpha u+\beta v) = \alpha\varphi(u)+\beta\varphi(v)$, for all $u, v \in \mathcal{B}$ and $\alpha, \beta \in \mathbb{R}$.

A linear functional $\varphi : \mathcal{B} \to \mathbb{R}$ is *continuous*, if for any $\varepsilon > 0$ and any $u_0 \in \mathcal{B}$, there exists $\delta > 0$ such that $\|v - u_0\| < \delta$ implies $|\varphi(v) - \varphi(u_0)| < \varepsilon$.

Continuity of a functional φ is equivalent to: there exists $M_\varphi > 0$ such that $|\varphi(u)| \le M_\varphi\|u\|$, for all $u \in \mathcal{B}$.

Definition 2.2. A *linear boundary condition* on $[a,b] \subseteq I$ is a condition of the form $\varphi(y) = r$, where φ is a linear functional on $C^{(n-1)}[a,b]$.

An example of a linear BVP on $[a,b] \subseteq I$ is the problem of finding a solution of

$$\begin{cases} y^{(n)} + \sum_{i=0}^{n-1} a_i(x)y^{(i)} = f(x). & (2.1) \\ \varphi_j(y) = r_j, \ \ 1 \le j \le n, & (2.8) \end{cases}$$

where each φ_j is a continuous linear functional on $C^{(n-1)}[a,b]$. Examples of the common types of linear BC's are:

1). Define $\varphi(y) = y(a)$;
2). Define $\varphi(y) = \sum_{j=0}^{n-1} \alpha_j y^{(j)}(x_0)$, where $a \le x_0 \le b$;

3). Define $\varphi(y) = \sum_{j=0}^{n-1} \left[\alpha_j y^{(j)}(a) + \beta_j y^{(j)}(b)\right]$;
4). Define $\varphi(y) = \int_a^b y(s)\,ds$.

Theorem 2.2. *Let φ_j, $1 \le j \le n$, be continuous linear functionals on $C^{(n-1)}[a,b]$, where $[a,b] \subseteq I$. Then the BVP*

$$\begin{cases} y^{(n)} + \sum_{i=0}^{n-1} a_i(x) y^{(i)} = f(x), & (2.1) \\ \varphi_j(y) = r_j, \ \ 1 \le j \le n, & (2.8) \end{cases}$$

has a solution (in fact, a unique solution) for every choice of $f(x)$ and any choice of boundary values $r_j \in \mathbb{R}$, $1 \le j \le n$, if and only if the corresponding homogeneous BVP

$$\begin{cases} y^{(n)} + \sum_{i=0}^{n-1} a_i(x) y^{(i)} = 0, & (2.9) \\ \varphi_j(y) = 0, \ \ 1 \le j \le n, & (2.10) \end{cases}$$

has only the solution, $y(x) \equiv 0$.

Proof. For $1 \le j \le n$, let $y_j(x)$ be the fundamental set of solutions of (2.9) satisfying the initial conditions, $y_j^{(k-1)}(a) = \delta_{jk}$. If we define $h(x) = \int_a^x u(x,s) f(s)\,ds$, where $u(x,s)$ is the Cauchy function for (2.9), then, as a consequence of Theorem 2.1, every solution of (2.1) is of the form $y(x) = \sum_{j=1}^n c_j y_j(x) + h(x)$. Hence the stated BVP (2.1), (2.8) has a solution if and only if the linear system of equations

$$\varphi_k(y) = \varphi_k\left(\sum_{j=1}^n c_j y_j(x) + h(x)\right) = \sum_{j=1}^n c_j \varphi_k(y_j) + \varphi_k(h) = r_k,$$

where $1 \le k \le n$, can be solved for $c_1, \ldots, c_n$; i.e., if and only if

$$\sum_{j=1}^n c_j \varphi_k(y_j) = r_k - \varphi_k(h), \ \ 1 \le k \le n,$$

can be solved for the n unknowns $c_1, c_2, \ldots, c_n$. This system has a solution for every choice of the r_k's and every choice of $f(x)$ if and only if $\det\left[\varphi_k(y_j)\right]_{j,k=1}^n \neq 0$; if and only if BVP (2.9), (2.10) has only the solution, $y(x) \equiv 0$. □

Corollary 2.1. *The BVP (2.1), (2.8) has a solution for every choice of $f(x)$ and every choice of boundary values, $r_k \in \mathbb{R}$, $1 \le k \le n$, if and only if solutions, when they exist, are unique.*

Exercise **20.** 1). Show that

$$\begin{cases} y'' + p(x)y' + q(x)y = f(x), & (2.11)\\ y(a) = A, \ y(b) = B, & (2.12) \end{cases}$$

has a unique solution, for every $f(x)$, A, and B, if and only if $u(b, a) \neq 0$, where $u(x, s)$ is the Cauchy function for $y'' + p(x)y' + q(x)y = 0$.

2). If $u(b, a) = 0$ and if $f(x)$ is chosen, then find a relation between A and B such that BVP (2.11), (2.12) will have a solution.

3). If $u(b, a) = 0$, show there exists a constant $k \neq 0$ such that $u(x, a) = ku(x, b)$.

Exercise **21.** 1). Show that the BVP

$$\begin{cases} y'' + y = f(x), \\ y(a) = A, \ y(b) = B, \end{cases}$$

has a unique solution if and only if $b - a$ is not an integer multiple of π.

2). Find a necessary and sufficient condition on $b-a$ such that the BVP $y'' + y = f(x)$,

$$y(a) = A, \quad \int_a^b y(s)\, ds = B,$$

has a unique solution.

3). Determine a condition on $\alpha_1, \alpha_2, \beta_1, \beta_2$ to insure that the BVP

$$\begin{cases} y'' = 0, \\ \alpha_1 y(a) + \beta_1 y'(a) = 0, \\ \alpha_2 y(b) + \beta_2 y'(b) = 0, \end{cases}$$

has only the solution, $y(x) \equiv 0$.

2.3 Green's Functions for Linear BVP's

As motivation for this section, we consider solutions of the BVP

$$\begin{cases} y'' = f(x), \\ y(a) = A, \ y(b) = B. \end{cases}$$

This BVP has a unique solution for every choice of $f(x)$ and $A, B \in \mathbb{R}$ if and only if the BVP $y'' = 0$, $y(a) = 0 = y(b)$ has only the solution, $y(x) \equiv 0$. This is indeed the case because of the disconjugacy of $y'' = 0$.

Thus, $y'' = f(x)$, $y(a) = A$, $y(b) = B$, does have a unique solution. In particular, we will be concerned with computing the solution of

$$\begin{cases} y'' = f(x), \\ y(a) = 0, \ y(b) = 0. \end{cases}$$

First the Cauchy function $u(x,s)$ for $y'' = 0$ is $u(x,s) = x - s$. Hence, all solutions of the differential equation $y'' = f(x)$ are of the form

$$y(x) = c_1 + c_2 x + \int_a^x (x-s) f(s)\, ds,$$

where 1 and x are linearly independent solutions of $y'' = 0$. So, we need to determine c_1 and c_2 in order that the BC's are satisfied:

$$y(a) = c_1 + c_2 a = 0,$$

$$y(b) = c_1 + c_2 b + \int_a^b (b-s) f(s)\, ds = 0,$$

which gives $c_1 = -ac_2$ and $c_2(b-a) = -\int_a^b (b-s) f(s)\, ds$. Hence, $c_2 = -\int_a^b \frac{(b-s)f(s)\, ds}{b-a}$ and $c_1 = \int_a^b \frac{a(b-s)f(s)}{b-a}\, ds$. Therefore, the unique solution is

$$\begin{aligned} y(x) &= \int_a^b \frac{a(b-s)}{b-a} f(s)\, ds + \int_a^b \frac{-x(b-s)}{b-a} f(s)\, ds + \int_a^x (x-s) f(s)\, ds. \\ &= \int_a^b \frac{(a-x)(b-s)}{b-a} f(s)\, ds + \int_a^x (x-s) f(s)\, ds, \end{aligned}$$

or

$$y(x) = \int_a^b G(x,s) f(s)\, ds,$$

where

$$G(x,s) = \begin{cases} \frac{(a-x)(b-s)}{b-a} + x - s, & a \le s \le x \le b, \\ \frac{(a-x)(b-s)}{b-a}, & a \le x \le s \le b. \end{cases}$$

Now, $\frac{(a-x)(b-s)}{b-a} + x - s = \frac{(a-x)(b-s)}{b-a}$. Therefore

$$G(x,s) = \begin{cases} \frac{(a-s)(b-x)}{b-a}, & a \le s \le x \le b, \\ \frac{(a-x)(b-s)}{b-a}, & a \le x \le s \le b. \end{cases}$$

It is customary to write

$$G(x,s) = \begin{cases} \frac{(a-x)(b-s)}{b-a}, & a \le x \le s \le b, \\ \frac{(a-s)(b-x)}{b-a}, & a \le s \le x \le b. \end{cases}$$

$G(x,s)$ is called the *Green's function* for the BVP $y'' = 0$, $y(a) = y(b) = 0$, and the unique solution of $y'' = f(x)$, $y(a) = y(b) = 0$ is

$$y(x) = \int_a^b G(x,s) f(s)\, ds.$$

2.4 Properties of the Green's Function

All of the four properties listed here are easily verifiable for our above constructed Green's function.

1) For each fixed s, as a function of x, $G(x,s)$ is a solution of $y''=0$ on $[a,s]$ and on $[s,b]$, (because for s fixed, $G(x,s)$ is just a constant multiple of $a-x$ on $[a,s]$ and a constant multiple of $b-x$ on $[s,b]$).
2) For each fixed s, as a function of x, $G(x,s)$ satisfies the BC's on $[a,s]$ and on $[s,b]$; i.e., $G(a,s)=0$, $G(b,s)=0$.
3) $G(x,s)$ is continuous on $[a,b]\times[a,b]$, and $G(x,s)$ is $C^{(1)}$ on the triangles $a\le x\le s\le b$ and on $a\le s\le x\le b$.

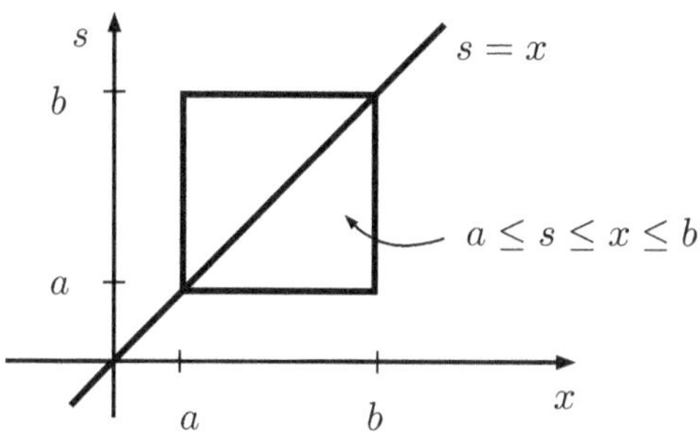

Fig. 2.1 Continuity and differentiability of $G(x,s)$.

4) $\frac{\partial G(s^+,s)}{\partial x}-\frac{\partial G(s^-,s)}{\partial x}=1.$

Let's check property 4) on our previously constructed Green's function. We have

$$\frac{-(a-s)}{b-a}-\frac{-(b-s)}{b-a}=\frac{-a+s+b-s}{b-a}=1.$$

These 4 properties say that the Green's function is "close in some sense" to a solution of the homogeneous BVP, which, of course, has only the trivial solution.

Theorem 2.3. *The properties* 1)–4) *listed above characterize the Green's function for the BVP*

$$\begin{cases} y''=0, \\ y(a)=y(b)=0. \end{cases}$$

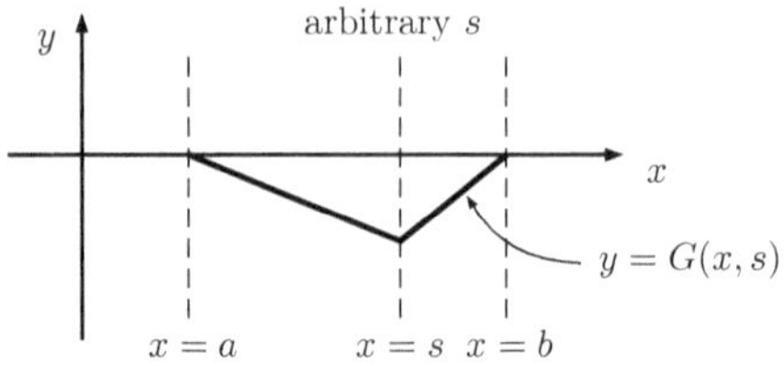

Fig. 2.2 The graph of $y = G(x, s)$ for any fixed $s \in (a, b)$.

Proof. Let $a < s < b$ be arbitrary but fixed, and assume that $G_1(x, s)$ and $G_2(x, s)$ both satisfy 1)–4). Then define $H(x, s) = G_1(x, s) - G_2(x, s)$. As a function of x, $H(x, s) \in C^{(1)}[a, s]$ and $H(x, s) \in C^{(1)}[s, b]$. Moreover, $\frac{\partial H(s^+, s)}{\partial x} - \frac{\partial H(s^-, s)}{\partial x} = 0$ and so $H(x, s) \in C^{(1)}[a, b]$. Since $G_1(x, s)$ and $G_2(x, s)$ are solutions of $y'' = 0$ and both vanish at $x = a$ and $x = b$, it follows that $H(x, s)$ is a solution of $y'' = 0$, $y(a) = y(b) = 0$. This BVP has only the trivial solution, hence $G_1(x, s) = G_2(x, s)$ and the proof is complete. □

We will now consider another method for construction of $G(x, s)$.

2.4.1 *Second method for constructing the Green's function*

Again we are concerned with a Green's function for $y'' = 0$, $y(a) = y(b) = 0$. From the above properties, the Green's function (for each fixed $s \in (a, b)$) is a solution of $y'' = 0$ on $[a, s]$ and on $[s, b]$. So,

$$G(x, s) = \begin{cases} c_1(s) + c_2(s)x, & a \le x \le s \le b, \\ c_3(s) + c_4(s)x, & a \le s \le x \le b. \end{cases}$$

Moreover, from properties 2), 3), 4),

$$G(a, s) = 0,$$

$$G(b, s) = 0,$$

$$G(s^+, s) = G(s^-, s),$$

$$\frac{\partial G(s^+, s)}{\partial x} - \frac{\partial G(s^-, s)}{\partial x} = 1.$$

Thus we solve these 4 equations for the 4 unknowns $c_1(s), \ldots, c_4(s)$.

Exercise **22.** Determine c_1, c_2, c_3, c_4 and then write down the Green's function using this method.

We next look at adapting the above four properties for construction of $G(x, s)$ for a more general second order BVP.

2.4.2 *Another approach using the properties*

We remark here that if $y(x) \equiv 0$ is the only solution of a linear homogeneous differential equation satisfying linear homogeneous BC's, then a Green's function exists for that BVP.

Let's now look at the linear second order equation

$$y'' + p(x)y' + q(x)y = 0 \tag{2.13}$$

with BC's

$$\alpha_1 y(a) + \beta_1 y'(a) = 0, \ \alpha_1^2 + \beta_1^2 \neq 0, \tag{2.14}$$

$$\alpha_2 y(b) + \beta_2 y'(b) = 0, \ \alpha_2^2 + \beta_2^2 \neq 0. \tag{2.15}$$

Now assume that BVP (2.13), (2.14), (2.15) has only the trivial solution $y(x) \equiv 0$, (Thus the corresponding nonhomogeneous BVP has a unique solution.)

Let us list the desired properties of the Green's function and see if we can construct a Green's function such that $y(x) = \int_a^b G(x,s)f(s)\,ds$ is a solution of

$$\begin{cases} y'' + p(x)y' + q(x)y = f(x), \\ \alpha_1 y(a) + \beta_1 y'(a) = 0, \\ \alpha_2 y(b) + \beta_2 y'(b) = 0. \end{cases}$$

1) For each fixed $a < s < b$, as a function of x, $G(x,s)$ is a solution of (2.13) on $[a,s]$ and on $[s,b]$.
2) For each fixed $a < s < b$, as a function of x, $G(x,s)$ satisfies BC (2.14) on $[a,s]$, $\alpha_1 G(a,s) + \beta_1 G'(a,s) = 0$, and BC (2.15) on $[s,b]$, $\alpha_2 G(b,s) + \beta_2 G'(b,s) = 0$.
3) $G(x,s)$ is continuous on $[a,b] \times [a,b]$ and $\frac{\partial G(x,s)}{\partial x}$ is continuous on $a \leq x \leq s \leq b$ and on $a \leq s \leq x \leq b$.
4) $\frac{\partial G(s^+,s)}{\partial x} - \frac{\partial G(s^-,s)}{\partial x} = 1.$

Construction of the Green's function:

Let $u(x)$ be a nontrivial solution of (2.13) satisfying BC (2.14), e.g., let $u(x)$ be a solution of the IVP for (2.13) with IC's

$$u(a) = \beta_1, \ u'(a) = \alpha_1.$$

Then $\alpha_1 u(a) + \beta_1 u'(a) = 0$.

Exercise **23.** If $u(x)$ and $z(x)$ are solutions of (2.13), both satisfying (2.14), then $u(x)$ and $z(x)$ are linearly dependent.

Similarly, let $v(x)$ be a nontrivial solution of (2.13) satisfying BC (2.15).

Claim. $u(x)$ and $v(x)$ are linearly independent solutions of (2.13).

Proof. If not, then there exists k such that $u(x) = kv(x)$. So, $u(x)$ satisfies Eq. (2.13) and conditions (2.14), (2.15). Hence $u(x) \equiv 0$, a contradiction. Thus, u and v are linearly independent. □

Now fix $a < s < b$ and let $G(x,s)$ satisfy the properties. Then by properties 1) and 2) and the above exercise

$$G(x,s) = \begin{cases} c_1(s)u(x), & a \le x \le s \le b, \\ c_2(s)v(x), & a \le s \le x \le b. \end{cases}$$

By property 3) and property 4),

$$c_1(s)u(s) - c_2(s)v(s) = 0, \tag{2.16}$$

$$-c_1(s)u'(s) + c_2(s)v'(s) = 1. \tag{2.17}$$

Solving for c_1, c_2, we have

$$c_1(s) = \frac{\begin{vmatrix} 0 & -v \\ 1 & v' \end{vmatrix}}{\begin{vmatrix} u & -v \\ -u' & v' \end{vmatrix}} = \frac{v(s)}{W(s;u,v)},$$

where $W(s;u,v)$ is the *Wronskian* of u and v. Similarly

$$c_2(s) = \frac{u(s)}{W(s;u,v)}.$$

Therefore,

$$G(x,s) = \begin{cases} \frac{v(s)u(x)}{W(s;u,v)}, & a \le x \le s \le b, \\ \frac{u(s)v(x)}{W(s;u,v)}, & a \le s \le x \le b. \end{cases} \tag{2.18}$$

As in Theorem 2.3, it can again be shown that properties 1)–4) characterize the Green's function, (i.e., $G(x,s)$ is unique).

Note: From (2.16) and (2.17), it follows that

$$u(x,s) = c_2(s)v(x) - c_1(s)u(x),$$

where $u(x,s)$ is the Cauchy function for (2.13) since $u(s,s) = 0$ and $u'(s,s) = 1$.

Example 2.2. Find the Green's function for

$$\begin{cases} y'' = 0, \\ y(a) = 0, \quad y(b) = 0, \end{cases}$$

using the above method.

Here $p = q = 0$, $\alpha_1 = \alpha_2 = 1$, $\beta_1 = \beta_2 = 0$. Take $u(x) = x - a$, $v(x) = x - b$, so that

$$W(s; u, v) = \begin{vmatrix} s-a & s-b \\ 1 & 1 \end{vmatrix} = b - a.$$

Therefore

$$G(x, s) = \begin{cases} \frac{(s-b)(x-a)}{b-a}, & a \le x \le s \le b, \\ \frac{(x-b)(s-a)}{b-a}, & a \le s \le x \le b. \end{cases}$$

Let us show that the Green's function for (2.13) is such that $\int_a^b G(x,s)f(s)\,ds$ is the solution of the BVP

$$\begin{cases} y'' + p(x)y' + q(x)y = f(x), & (2.19) \\ \alpha_1 y(a) + \beta_1 y'(a) = 0, & (2.14) \\ \alpha_2 y(b) + \beta_2 y'(b) = 0. & (2.15) \end{cases}$$

Let

$$\begin{aligned} y(x) &= \int_a^b G(x,s)f(s)\,ds \\ &= v(x)\int_a^x \frac{u(s)f(s)}{W(s;u,v)}\,ds + u(x)\int_x^b \frac{v(s)f(s)}{W(s;u,v)}\,ds. \end{aligned}$$

Then

$$y'(x) = v'(x)\int_a^x \frac{u(s)f(s)}{W(s;u,v)}\,ds + u'(x)\int_x^b \frac{v(s)f(s)}{W(s;u,v)}\,ds$$

and

$$\begin{aligned} y''(x) &= v''(x)\int_a^x \frac{u(s)f(s)}{W(s;u,v)}\,ds + u''(x)\int_x^b \frac{v(s)f(s)}{W(s;u,v)}\,ds \\ &\quad + \left[\frac{v'(x)u(x) - u'(x)v(x)}{W(x;u,v)}\right] f(s) \\ &= v''(x)\int_a^x \frac{u(s)f(s)}{W(s;u,v)}\,ds + u''(x)\int_x^b \frac{v(s)f(s)}{W(s;u,v)}\,ds + f(x). \end{aligned}$$

It follows easily that $y(x)$ is a solution of (2.19). Moreover, $y(a) = u(a)\int_a^b \frac{v(s)f(s)}{W(s;u,v)}$, (a constant multiple of $u(a)$), and $y'(a) = u'(a)\int_a^b \frac{v(s)f(s)}{W(s;u,v)}$, (the same constant multiple of $u'(a)$). Thus $\alpha_1 y(a) + \beta_1 y'(a) = 0$. Similarly, y satisfies BC (2.15); therefore, $y(x) = \int_a^b G(x,s)f(s)\,ds$ is the unique solution of BVP (2.19), (2.14), (2.15).

Exercise **24.** Show that $y(x) = z(x) + \int_a^b G(x,s)f(s)\,ds$ is the unique solution of the BVP

$$\begin{cases} y'' + p(x)y' + q(x)y = f(x), \\ \alpha_1 y(a) + \beta_1 y'(a) = A, \\ \alpha_2 y(b) + \beta_2 y'(b) = B, \end{cases}$$

where $z(x)$ is the solution of

$$\begin{cases} y'' + p(x)y' + q(x)y = 0, \\ \alpha_1 y(a) + \beta_1 y'(a) = A, \\ \alpha_2 y(b) + \beta_2 y'(b) = B. \end{cases}$$

Thus, we now have an expression for a rather general class of linear BVP's for second order equations.

2.4.3 *Green's functions for general linear homogeneous BVP's*

In the subsection we are interested in Green's functions for

$$\begin{cases} L[y] = y^{(n)} + \sum_{i=0}^{n-1} a_i(x) y^{(i)} = 0, & (2.20) \\ \varphi_k(y) = 0, \ 1 \le k \le n, & (2.21) \end{cases}$$

where φ_k, $k = 1, 2, \ldots n$, are linear functionals on $C^{(n-1)}[a,b]$.

Assume that the only solution of (2.20), (2.21) is $y \equiv 0$. Then there is a Green's function associated with this BVP.

Let $u(x,s)$ be the Cauchy function for $L[y] = 0$, and for a fixed $s \in [a,b]$, define

$$\psi_k(s) \equiv \varphi_k[u(\cdot, s)], \ 1 \le k \le n, \quad \text{(i.e., apply } \varphi_k \text{ to } u(x,s)).$$

Let $y_k(x)$ denote the solution of $L[y] = 0$ satisfying the BC's $\varphi_k(y_k) = 1$, $\varphi_i(y_k) = 0$, $1 \le i \le n$, $i \ne k$.

Exercise **25.** Show that the $y_k(x)$, $k = 1, 2, \ldots, n$, are linearly independent solutions of $L[y] = 0$. Show also that $u(x,s) = \sum_{k=1}^{n} \psi_k(s) y_k(x)$.

In constructing our Green's function, let's assume that each φ_j is of the form $\varphi_j(y) = \sum_{i=0}^{n-1} \alpha_{ji} y^{(i)}(x_j)$, where $x_j \in [a,b]$, $\alpha_{ji} \in \mathbb{R}$. Moreover, assume that the φ_j's are numbered so that $j < k$. Hence, $x_j \le x_k$. Also, we assume that $x_1 = a$ and $x_n = b$, otherwise, we can simply shorten our interval.

Now some of our functionals may be evaluated at the same point, e.g., we might have $\varphi_1(y) = y(a)$, $\varphi_2(y) = y'(a)$. We want to identify these together:

Let the integers, $1 \le j_1 < j_2 < \cdots < j_r = n$ be such that for

$$\begin{array}{ll} 1 \le i \le j_1, & x_i = a, \\ j_{r-1}+1 \le i \le n, & x_i = b, \\ j_{p-1}+1 \le i_1, i_2 \le j_p, & x_{i_1} = x_{i_2}, \text{ for } 2 \le p \le r-1, \\ & x_{j_p} < x_{j_{p+1}}, \text{ for } 1 \le p \le r-1. \end{array}$$

We now define the Green's function:

For $x_{j_i} \le s < x_{j_{i+1}}$, $1 \le i \le r-2$,
(A)

$$G(x,s) = \begin{cases} -\sum_{k=j_i+1}^{n} \psi_k(s) y_k(x), & a \le x \le s \le b, \\ +\sum_{k=1}^{j_i} \psi_k(s) y_k(x), & a \le s \le x \le b. \end{cases}$$

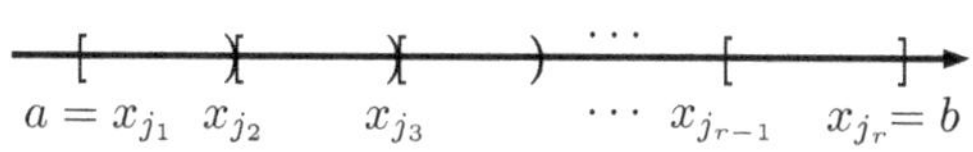

Fig. 2.3 Partition of the interval $[a, b]$.

For $x_{j_{r-1}} \le s \le x_{j_r} = b$,
(B)

$$G(x,s) = \begin{cases} -\sum_{k=j_{r-1}+1}^{n} \psi_k(s) y_k(x), & a \le x \le s \le b, \\ +\sum_{k=1}^{j_{r-1}} \psi_k(s) y_k(x), & a \le s \le x \le b. \end{cases}$$

Or if we let $x_{j_i} \le x < x_{j_{i+1}}$, then

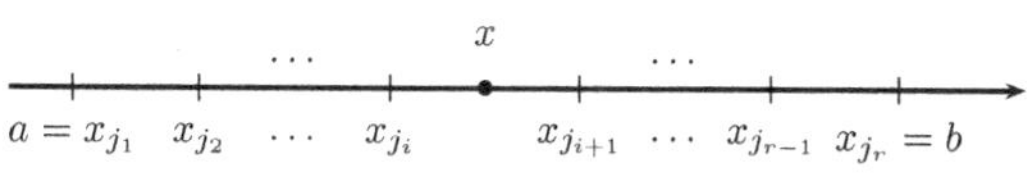

Fig. 2.4 $x_{j_i} \le x < x_{j_{i+1}}$.

(C)

$$G(x,s) = \begin{cases} +\sum_{k=1}^{j_p} \psi_k(s) y_k(x), & x_{j_p} \le s < x_{j_{p+1}},\ s \le x,\ 1 \le p \le i, \\ -\sum_{k=j_p+1}^{n} \psi_k(s) y_k(x), & x_{j_p} \le s < x_{j_{p+1}},\ s \ge x,\ i \le p \le r-1. \end{cases}$$

Now properties of the Green's function labeled (A) and (B) are:

1). For each fixed $a < s < b$, as a function of x, $G(x,s)$ is a solution of (2.20) on $[a,s]$ and $[s,b]$.
2). For each fixed $a < s < b$, as a function of x, $G(x,s)$ satisfies the BC's (2.21) on $[a,s]$ and on $[s,b]$.
3). As a function of x, $G(x,s) \in C^{(n-1)}$ on $[a,s]$ and on $[s,b]$, and $G(x,s) \in C^{(n-2)}([a,b] \times [a,b])$; $n-2$ continuous derivatives in (x,s) on the square.
4).

$$\frac{\partial^{(k)} G(s^+,s)}{\partial x^{(k)}} - \frac{\partial^{(k)} G(s^-,s)}{\partial x^{(k)}} = u^{(k)}(s,s) = \begin{cases} 0, & 0 \le k \le n-2, \\ 1, & k = n-1. \end{cases}$$

As before, these properties completely characterize $G(x,s)$ in the sense that if $G_1(x,s)$ and $G_2(x,s)$ both satisfy 1)–4), then $G_1(x,s) \equiv G_2(x,s)$ on $[a,b] \times [a,b]$.

We claim that $y(x) = \int_a^b G(x,s) f(s)\, ds$ is the solution of the nonhomogeneous BVP

$$\begin{cases} L[y] = f(x), & (2.22) \\ \varphi_k(y) = 0,\ \ 1 \le k \le n. & (2.21) \end{cases}$$

Proof. To begin with, the solution is of the form

$$y(x) = \sum_{k=1}^{n} c_k y_k(x) + \underbrace{\int_a^x u(x,s) f(s)\, ds}_{h(x)}.$$

Applying φ_j to $y(x)$, we have

$$\varphi_j(y) = c_j + \varphi_j(h), \quad \text{since } \varphi_j(y_k) = \delta_{jk}.$$

In order for $y(x)$ to be a solution, we want to satisfy the BC's; that is,

$$\varphi_j(y) = c_j + \varphi_j(h) = 0, \quad \varphi_j(h) = \sum_{i=0}^{n-1} \varphi_{ji} h^{(i)}(x_j).$$

Now we know

$$h'(x) = \int_a^x u'(x,s)f(s)\,ds,$$
$$h''(x) = \int_a^x u''(x,s)f(s)\,ds,$$
$$\vdots$$
$$h^{(n-1)}(x) = \int_a^x u^{(n-1)}(x,s)f(s)\,ds,$$

and so

$$\begin{aligned}\varphi_j(h) &= \int_a^{x_j} \sum_{i=0}^{n-1} \alpha_{ji} u^{(i)}(x_j,s)f(s)\,ds = \int_a^{x_j} \varphi_j[u(x,s)]f(s)\,ds \\ &= \int_a^{x_j} \psi_j(s)f(s)\,ds = -c_j.\end{aligned}$$

Hence

$$y(x) = \sum_{k=1}^{n} \int_a^{x_k} -\psi_k(s)y_k(x)f(s)\,ds + \int_a^x u(x,s)f(s)\,ds. \tag{2.23}$$

Now for $1 \le k \le j_1$, $x_k = a$. The right side of (2.23) in the first interval takes the form (i.e. for s ranging over $[a, x_{j_2}]$.)

$$\begin{aligned}&\int_a^{x_{j_2}} \left\{ u(x,s) - \sum_{k=j_1+1}^{n} \psi_k(s)y_k(x) \right\} f(s)\,ds \\ &= \int_a^{x_{j_2}} \underbrace{\sum_{k=1}^{j_1} \psi_k(s)y_k(x)}\, f(s)\,ds.\end{aligned}$$

This is the Green's function on $[a, x_{j_2}]$;
i.e., $a \le s \le x$ and say $x_{j_2} \le x$.

Continue this in the next interval, and so on, and it eventually follows that $\int_a^b G(x,s)f(s)\,ds$ is the solution of (2.22), (2.21). □

Example 2.3. 1). Using the method outlined above, find the Green's function for

$$\begin{cases} y'' = 0, \\ y(a) = y(b) = 0. \end{cases}$$

A Green's function exists, since the solution of the BVP is $y(x) \equiv 0$. First, since we are interested in conditions $y(a)$ and $y(b)$, we have

$$\varphi_1(y) = y(a), \ \varphi_2(y) = y(b).$$

Also $u(x,s) = x - s$. Hence $\psi_1(s) = \varphi_1[u(x,s)] = u(a,s) = a - s$ and $\psi_2(s) = \varphi_2[u(x,s)] = u(b,s) = b - s$.

Thus, if $y_1(x)$, $y_2(x)$ are solutions of $y'' = 0$ satisfying

$$\begin{cases} \varphi_1(y_1) = 1, \\ \varphi_2(y_1) = 0, \end{cases} \quad \text{and} \quad \begin{cases} \varphi_1(y_2) = 0, \\ \varphi_2(y_2) = 1, \end{cases}$$

and since solutions are $c_1 + c_2 x$, it follows that $y_1(x) = \frac{x-b}{a-b}$ and $y_2(x) = \frac{x-a}{b-a}$. Thus for fixed $a < s < b$,

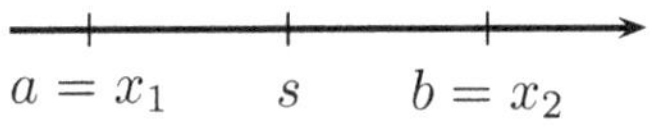

Fig. 2.5 $s \in [a,b]$.

$$G(x,s) = \begin{cases} -\psi_2(s) y_2(x), & a \le x \le s \le b, \\ \psi_1(s) y_1(x), & a \le s \le x \le b, \end{cases}$$

$$= \begin{cases} -\frac{(b-s)(x-a)}{b-a}, & a \le x \le s \le b, \\ +\frac{(a-s)(x-b)}{b-a}, & a \le s \le x \le b. \end{cases}$$

2). Find the Green's function, using this latter method, for

$$\begin{cases} y''' = 0, \\ y(a) = y'(a) = y(b) = 0. \end{cases}$$

The disconjugacy of $y''' = 0$ implies a Green's function exists. Moreover, solutions are of the form: $y(x) = c_1 + c_2(x-a) + c_3(x-a)^2$. The BC's indicate that

$$\varphi_1(y) = y(a), \ \varphi_2(y) = y'(a), \ \text{ and } \varphi_3(y) = y(b).$$

The Cauchy function is given by $u(x,s) = \frac{1}{2}(x-s)^2$, so that

$$\psi_1(s) = \varphi_1[u(x,s)] = u(a,s) = \frac{1}{2}(a-s)^2,$$
$$\psi_2(s) = \varphi_2[u(x,s)] = u'(a,s) = (a-s),$$
$$\psi_3(s) = \varphi_3[u(x,s)] = u(b,s) = \frac{1}{2}(b-s)^2.$$

If y_1, y_2, y_3 are the linearly independent, and since each is also of the form $c_1 + c_2(x-a) + c_3(x-a)^2$, then

$$\varphi_1(y_1) = 1 = y_1(a) = c_1,$$
$$\varphi_2(y_1) = 0 = y_1'(a) = c_2,$$
$$\varphi_3(y_1) = 0 = y_1(b) = 1 + c_3(b-a)^2, \implies c_3 = -\frac{1}{(b-a)^2}.$$

So

$$y_1 = 1 - \frac{(x-a)^2}{(b-a)^2}.$$

Also,

$$\varphi_1(y_2) = 0 = y_2(a) = c_1,$$
$$\varphi_2(y_2) = 1 = y_2'(a) = c_2,$$
$$\varphi_3(y_2) = 0 = y_2(b) = (b-a) + c_3(b-a)^2, \implies c_3 = -\frac{1}{b-a}.$$

So

$$y_2 = x - a - \frac{(x-a)^2}{b-a}.$$

Similarly,

$$\varphi_1(y_3) = 0 = y_3(a) = c_1,$$
$$\varphi_2(y_3) = 0 = y_3'(a) = c_2,$$
$$\varphi_3(y_3) = 1 = y_3(b) = c_3(b-a)^2 \implies c_3 = \frac{1}{(b-a)^2}.$$

So

$$y_3 = \frac{(x-a)^2}{(b-a)^2}.$$

Hence, for fixed $a < s < b$,

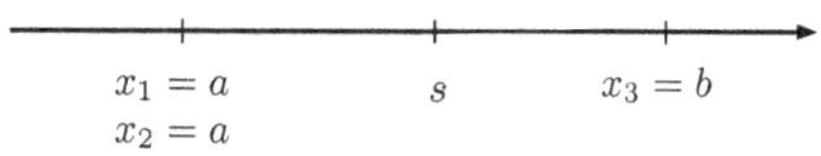

Fig. 2.6 $s \in [a, b]$.

$$G(x,s) = \begin{cases} -\psi_3(s)y_3(x), & a \le x \le s \le b, \\ \psi_1(s)y_1(x) + \psi_2(s)y_2(x), & a \le s \le x \le b, \end{cases}$$

$$= \begin{cases} -\frac{\frac{1}{2}(b-s)^2(x-a)^2}{(b-a)^2}, & a \le x \le s \le b, \\ \frac{1}{2}(a-s)^2\left[1 - \frac{(x-a)^2}{(b-a)^2}\right] & \\ +(a-s)\left[(x-a) - \frac{(x-a)^2}{b-a}\right], & a \le s \le x \le b. \end{cases}$$

Thus, we have illustrated three ways in which to find the Green's function. In particular, for the above problem

$$\begin{cases} y''' = 0, \\ y(a) = y'(a) = y(b) = 0, \end{cases}$$

we can

a) Proceed as above;
b) Use the 4 properties of the Green's function;
c) Use the form of the solution

$$y(x) = c_1 + c_2(x-a) + c_3(x-a)^2 + \int_a^x u(x,s)f(s)\,ds$$

and determine $c_1,\ c_2,\ c_3$ to satisfy the BC's.

Exercise **26.** 1). Determine the Green's function for the BVP

$$\begin{cases} y''' = 0, \\ y(a) = y(b) = y(c) = 0. \end{cases}$$

Then, $j_1 = 1,\ j_2 = 2,\ j_3 = 3$. You should obtain two expressions: one for

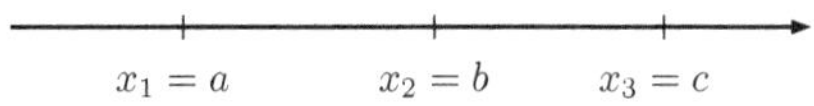

Fig. 2.7 $a < b < c$.

$a < s < b$ and one for $b < s < c$.

2). For the Green's function $G(x,s)$ of the BVP $y'' = 0$, $y(a) = y(b) = 0$,

$$G(x,s) = \begin{cases} -\frac{(b-s)(x-a)}{b-a}, & a \le x \le s \le b, \\ \frac{(a-s)(x-b)}{a-b}, & a \le s \le x \le b, \end{cases}$$

show that

$$\max_{a\le x\le b} \int_a^b |G(x,s)|\,ds = \frac{(b-a)^2}{8}, \quad \text{and} \quad \max_{a\le x\le b} \int_a^b \left|\frac{\partial G(x,s)}{\partial x}\right| ds = \frac{b-a}{2}.$$

3). Let $G(x,s)$ be the Green's function for $y''' = 0$, $y(a) = y'(a) = y(b) = 0$. Show that

$$\max_{a\le x\le b} \int_a^b |G(x,s)|\,ds = \frac{2(b-a)^3}{81},$$

$$\max_{a\le x\le b} \int_a^b \left|\frac{\partial G(x,s)}{\partial x}\right| ds \le \frac{1}{6}(b-a)^2,$$

$$\max_{a\le x\le b} \int_a^b \left|\frac{\partial^2 G(x,s)}{\partial x^2}\right| ds = \frac{2}{3}(b-a).$$

Bibliography

Agarwal R. P. (1986). *Boundary Value Problems for Higher Order Differential Equations* (World Scientific Publishing Co., Inc., Teaneck, NJ).

Beesack P. R. (1962). On the Green's function of an N-point boundary value problem, *Pacific J. Math.* **12**, pp. 801–812.

Cabada A. (2014). *Green's Functions in the Theory of Ordinary Differential Equations* (Springer Briefs in Mathematics, Singapore).

Coddington E. A. and Levinson N. (1955). *Theory of Ordinary Differential Equations* (McGraw-Hill Book Company, Inc., New York).

Das K. M. and Vatsala A. S. (1973). On the Green's function of an N-point boundary value problem, *Trans. Amer. Math. Soc.* **182**, pp. 469–480.

Hartman P. (1964). *Ordinary Differential Equations* (Wiley, New York).

Jackson L. K. (1977). Boundary value problems for ordinary differential equations, *Studies in Ordinary Differential Equations*, Stud. in Math., Vol. 14 (Editor: J. K. Hale), Math. Association of America, Washington, D.C., pp. 93–127.

Chapter 3

Existence of Solutions I

For conjugate 2-point boundary value problems for second order nonlinear ODE's, local existence results are obtained by fixed point theorems and *a priori* bounds methods. Then a "uniqueness implies existence" result is proven, followed by an application for Lipschitz differential equations. The existence of solutions for a higher order BVP is discussed in the last section.

3.1 Local Existence Theorems for Solutions of BVP's for the Nonlinear Equation, $y'' = f(x, y, y')$

The first result establishes the equivalence of solutions of BVP's associated with $y'' = f(x, y, y')$ with solutions of an integral equation.

Theorem 3.1. *Assume that $f(x, y_1, y_2) : [a, b] \times \mathbb{R}^2 \to \mathbb{R}$ is continuous. Then a function $y(x)$ is of class $C^{(2)}[a, b]$ and is a solution of the BVP*

$$\begin{cases} y'' = f(x, y, y'), \\ y(a) = A, \; y(b) = B \end{cases}$$

if and only if $y(x)$ is of class $C^{(1)}[a, b]$ and is a solution of the integral equation,

$$y(x) = w(x) + \int_a^b G(x, s) f(s, y(s), y'(s))\, ds,$$

on $[a, b]$ where $G(x, s)$ is the Green's function for $y'' = 0$, $y(a) = y(b) = 0$, and $w(x)$ is the solution of $y'' = 0$, $y(a) = A$, $y(b) = B$.
[Note: Solutions of $y'' = 0$ are of the form $c_1 + c_2 x$. Thus $w(x)$ is simply the line determined by (a, A) and (b, B). So, $w(x) = \frac{(x-a)B}{b-a} + \frac{(b-x)A}{b-a}$.]

Proof. First assume that $y(x) \in C^{(2)}[a, b]$ and is a solution of the stated BVP. Then $y(x) \in C^{(1)}[a, b]$ and $y''(x) = f(x, y(x), y'(x))$ and satisfies $y(a) = A$, $y(b) = B$. Define $h(x) := f(x, y(x), y'(x)) \in C[a, b]$.

Thus, we are considering a solution of

$$\begin{cases} y'' = h(x), \\ y(a) = A, \ y(b) = B. \end{cases}$$

From Exercise 24 of Section 2.4.2, it follows that

$$\begin{aligned} y(x) &= w(x) + \int_a^b G(x,s)h(s)\,ds \\ &= w(x) + \int_a^b G(x,s)f(s,y(s)),y'(s))\,ds. \end{aligned}$$

Conversely, let's suppose now that $y(x) \in C^{(1)}[a,b]$ and satisfies the specified integral equation on $[a,b]$. Now using properties of $w(x)$ and the role of $G(x,s)$ in solutions of BVP's in Exercise 24, it follows that $y''(x) = 0 + f(x,y(x),y'(x))$ on $[a,b]$. So, $y \in C^{(2)}[a,b]$, and moreover, from the properties of $w(x)$ and $G(x,s)$, we also have $y(a) = A + 0 = A$, $y(b) = B + 0 = B$. Therefore $y(x)$ is a solution of the BVP. □

We next impose restrictions on f and on the length of the interval $[a,b]$ which are sufficient for the existence of unique solutions of $y'' = f(x,y,y')$, $y(a) = A$, $y(b) = B$. Use will be made of the bounds found for $\int_a^b |G(x,s)|\,ds$ and $\int_a^b \left|\frac{\partial G(x,s)}{\partial x}\right| ds$ in Exercise 26 in Section 2.4.3, along with the fixed point theorem knows as the Contraction Mapping Principle.

Theorem 3.2 (Contraction Mapping Principle). *Let $\langle \mathcal{M}, d\rangle$ be a complete metric space and let $T : \mathcal{M} \to \mathcal{M}$ be such that there exists $0 \le \alpha < 1$, with $d(T(x),T(y)) \le \alpha d(x,y)$, for all $x,y \in \mathcal{M}$. Then T has a unique fixed point in $\mathcal{M}$ (i.e., there exists a unique $x_0 \in M$ such that $T(x_0) = x_0$). Moreover, if $x_1 \in \mathcal{M}$, the sequence $\{x_1, T(x_1), T[T(x_1)], \ldots\}$ converges to the unique fixed point.*

Exercise 27. Let $\mathcal{M} = C^{(1)}[a,b]$. For $h \in \mathcal{M}$, define

$$|h|_\infty = \max_{a\le x\le b} |h(x)| \ \text{ and } \ |h'|_\infty = \max_{a \le x \le b} |h'(x)|.$$

Let $K > 0$, $L > 0$ be fixed constants, then define $\|h\| = K|h|_\infty + L|h'|_\infty$. Show that $\|\cdot\|$ is a norm on $\mathcal{M}$, and $\mathcal{M}$ is a complete metric space with respect to the metric $d(h,g) = \|h - g\|$.

Our first local existence result puts a restriction on the interval length $b - a$.

Theorem 3.3. *Let K, $L \geq 0$ and let $f(x, y_1, y_2) : [a, b] \times \mathbb{R}^2 \to \mathbb{R}$ be continuous and satisfy a Lipschitz condition*

$$|f(x, y_1, y_2) - f(x, z_1, z_2)| \leq K|y_1 - z_1| + L|y_2 - z_2|$$

on $[a, b] \times \mathbb{R}^2$. If

$$\frac{K(b-a)^2}{8} + \frac{L(b-a)}{2} < 1,$$

then, for any $A, B \in \mathbb{R}$, the BVP $y'' = f(x, y, y')$, $y(a) = A$, $y(b) = B$ has a unique solution.

Proof. First define a mapping $T : C^{(1)}[a, b] \to C^{(1)}[a, b]$ by

$$(Th)(x) = w(x) + \int_a^b G(x, s) f(s, h(s), h'(s))\, ds,$$

for $a \leq x \leq b$, $h \in C^{(1)}[a, b]$, where $G(x, s)$ is the Green's function for $y'' = 0$, $y(a) = y(b) = 0$, and $w(x)$ is the solution of $y'' = 0$, $y(a) = A$, $y(b) = B$.

We shall show that T is a contraction mapping with respect to the metric, $d(h, g) = \|h - g\| = K|h - g|_\infty + L|h' - g'|_\infty$.

So, let $h, g \in C^{(1)}[a, b]$. Then, by Exercise 26, for $a \leq x \leq b$,

$$\begin{aligned}
|(Th)(x) - (Tg)(x)| &\leq \int_a^b |G(x, s)||f(s, h(s), h'(s)) - f(s, g(s), g'(s))|\, ds \\
&\leq \int_a^b |G(x, s)|\big[K|h(s) - g(s)| + L|h'(s) - g'(s)|\big]\, ds \\
&\leq \int_a^b |G(x, s)|\big[K|h - g|_\infty + L|h' - g'|_\infty\big]\, ds \\
&= \int_a^b |G(x, s)| \underbrace{\|h - g\|}_{\text{constant}}\, ds \\
&\leq \frac{(b-a)^2}{8}\|h - g\|.
\end{aligned}$$

Similarly, by Exercise 26, for $a \leq x \leq b$,

$$|(Th)'(x) - (Tg)'(x)| \leq \|h - g\| \int_a^b \left|\frac{\partial G(x, s)}{\partial x}\right| ds \leq \frac{(b-a)}{2}\|h - g\|.$$

Each of these bounds are independent of x, hence

$$|Th - Tg|_\infty \leq \frac{(b-a)^2}{8}\|h - g\| \quad \text{and} \quad |(Th)' - (Tg)'|_\infty \leq \frac{(b-a)}{2}\|h - g\|.$$

Consequently,

$$\begin{aligned}\|Th - Tg\| &= K|Th - Tg|_\infty + L|(Th)' - (Tg)'|_\infty \\ &\le \left(\frac{K(b-a)^2}{8} + \frac{L(b-a)}{2}\right)\|h - g\|.\end{aligned}$$

Therefore, if $\frac{k(b-a)^2}{8} + \frac{L(b-a)}{2} < 1$, then T is a contraction mapping, and by Theorem 3.2, T has a unique fixed point $y(x) \in C^{(1)}[a,b]$. In particular, there is a unique $y(x) \in C^{(1)}[a,b]$ satisfying

$$y(x) = (Ty)(x) = w(x) + \int_a^b G(x,s)f(s,y(s),y'(s))\,ds,$$

and by Theorem 3.1, $y(x)$ is the unique solution of

$$\begin{cases} y'' = f(x,y,y'), \\ y(a) = A, \ y(b) = B. \end{cases}$$

This completes the proof. □

Exercise **28.** 1). Show that $f(x,y,y') = 2|y| + 5y'$ satisfies a Lipschitz condition on $\mathbb{R} \times \mathbb{R}^2$ of the type in Theorem 3.3. Determine $h > 0$ such that the BVP as in Theorem 3.3 has a unique solution for any interval $[a,b]$ with $b - a < h$.

2). Same as the previous exercise for $[a,b] \subseteq [0,5]$ and $y'' + 2xy' + x^2y = \sin x$.

In the event that $f(x,y_1,y_2)$ is not Lipschitz, local solvability can still be established under weaker hypotheses on f. For this the following fixed point theorem is often applied.

Theorem 3.4 (Schauder-Tychonoff Fixed Point Theorem). *Let K be a closed, bounded, convex subset of a Banach space, $\mathcal{B}$. Let $T : K \to K$ be continuous on K and be such that for any sequence $\{y_k\}_{k=1}^\infty \subseteq K$, the sequence $\{T(y_k)\}_{k=1}^\infty$ contains a convergent subsequence, (this is equivalent to saying $T(K)$ is "compact" or "completely continuous"). Then T has a fixed point in K, (not necessarily unique).*

In what follows, let $\mathcal{B} = C^{(1)}[x_1,x_2]$, $x_1, x_2 \in \mathbb{R}$. Let $M > 0$, $N > 0$ be given and define

$$K := \{h \in \mathcal{B} \mid |h|_\infty \le 2M,\ |h'|_\infty \le 2N\},$$

where for $h \in \mathcal{B}$, we define $\|h\| \equiv \max\{|h|_\infty, |h'|_\infty\}$.

We claim that the hypotheses of the Schauder-Tychonoff Theorem are satisfied by the definitions of $\mathcal{B}$ and K.

First, K is closed: To see this, let $\{h_n\}_{n=1}^{\infty} \subseteq K$ and let $h_0 \in \mathcal{B}$ be such that $\|h_n - h_0\| \to 0$, as $n \to \infty$. Hence, $h_n \to h_0$ and $h_n' \to h_0'$ on $[x_1,\ x_2]$. Thus $|h_0(x)| \le 2M$ and $|h_0'(x)| \le 2N$ on $[x_1,\ x_2]$, which imply $h_0 \in K$ and therefore K is closed.

Second, K is bounded: This is obvious by the definition of K.

Third, K is convex: Now, let $h, g \in K$ and consider $\lambda h + (1-\lambda)g$, $0 \le \lambda \le 1$. Then, for all $x \in [x_1, x_2]$,

$$\begin{aligned} |\lambda h(x) + (1-\lambda)g(x)| &\le \lambda|h(x)| + (1-\lambda)|g(x)| \\ &\le \lambda 2M + (1-\lambda)2M = 2M. \end{aligned}$$

Similarly $|\lambda h'(x) + (1-\lambda)g'(x)| \le 2N$, for all $x \in [x_1, x_2]$. Therefore, $\lambda h + (1-\lambda)g \in K$, and thus K is convex.

Therefore, the claim is verified; that is, the hypotheses of the Schauder-Tychonoff Theorem are satisfied.

Theorem 3.5. *Let $f(x, u_1, u_2) : [a, b] \times \mathbb{R}^2 \to \mathbb{R}$ be continuous and let $M > 0$, $N > 0$ be given. Let $Q = \max\{|f(x, u_1, u_2)| \,|\, a \le x \le b,\ |u_1| \le 2M,\ |u_2| \le 2N\}$. Then for any $[x_1, x_2] \subseteq [a, b]$, the BVP*

$$\begin{cases} y'' = f(x, y, y'), \\ y(x_1) = y_1,\ y(x_2) = y_2 \end{cases}$$

has a solution, provided $\max\{|y_1|, |y_2|\} \le M$ and $\left|\frac{y_1 - y_2}{x_1 - x_2}\right| \le N$, and $x_2 - x_1 \le \delta(M, N)$, where $\delta(M, N) = \min\left\{\sqrt{\frac{8M}{Q}}, \frac{2N}{Q}\right\}$.

Proof. First note that $\frac{y_1 - y_2}{x_1 - x_2}$ is the slope of the line determined by the points (x_1, y_1), (x_2, y_2).

Let $a \le x_1 < x_2 \le b$ and $\mathcal{B}$ be the Banach space $C^{(1)}[x_1, x_2]$, with $\|h\| = \max\{|h|_\infty, |h'|_\infty\}$, for $h \in \mathcal{B}$, and let $K := \{h \in \mathcal{B} \,|\, |h|_\infty \le 2M,\ |h'|_\infty \le 2N\}$. Define the mapping $T : \mathcal{B} \to \mathcal{B}$ via

$$(Th)(x) = w(x) + \int_{x_1}^{x_2} G(x, s) f(s, h(s), h'(s))\, ds, \quad \text{for } h \in \mathcal{B},$$

where $G(x, s)$ is the Green's function for $y'' = 0$, $y(x_1) = y(x_2) = 0$, and $w(x)$ is the solution of $y'' = 0$, $y(x_1) = y_1$, $y(x_2) = y_2$. Observe that $w(x)$ is the straight line joining the points (x_1, y_1), (x_2, y_2).

Claim 1: T maps K into K: Assume $h \in K$. From the hypotheses, $|w(x)| \le M$ and $|w'(x)| \le N$ on $[x_1, x_2]$. Consequently, on $[x_1, x_2]$,

$$|(Th)(x)| \le M + \int_{x_1}^{x_2} |G(x,s)| Q\, ds \le M + Q\left(\frac{(x_2 - x_1)^2}{8}\right),$$

$$|(Th)'(x)| \le N + \int_{x_1}^{x_2} \left|\frac{\partial G(x,s)}{\partial x}\right| Q\, ds \le N + Q\left(\frac{x_2 - x_1}{2}\right).$$

Thus, we conclude that if $x_2 - x_1 \le \delta(M, N)$, then $|Th|_\infty \le 2M$, $|(Th)'|_\infty \le 2N$. Hence, when $|x_2 - x_1| \le \delta(M, N)$, T maps K into K.

Claim 2: T is continuous on K: First note that $H = [a,b] \times [-2M, 2M] \times [-2N, 2N]$ is a compact subset of $[a,b] \times \mathbb{R}^2$. Hence, given $\varepsilon > 0$, there exists $\delta(\varepsilon) > 0$ such that for each (x_1, u_1, u_2), $(x_2, \hat{u_1}, \hat{u_2}) \in H$ with $|x_1 - x_2| < \delta$, $|u_1 - \hat{u_1}| < \delta$, $|u_2 - \hat{u_2}| < \delta$, it follows that $|f(x_1, u_1, u_2) - f(x_2, \hat{u_1}, \hat{u_2})| < \varepsilon$. Let $h, g \in K$ with $\|h - g\| < \delta$. Then $|h - g|_\infty < \delta$ and $|h' - g'|_\infty < \delta$. Arguments similar to those used in Claim 1 show that

$$|(Th)(x) - (Tg)(x)| < \frac{\varepsilon(x_2 - x_1)^2}{8},$$

$$|(Th)'(x) - (Tg)'(x)| < \frac{\varepsilon(x_2 - x_1)}{2},$$

for all $x \in [x_1, x_2]$. Hence,

$$\|Th - Tg\| < \varepsilon\left[\frac{(x_2 - x_1)^2}{8} + \frac{(x_2 - x_1)}{2}\right],$$

whenever $\|h - g\| < \delta$, which yields that T is continuous on K.

Now let $\{h_k\}_{k=1}^\infty \subseteq K$ and consider the sequence $\{Th_k\}_{k=1}^\infty \subseteq K$. (Note: Convergence of $\{h_k\}$ in norm, $\|\cdot\|$, is satisfied if and only if both $\{h_k\}$ and $\{h_k'\}$ converge uniformly with respect to the sup norm, $|\cdot|_\infty$.)

For all k, $(Th_k)''(x) = f(x, h_k(x), h_k'(x))$ implies $|(Th_k)''(x)| \le Q$ for all $x \in [x_1, x_2]$ and for all k. Consequently, for $x, y \in [x_1, x_2]$, $|(Th_k)'(x) - (Th_k)'(y)| \le Q|x - y|$, for all k and so $\{(Th_k)'\}$ is an equicontinuous family of functions. Since $|(Th_k)'|_\infty \le 2N$, for all k, it follows from the Ascoli-Arzelá Theorem that there exists a subsequence $\{(Th_{k_j})'\}$ which is uniformly convergent on $[x_1, x_2]$. By the same token, $|(Th_{k_j})'|_\infty \le 2N$ implies that $\{Th_{k_j}\}$ is an equicontinuous family of functions and this, coupled with $|Th_{k_j}| \le 2M$ and the Ascoli-Arzela Theorem, implies there exists a further subsequence $\{Th_{k_{j_i}}\}$ which is uniformly convergent on $[x_1, x_2]$. Consequently $\{Th_{k_{j_i}}\}$ converges in the norm of $\mathcal{B}$.

Thus, the hypotheses of the Schauder-Tychonoff Theorem (Theorem 3.4) are satisfied; hence T has a fixed point $y(x) \in K$ and the fixed point is a solution of the BVP. □

Corollary 3.1. *Let $f(x, u_1, u_2) : [a, b] \times \mathbb{R}^2 \to \mathbb{R}$ be continuous and let $\varphi \in C^{(1)}[a, b]$. Then there exists $\delta > 0$ such that for any x_1, x_2 with $a \le x_1 < x_2 \le b$ and $x_2 - x_1 \le \delta$, the BVP*

$$\begin{cases} y'' = f(x, y, y'), \\ y(x_1) = \varphi(x_1), \; y(x_2) = \varphi(x_2), \end{cases}$$

has a solution.

Proof. Take $\delta = \delta(M, N)$ of Theorem 3.5 with $M = \max_{a \le x \le b} |\varphi(x)|$ and $N = \max_{a \le x \le b} |\varphi'(x)|$. The conclusion then follows, for by the Mean Value Theorem,

$$\left| \frac{\varphi(x_1) - \varphi(x_2)}{x_1 - x_2} \right| = |\varphi'(x_0)| \le N, \text{ for some } x_0 \in (x_1, x_2).$$

Then Theorem 3.5 is applied. □

Corollary 3.2. *Let $f(x, u_1, u_2)$ be continuous and bounded on $[a, b] \times \mathbb{R}^2$. Then for any $A, B \in \mathbb{R}$, the BVP*

$$\begin{cases} y'' = f(x, y, y'), \\ y(a) = A, \; y(b) = B, \end{cases}$$

has a solution.

Proof. Let $Q = \sup\{|f(x, u_1, u_2)| \; : \; a \le x \le b, \; |u_1| < \infty, \; |u_2| < \infty\}$. Choose $M, \; N > 0$ such that $\max\{|A|, |B|\} \le M$, $\left|\frac{A-B}{a-b}\right| \le N$ and $\sqrt{\frac{8M}{Q}} \ge b - a$, $\frac{2N}{Q} \ge b - a$. So, $b - a \le \delta(M, N)$ of Theorem 3.5, and then Theorem 3.5 can be applied for the conclusion. □

We remark that Corollary 3.2 is a "global" existence result.

Our next results are concerned with existence where *a priori* bounds are given to solutions and their derivatives.

Definition 3.1. Let $\alpha(x), \; \beta(x) \in C[a, b]$, with $\alpha(x) \le \beta(x)$ on $[a, b]$. Then $y \in C^{(2)}(a, b) \cap C^{(1)}[a, b]$ is said to satisfy a *Nagumo condition* on $[a, b]$ with respect to the pair $\alpha(x)$ and $\beta(x)$ in case

$$\alpha(x) \le y(x) \le \beta(x) \quad \text{on } [a, b],$$

and

$$|y''(x)| \le \varphi(|y'(x)|) \quad \text{on } (a, b),$$

where $\varphi : [0, +\infty) \to (0, +\infty)$ is continuous and satisfies

$$\int_{\lambda}^{+\infty} \frac{s}{\varphi(s)} ds > \max_{a \le x \le b} \beta(x) - \min_{a \le x \le b} \alpha(x),$$

and where

$$\lambda = \max \left\{ \frac{|\alpha(b) - \beta(a)|}{b - a}, \frac{|\alpha(a) - \beta(b)|}{b - a} \right\}.$$

Lemma 3.1. *Assume that $y(x)$ satisfies a Nagumo condition on $[a, b]$ with respect to the pair $\alpha(x)$ and $\beta(x)$. Then there exists $N > 0$, depending on α, β, φ, such that $|y'(x)| \le N$ on $[a, b]$.*

Proof. From the hypotheses concerning φ, we can choose $N_0 > 0$ such that $\int_{\lambda}^{N_0} \frac{s\,ds}{\varphi(s)} > \max \beta(x) - \min \alpha(x)$.

We will show that N_0 is the "$N(\alpha, \beta, \varphi)$" in the statement of the lemma.

First by the Mean Value Theorem, $\frac{|y(b)-y(a)|}{b-a} = |y'(x_0)|$, some $x_0 \in (a, b)$. We claim that $|y'(x_0)| \le \lambda$. To see this, observe first that $\alpha(b) \le y(b) \le \beta(b)$, since $y(x)$ satisfies a Nagumo condition. Also, $-\beta(a) \le -y(a) \le -\alpha(a)$, and so $\alpha(b) - \beta(a) \le y(b) - y(a) \le \beta(b) - \alpha(a)$. It follows from the above definition of λ, that $|y'(x_0)| \le \lambda$.

For the remainder of the proof, we assume that

$$|y'(x)| > N_0 > \lambda \text{ for some } x = x_2 \in [a, b]. \tag{3.1}$$

Without loss of generality, assume $x_2 < x_0$ and that $y'(x_2)$ is negative, so that $y'(x_2) < -N_0$. By continuity of $y'(x)$, there exist points x_3 and x_1, with $x_2 < x_3 < x_1 \le x_0$, such that $y'(x_3) = -N_0$, $y'(x_1) = -\lambda$, and $-N_0 < y'(x) < -\lambda$ on (x_3, x_1). Hence,

$$\frac{-y'(x)y''}{\varphi(-y'(x))} \le \frac{-y(x)|y''(x)|}{\varphi(-y'(x))} \le -y'(x).$$

Integrating the extremes here over $[x_3, x_1]$, we have

$$\int_{x_3}^{x_1} \frac{-y'(x)y''}{\varphi(-y'(x))} dx \le \int_{x_3}^{x_1} -y'(x) dx = y(x_3) - y(x_1).$$

Let $s = -y'(x)$, so that $ds = -y''(x)dx$ and

$$\int_{-y'(x_3)}^{-y'(x_1)} \frac{-s}{\varphi(s)} ds = \int_{N_0}^{\lambda} \frac{-s}{\varphi(s)} ds = \int_{\lambda}^{N_0} \frac{s}{\varphi(s)} ds > \max \beta(x) - \min \alpha(x).$$

Hence, $y(x_3) - y(x_1) > \max \beta(x) - \min \alpha(x)$. But this is a contradiction, since $y(x)$ is trapped between $\alpha(x)$ and $\beta(x)$. Thus (3.1) is false and the conclusion of the lemma holds with $N = N_0$. □

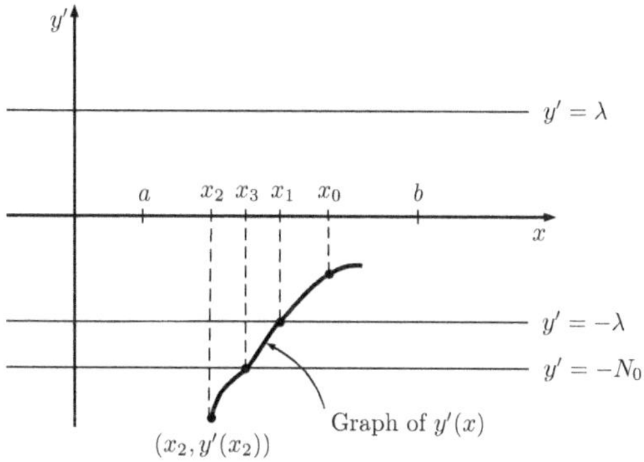

Fig. 3.1 Graph of $y'(x)$.

Definition 3.2. The differential equation $y'' = f(x, y, y')$ is said to satisfy a *Nagumo condition* on $[a, b]$ with respect to the pair $\alpha(x)$, $\beta(x) \in C[a, b]$ with $\alpha(x) \leq \beta(x)$ on $[a, b]$ in case

1). $f(x, u_1, u_2)$ is continuous on $[a, b] \times \mathbb{R}^2$, and
2). $|f(x, y, y')| \leq \varphi(|y'|)$, for all (x, y, y') with $a \leq x \leq b$ and $\alpha(x) \leq y \leq \beta(x)$, where $\varphi : [0, +\infty) \to (0, +\infty)$ is continuous, and

$$\int_{\lambda}^{+\infty} \frac{sds}{\varphi(s)} > \max_{a \leq x \leq b} \beta(x) - \min_{a \leq x \leq b} \alpha(x),$$

where $\lambda = \max\left\{\frac{|\alpha(a)-\beta(b)|}{b-a}, \frac{|\alpha(b)-\beta(a)|}{b-a}\right\}$.

Definition 3.3. The functions $\gamma(x)$ and $\psi(x)$ are defined to be, respectively, *lower and upper solutions* of $y'' = f(x, y, y')$ on $[a, b]$, in case

1). $\gamma(x),\ \psi(x) \in C^{(1)}[a, b] \cap C^{(2)}(a, b)$,
2). $\psi'' \leq f(x, \psi(x), \psi'(x))$ on (a, b), and
3). $\gamma''(x) \geq f(x, \gamma(x), \gamma'(x))$ on (a, b).

Our next local existence result makes application of an upper solution and a lower solution with which the nonlinearity also satisfies a Nagumo condition.

Theorem 3.6. *Assume that $f(x, u_1, u_2)$ is continuous on $[a, b] \times \mathbb{R}^2$, that $\alpha(x)$ and $\beta(x)$ are respectively lower and upper solutions of $y'' = f(x, y, y')$ on $[a, b]$ with $\alpha(x) \leq \beta(x)$ on $[a, b]$, and assume $y'' = f(x, y, y')$ satisfies a*

Nagumo condition on $[a,b]$ with respect to the pair $\alpha(x)$ and $\beta(x)$. Then for any A and B, with $\alpha(a) \le A \le \beta(a)$ and $\alpha(b) \le B \le \beta(b)$, the BVP

$$\begin{cases} y'' = f(x,y,y'), \\ y(a) = A, \ y(b) = B, \end{cases}$$

has a solution $y(x)$, such that $\alpha(x) \le y(x) \le \beta(x)$ on $[a,b]$ and $|y'(x)| \le N_0$ on $[a,b]$, where

$$\int_\lambda^{N_0} \frac{s}{\varphi(s)}\, ds = \max_{a\le x\le b} \beta(x) - \min_{a\le x\le b} \alpha(x).$$

Proof. Let $N_1 = \max_{a\le x\le b} |\alpha'(x)|$, $N_2 = \max_{a\le x\le b} |\beta'(x)|$, $N = \max\{N_0, N_1, N_2\} + 1$. Define

$$F^*(x,y,y') = \begin{cases} f(x,y,+N), & y' > N, \\ f(x,y,y'), & -N \le y' \le N, \\ f(x,y,-N), & y' < -N. \end{cases}$$

Then, F^* is continuous on $[a,b] \times \mathbb{R}^2$.

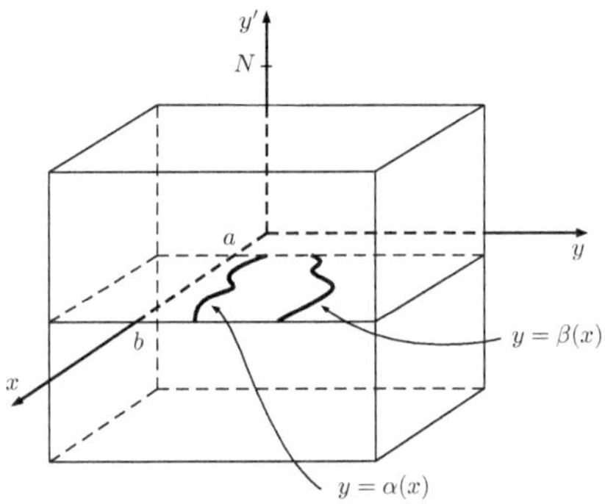

Fig. 3.2 The (x,y,y')-space.

Now define

$$F(x,y,y') = \begin{cases} F^*(x,\beta(x),y') + \frac{y-\beta(x)}{1+|y-\beta(x)|}, & y > \beta(x), \\ F^*(x,y,y'), & \alpha(x) \le y \le \beta(x), \\ F^*(x,\alpha(x),y') + \frac{y-\alpha(x)}{1+|y-\alpha(x)|}, & y < \alpha(x). \end{cases}$$

Thus defined, $F(x,y,y')$ is continuous and bounded on $[a,b] \times \mathbb{R}^2$. Hence, from Corollary 3.2, there exists a solution $y(x)$ of

$$\begin{cases} y'' = F(x,y,y'), \\ y(a) = A, \quad y(b) = B. \end{cases}$$

We claim that this solution $y(x)$ possesses all of the properties specified in the statement of the theorem.

First, we show $\alpha(x) \le y(x) \le \beta(x)$ on $[a,b]$. We prove only that $y(x) \le \beta(x)$ on $[a,b]$. Assume on the contrary that $y(x) > \beta(x)$ at some points in $[a,b]$. Since $y(a) \le A \le \beta(a)$ and $y(b) \le B \le \beta(b)$ it must be the case that $y(x) - \beta(x)$ has a positive maximum at some point $x_0 \in (a,b)$. This implies that $y(x_0) > \beta(x_0)$, $y'(x_0) = \beta'(x_0)$ and $y''(x_0) \le \beta''(x_0)$. But, since $\beta(x)$ is an upper solution,

$$\begin{aligned} y''(x_0) &= F(x_0, y(x_0), y'(x_0)) \\ &= F^*(x_0, \beta(x_0), y'(x_0)) + \frac{y(x_0) - \beta(x_0)}{1 + |y(x_0) - \beta(x_0)|} \\ &= F^*(x_0, \beta(x_0), \beta'(x_0)) + \frac{y(x_0) - \beta(x_0)}{1 + |y(x_0) - \beta(x_0)|} \\ &\ge \beta''(x_0) + \frac{y(x_0) - \beta(x_0)}{1 + |y(x_0) - \beta(x_0)|} \\ &> \beta''(x_0), \end{aligned}$$

which is contrary to $y''(x_0) \le \beta''(x_0)$. Thus $y(x) \le \beta(x)$.

Exercise **29.** Show that $\alpha(x) \le y(x)$ on $[a,b]$.

Consequently, $\alpha(x) \le y(x) \le \beta(x)$ on $[a,b]$; hence $y(x)$ is a solution of $y'' = F^*(x, y, y')$, $y(a) = A$, $y(b) = B$.

By mimicking the proof of Lemma 3.1, we shall show that $|y'(x)| \le N$ on $[a,b]$. First, since $\alpha(x) \le y(x) \le \beta(x)$ on $[a,b]$, there exists an $x_0 \in (a,b)$ by the Mean Value Theorem such that

$$\begin{aligned} \left|\frac{y(a) - y(b)}{a - b}\right| &= |y'(x_0)| \\ &\le \max\left\{\frac{|\alpha(b) - \beta(a)|}{b - a}, \frac{|\alpha(a) - \beta(b)|}{b - a}\right\} \\ &= \lambda < N_0 < N. \end{aligned}$$

Now if we assume that $|y'(x)| > N$, for some $x = x_2 \in [a,b]$, then as in Lemma 3.1, we may assume without loss of generality that $x_2 < x_0$ and that $y'(x_2)$ is negative. This implies $y'(x_2) < -N$. Then there exist points $x_2 < x_3 < x_1 \le x_0$ such that, $y'(x_3) = -N$, $y'(x_1) = -\lambda$, and $-N < y'(x) < -\lambda$ on (x_3, x_1). Then on $[x_3, x_1]$, $y(x)$ satisfies

$$y''(x) = f(x, y(x), y'(x)),$$

and by the Nagumo condition, $|y''| \le \varphi(|y'|)$ on $[x_3, x_1]$. As in the proof of Lemma 3.1,

$$y(x_3) - y(x_1) = \int_\lambda^N \frac{s}{\varphi(s)} ds > \int_\lambda^{N_0} \frac{s}{\varphi(s)} ds = \max \beta(x) - \min \alpha(x).$$

This is a contradiction, since $y(x)$ is trapped between $\alpha(x)$ and $\beta(x)$.

Therefore, $|y'(x)| \le N$ on $[a, b]$, and consequently from the manner in which F^* was defined, $y(x)$ is a solution of

$$\begin{cases} y'' = f(x, y, y'), \\ y(a) = A, \;\; y(b) = B. \end{cases}$$

Moreover, only a slight modification of the last argument above shows that $|y'(x)| \le N_0$ on $[a, b]$. □

Exercise 30. 1). Prove that for any A, B with $0 \le A, B \le 1$, the BVP

$$\begin{cases} y'' = y^3 - y, \\ y(a) = A, \;\; y(b) = B, \end{cases}$$

has a solution. Hint: Find functions α, β, φ such that Theorem 3.6 is satisfied.

2). Using Theorem 3.6 and Exercise 1 of Section 1.1, prove that there *does not exist* any function $\varphi \in C^{(1)}[0, \pi] \cap C^{(2)}(0, \pi)$ such that $\varphi''(x) \ge 1 + (\varphi'(x))^2$ on $(0, \pi)$.

Exercise 31. Assume that $f(x, u_1, u_2)$ is continuous on $[a, b] \times \mathbb{R}^2$ and that $|f(x, y, y')| \le \varphi(|y'|)$ on $[a, b] \times \mathbb{R}^2$, where $\varphi : [0, +\infty) \to (0, +\infty)$ is continuous and $\int_0^{+\infty} \frac{s\,ds}{\varphi(s)} = +\infty$. Assume, moreover, that the differential equations

$$y'' = -\varphi(|y'|), \tag{3.2}$$

$$y'' = \varphi(|y'|), \tag{3.3}$$

both have solutions on $[a, b]$.

Then prove that the BVP

$$\begin{cases} y'' = f(x, y, y'), \\ y(a) = A, \;\; y(b) = B, \end{cases}$$

has a solution, for all A, B.

Hint: If (3.2) and (3.3) have solutions, then those solutions plus constants are also solutions of the respective equations. In that manner, you can obtain $\alpha(x)$ and $\beta(x)$ and then apply Theorem 3.6.

The next result establishes that if IVP's are unique, then lower and upper solutions cannot touch tangentially.

Theorem 3.7. *Assume that $f(x, u_1, u_2)$ is continuous on $[a, b] \times \mathbb{R}^2$, that solutions of IVP's for $y'' = f(x, y, y')$ are unique, and that $\alpha(x)$ and $\beta(x)$ are respectively lower and upper solutions of the differential equation, with $\alpha(x) \le \beta(x)$ on $[a, b]$ such that there exists $x_0 \in [a, b]$ with $\alpha(x_0) = \beta(x_0)$ and $\alpha'(x_0) = \beta'(x_0)$. Then $\alpha(x) \equiv \beta(x)$ on $[a, b]$.*

Proof. Let's assume the hypotheses of the theorem, but that on the contrary $\alpha(x) \not\equiv \beta(x)$ on $[a, b]$. To consider a specific case, assume there exists $a \le x_0 < b$ such that $\alpha(x) < \beta(x)$ at some points in $(x_0, b]$. Without loss of generality, we may assume that we have a subinterval $[x_0, x_1] \subseteq [a, b]$ such that $\alpha(x_0) = \beta(x_0)$, $\alpha'(x_0) = \beta'(x_0)$, $\alpha(x) < \beta(x)$ on $(x_0, x_1]$.

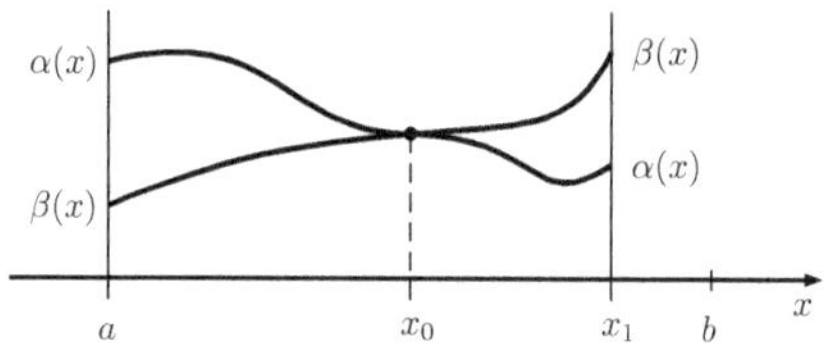

Fig. 3.3 $\alpha(x)$ and $\beta(x)$ touch tangentially.

Now choose $N = \max\{N_1, N_2\} + 1$, where $N_1 = \max_{a \le x \le b} |\alpha'(x)|$, $N_2 = \max_{a \le x \le b} |\beta'(x)|$. Then let $\alpha(x_1) < B < \beta(x_1)$. As in the proof of Theorem 3.6, the BVP

$$\begin{cases} y'' = F^*(x, y, y'), \\ y(x_0) = \alpha(x_0), \ y(x_1) = B, \end{cases}$$

has a solution $y(x)$ with $\alpha(x) \le y(x) \le \beta(x)$ on $[x_0, x_1]$, where

$$F^*(x, y, y') = \begin{cases} f(x, y, N), & y' > N, \\ f(x, y, y'), & |y'| \le N, \\ f(x, y, -N), & y' < -N. \end{cases}$$

Since $\alpha(x) \le y(x) \le \beta(x)$ on $[x_0, x_1]$, we must have $\alpha(x_0) = y(x_0) = \beta(x_0)$ and $\alpha'(x_0) = y'(x_0) = \beta'(x_0)$. So, $|y'(x_0)| < N$. Hence, by continuity there exists $x_0 < x_2 \le x_1$ such that $|y'(x) \le N$ on $[x_0, x_2]$. Assume that $[x_0, x_2]$ is the longest subinterval of $[x_0, x_1]$ on which $|y'(x) \le N$.

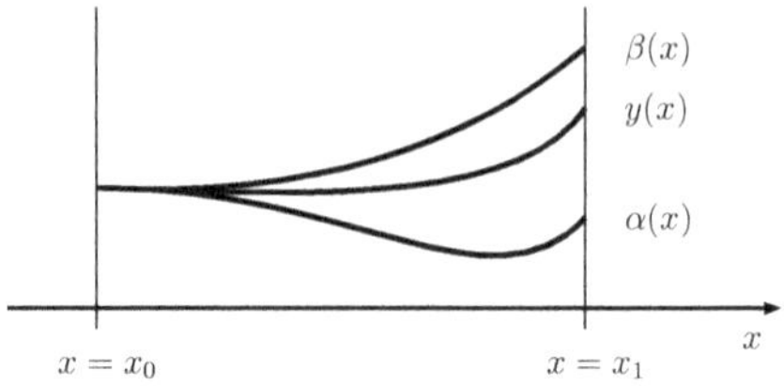

Fig. 3.4 $\alpha(x) \le y(x) \le \beta(x)$.

Then $y(x)$ is a solution of $y'' = f(x, y, y')$, and we claim moreover that $\alpha(x_2) < y(x_2) < \beta(x_2)$. There are two cases for the claim: i) If $x_2 = x_1$, then since $y(x_1) = B$ and $\alpha(x_1) < B < \beta(x_1)$, the claim is true. ii) If $x_0 < x_2 < x_1$, then $|y'(x_2)| = N$; otherwise $[x_0, x_2]$ would not be the longest subinterval. Then $\alpha(x_2) < y(x_2) < \beta(x_2)$, since $N > N_1 = \max|\alpha'(x)|$ and $N > N_2 = \max|\beta'(x)|$, and $\alpha(x) \le y(x) \le \beta(x)$ on $[x_0, x_1]$. Thus, the claim that $\alpha(x_2) < y(x_2) < \beta(x_2)$ is verified.

Let us now replace $\beta(x)$ by $y(x)$ and consider the situation we have. Certainly $y(x)$ is an upper solution on $[x_0, x_2]$ and $\alpha(x)$ is still a lower solution. Moreover, $\alpha(x_0) = y(x_0)$, $\alpha'(x_0) = y'(x_0)$, $\alpha(x) \le y(x)$ on $[x_0, x_2]$, and $\alpha(x_2) < y(x_2)$. Thus, we can repeat the above argument using $\alpha(x)$, $y(x)$, and $[x_0, x_2]$ in place of $\alpha(x)$, $\beta(x)$, and $[a, b]$.

Proceed as follows: there exists $x_0 \le \tau_0 < x_1$ such that $\alpha(\tau_0) = y(\tau_0)$, $\alpha'(\tau_0) = y'(\tau_0)$ and $\alpha(x) < y(x)$ on $(\tau_0, x_2]$. Repeating the above arguments, there exists $\tau_0 < \tau_2 \le x_2$ and a solution $z(x)$ of $y'' = f(x, y, y')$ such that, $\alpha(x) \le z(x) \le y(x)$ on $[\tau_0, \tau_2]$ and $\alpha(\tau_2) < z(\tau_2) < y(\tau_2)$. Hence, $z(\tau_0) = y(\tau_0)$ and $z'(\tau_0) = y'(\tau_0)$. By uniqueness of solutions of IVP's, $y(x) \equiv z(x)$ on $[\tau_0, \tau_2]$, however $z(\tau_2) < y(\tau_2)$; a contradiction.

Therefore the conclusion of the theorem is true. □

Corollary 3.3. *Assume $\beta(x) \in C^{(1)}[a, b] \cap C^{(2)}(a, b)$, that $\beta'' + p(x)\beta' + q(x)\beta \le 0$ on (a, b), and that $\beta(x) \ge 0$ on $[a, b]$. Then $\beta(x) \equiv 0$ on $[a, b]$ or $\beta(x) > 0$ on (a, b). If $\beta(a) = 0$ and $\beta(x) \not\equiv 0$ on (a, b), then $\beta'(a) > 0$. If $\beta(x) \not\equiv 0$ on (a, b) and $\beta(b) = 0$, then $\beta'(b) < 0$.*

Exercise 32. Prove Corollary 3.3.

3.2 Linear Second Order Inequalities

We are initially interested in this section with the "connection between functions that satisfy $\varphi'' \geq f(x, \varphi, \varphi')$ and the corresponding differential equation, $y'' = f(x, y, y')$." The simplest case, of course, is: $\varphi'' \geq 0$ and the corresponding differential equation, $y'' = 0$.

Definition 3.4. A real-valued function $\varphi(x)$ defined on an interval $I \subseteq \mathbb{R}$ is said to be *convex* on I in case, given any $[x_1, x_2] \subseteq I$ and any linear function $w(x) = \alpha x + \beta$, with $w(x_1) \geq \varphi(x_1)$, $w(x_2) \geq \varphi(x_2)$, it follows that $w(x) \geq \varphi(x)$ on $[x_1, x_2]$. (I.e. Graph$[w(x)]$ stays above the Graph$[\varphi(x)]$ on $[x_1, x_2]$.)

Some properties of convex functions:

1). A real-valued function $\varphi(x)$ is convex on $I \subseteq \mathbb{R}$ if and only if for any $[x_1, x_2] \subseteq I$ and any $0 \leq \lambda \leq 1$, we have $\varphi(\lambda x_1 + (1-\lambda)x_2) \leq \lambda\varphi(x_1) + (1-\lambda)\varphi(x_2)$.

2). Let $\varphi(x)$ be convex on I, let $[x_1, x_2] \subseteq I$, and let $w(x)$ be the linear function with $w(x_i) = \varphi(x_i)$, $i = 1, 2$. Then $\varphi(x) \geq w(x)$ on $I \setminus [x_1, x_2]$.

3). If $\varphi(x)$ is convex on I, then φ is continuous on the interior of the interval, I°. Geometric proof of 3):

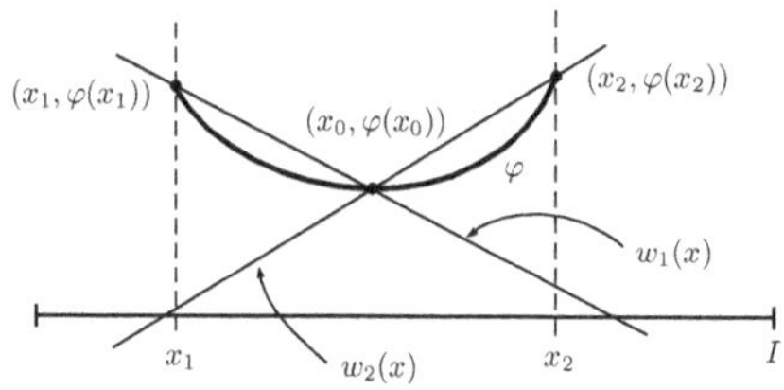

Fig. 3.5 Geometric proof of the fact that $\varphi(x) \to \varphi(x_0)$.

Show φ is continuous at x_0. Choose x_1 and x_2 in I°. By property 2), the Graph$[\varphi]$ is trapped between the lines w_1 and w_2. As one approaches x_0 from the right and the left, it is easily argued that $\varphi(x) \to \varphi(x_0)$, hence φ is continuous at x_0.

4). If φ is convex on I, then for all $x_0 \in I^\circ$, $\varphi'_+(x_0)$ and $\varphi'_-(x_0)$ exist, are finite and $\varphi'_-(x_0) \leq \varphi'_+(x_0)$.

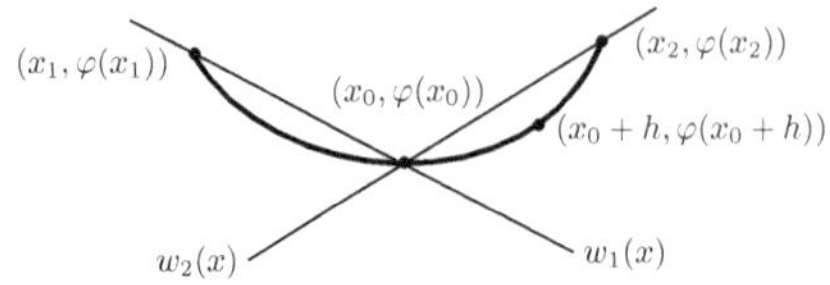

Fig. 3.6 The functions $\varphi(x)$, $w_1(x)$, and $w_2(x)$.

Proof: Again by picture. Consider the difference quotients and inequalities due to property 2),

$$\frac{w_1(x_0+h) - w_1(x_0)}{h} \le \frac{\varphi(x_0+h) - \varphi(x_0)}{h} \le \frac{w_2(x_0+h) - w_2(x_0)}{h}$$

Now, for all $h \neq 0$, the extremes on the inequalities are finite. If we let $h \to 0^+$, it may be the case that the limit of the middle quotient doesn't exist. If this is the case though, then

$$\varliminf_{h\to 0^+} \frac{\varphi(x_0+h) - \varphi(x_0)}{h} < \varlimsup_{h\to 0^+} \frac{\varphi(x_0+h) - \varphi(x_0)}{h}, \tag{3.4}$$

where

$$\varliminf_{h\to 0^+} \frac{\varphi(x_0+h) - \varphi(x_0)}{h} = \sup_{s\downarrow 0} \inf_{0<h<\delta} \frac{\varphi(x_0+h) - \varphi(x_0)}{h},$$

$$\varlimsup_{h\to 0^+} \frac{\varphi(x_0+h) - \varphi(x_0)}{h} = \inf_{\delta\downarrow 0} \sup_{0<h<\delta} \frac{\varphi(x_0+h) - \varphi(x_0)}{h}.$$

Now pick m such that

$$\varliminf_{h\to 0^+} \frac{\varphi(x_0+h) - \varphi(x_0)}{h} < m < \varlimsup_{h\to 0^+} \frac{\varphi(x_0+h) - \varphi(x_0)}{h}$$

and let $w(x) = m(x - x_0) + \varphi(x_0)$. From properties of "$\underline{\lim}$" and "$\overline{\lim}$",

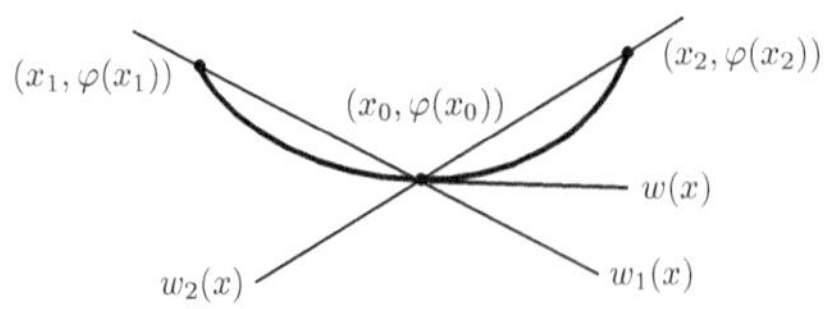

Fig. 3.7 $w(x) = m(x - x_0) + \varphi(x_0)$.

we can choose interlaced sequences $\{h_i\}$, $\{k_i\}$ such that

$$0 < \cdots < k_i < h_i < k_{i-1} < h_{i-1} < \cdots < k_3 < h_3 < k_2 < h_2 < k_1 < h_1$$

and $k_i \to 0^+$, $h_i \to 0^+$, as $i \to +\infty$, and such that

$$\lim_{h_i \to 0^+} \frac{\varphi(x_0 + h_i) - \varphi(x_0)}{h_i} = \underline{\lim}_{h \to 0^+} \frac{\varphi(x_0 + h) - \varphi(x_0)}{h},$$

$$\lim_{h_i \to 0^+} \frac{\varphi(x_0 + k_i) - \varphi(x_0)}{k_i} = \overline{\lim}_{h \to 0^+} \frac{\varphi(x_0 + h) - \varphi(x_0)}{h},$$

where the convergence is monotonic.

Since $\{h_i\}$ and $\{k_i\}$ are interlaced sequences, it follows that infinitely many quotients are such that

$$\frac{\varphi(x_0 + k_i) - \varphi(x_0)}{k_i} < m \quad \text{and} \quad \frac{\varphi(x_0 + h_i) - \varphi(x_0)}{h_i} > m.$$

But

$$\underline{\lim}_{h \to 0^+} \frac{\varphi(x_0 + h) - \varphi(x_0)}{h} < m < \overline{\lim}_{h \to 0^+} \frac{\varphi(x_0 + h) - \varphi(x_0)}{h},$$

yet at the same time we have sequences greater than m and less than m converging respectively to the values for "$\underline{\lim}$" and "$\overline{\lim}$"; a contradiction. Hence (3.4) is not a strict inequality; i.e.

$$\underline{\lim}_{h \to 0^+} \frac{\varphi(x_0 + h) - \varphi(x_0)}{h} = \overline{\lim}_{h \to 0^+} \frac{\varphi(x_0 + h) - \varphi(x_0)}{h},$$

which says that the limit (as $h \to 0^+$) exists, and so in particular, $\varphi'_+(x_0)$ exists.

Similarly $\varphi'_-(x_0)$ exists and it is not difficult to argue that $\varphi'_-(x_0) \le \varphi'_-(x_0)$.

5). If $\varphi \in C(I) \cap C^{(2)}(I^\circ)$, then φ is convex on I if and only if $\varphi'' \ge 0$ on I°.

Thus, in the simplest setting concerning $\varphi'' \ge 0$ and the differential equation, $y'' = 0$, φ behaves like a sublinear function; that is, its graph remains below some dominating line over an interval. Note, of course, that solutions of $y'' = 0$ are linear functions, $\alpha x + \beta$. With this as motivation, we make the following definition.

Definition 3.5. Assume $f(x, u_1, u_2)$ is continuous on $I \times \mathbb{R}^2$. A real-valued function $\varphi(x)$ defined on an interval I is said to be a *subfunction* on I with respect to solutions of $y'' = f(x, y, y')$, in case for any $[x_1, x_2] \subseteq I$ and any solution $y(x)$ of $y'' = f(x, y, y')$ on $[x_1, x_2]$ with $\varphi(x_1) \le y(x_1)$ and $\varphi(x_2) \le y(x_2)$, it follows that $\varphi(x) \le y(x)$ on $[x_1, x_2]$.

Before proving some of the basic results, we consider some *possible* properties which in turn will be the motivation for our first theorem in this section.

Suppose we have $\varphi \in C^{(1)}(I) \cap C^{(2)}(I^\circ)$ and suppose for the moment that φ is a subfunction on I wrt solutions of $y'' = f(x, y, y')$ if and only if $\varphi'' \geq f(x, \varphi, \varphi')$ on I°. Assume also that $y(x)$ and $z(x)$ are solutions of $y'' = f(x, y, y')$ on $[x_1, x_2] \subseteq I$ with $y(x_i) = z(x_i)$, $i = 1, 2$.

Since solutions themselves are subfunctions, one concludes that $y(x) \leq z(x)$ and $z(x) \leq y(x)$ on $[x_1, x_2]$. Hence, $y(x) \equiv z(x)$ on $[x_1, x_2]$. This says that, under these assumptions, the BVP

$$\begin{cases} y'' = f(x, y, y'), \\ y(x_1) = y_1, \ y(x_2) = y_2, \end{cases}$$

has at *most one* solution, where $[x_1, x_2] \subseteq I$ and $y_1, y_2 \in \mathbb{R}$.

Theorem 3.8. *Let $f(x, u_1, u_2)$ be continuous on $J \times \mathbb{R}^2$, where J is an interval. Assume that solutions of*

$$y'' = f(x, y, y') \tag{3.5}$$

extend to J, and assume that solutions of BVP's for (3.5) *satisfying $y(x_1) = y_1$, $y(x_2) = y_2$, $[x_1, x_2] \subseteq J$, are unique on J, when they exist. Let $I \subseteq J$ and assume that $\varphi \in C^{(1)}(I) \cap C^{(2)}(I^0\circ)$. Then φ is a subfunction on I with respect to solutions of* (3.5) *if and only if $\varphi''(x) \geq f(x, \varphi(x), \varphi'(x))$ on I°.*

Proof. (Sufficiency) This part of the proof is by contradiction. So assume that $\varphi''(x) \geq f(x, \varphi(x), \varphi'(x))$ on I°, but assume that φ is not a subfunction on I. Then there is a subinterval $[x_0, x_1] \subseteq I$ and a solution $y_0(x)$ of (3.5) such that $y_0(x_i) = \varphi(x_i)$, $i = 0, 1$, and $y_0(x) < \varphi(x)$ on (x_0, x_1).
Claim 1. On $[x_0, x_1]$, $\varphi(x)$ is a subfunction "in the small." (i.e., there exists $\delta > 0$ such that φ is a subfunction on subintervals of length at most δ).
Proof of Claim 1. Define

$$F(x, y, y') = \begin{cases} f(x, y, y'), & \text{for } y \geq \varphi(x), \\ f(x, \varphi(x), y') - (\varphi(x) - y), & \text{for } y < \varphi(x). \end{cases}$$

Then $F(x, y, y')$ is continuous on $[x_0, x_1] \times \mathbb{R}^2$ and recall that $\varphi \in C^{(1)}[x_0, x_1]$. By Corollary 3.1, there exists $\delta > 0$ such that, for any

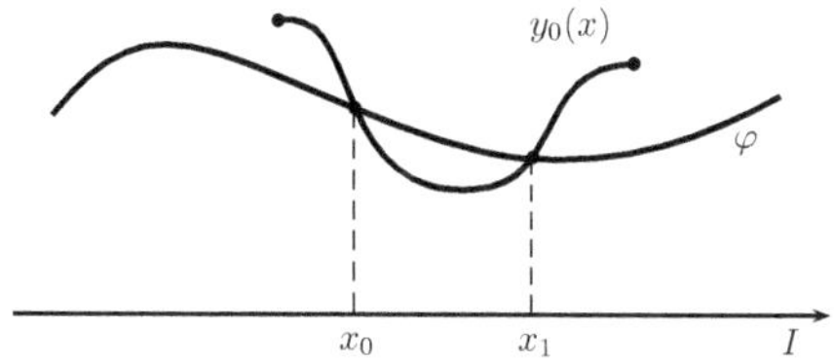

Fig. 3.8 $y_0(x_i) = \varphi(x_i)$, $i = 0, 1$, and $y_0(x) < \varphi(x)$ on (x_0, x_1).

$x_0 \leq t_1 < t_2 \leq x_1$, with $t_2 - t_1 \leq \delta$, the BVP

$$\begin{cases} y'' = F(x, y, y'), \\ y(t_1) = \varphi(t_1), \; y(t_2) = \varphi(t_2), \end{cases}$$

has a solution $y(x)$. Is it possible that $y(x) < \varphi(x)$ at some points $x \in (t_1, t_2)$? Suppose it is. Then $\varphi(x) - y(x)$ has a positive maximum at some $t_0 \in (t_1, t_2)$. Hence, at $x = t_0$, we have $\varphi(t_0) - y(t_0) > 0$, $\varphi'(t_0) - y'(t_0) = 0$, and $\varphi''(t_0) - y''(t_0) \leq 0$.

However, in contradiction to this, we have

$$\begin{aligned} &\varphi''(t_0) - y''(t_0) \\ &\quad \geq f(t_0, \varphi(t_0), \varphi''(t_0)) - F(t_0, y(t_0), y'(t_0)) \\ &\quad = f(t_0, \varphi(t_0), \varphi'(t_0)) - f(t_0, \varphi(t_0), \varphi'(t_0)) + \varphi(t_0) - y(t_0) \\ &\quad > 0. \end{aligned}$$

So, $y(x) \geq \varphi(x)$ on $[t_1, t_2]$. But from the manner in which F is defined, $y(x)$ is a solution of (3.5), $y'' = f(x, y, y')$, on $[t_1, t_2]$, $\varphi(t_1) = y(t_1)$, $\varphi(t_2) = y(t_2)$, and $\varphi(x) \leq y(x)$ on $[t_1, t_2]$. Hence, by definition, φ is a subfunction on $[t_1, t_2]$, and hence, by definition, φ is a subfunction "in the small" on $[x_0, x_1]$. Note here that if $x_1 - x_0 \leq \delta$, then this part of the proof is complete, for it would follow that $\varphi(x) \leq y_0(x)$ which contradicts $y_0(x) < \varphi(x)$ on (x_0, x_1). Hence, Claim 1 is proven. □

For the remainder of the sufficiency part of the proof, we assume without loss of generality that $x_1 - x_0 > \delta$.

Claim 2. Our next claim is that under the assumptions made on $y_0(x)$, there exists a subinterval $[s_0, s_1] \subseteq [x_0, x_1]$ with $s_1 - s_0 \leq x_1 - x_0 - \delta$ and there exists a solution $z_0(x)$ of (3.5) with $z_0(s_i) = \varphi(s_i)$, $i = 0, 1$, and $z_0(x) < \varphi(x)$ on (s_0, s_1).

Proof of Claim 2. Suppose no such interval exists.

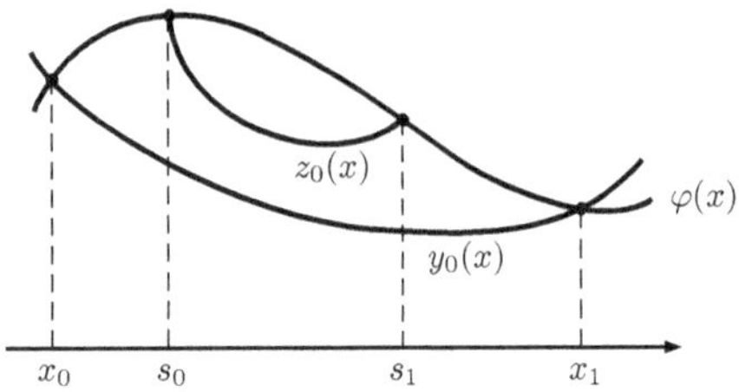

Fig. 3.9 $z_0(s_i) = \varphi(s_i)$, $i = 0, 1$, and $z_0(x) < \varphi(x)$ on (s_0, s_1).

Partition $[x_0, x_1]$ as follows: $x_0 < x_0+\delta < x_0+2\delta < \cdots < x_0+k\delta < x_1$, where $x_1 - (x_0 + k\delta) \leq \delta$. By Corollary 3.1, there exists a solution $y_1(x)$ of (3.5) satisfying $y(x_0) = \varphi(x)$, $y(x_0 + \delta) = \varphi(x_0 + \delta)$, and since φ is a subfunction "in the small" on $[x_0, x_1]$, $\phi(x) \leq y_1(x)$ on $[x_0,\ x + \delta]$. By the extendibility hypotheses in the theorem, $y_1(x)$ exists on all of J. Moreover, from the uniqueness assumption in the hypotheses of the theorem concerning two point BVP's and from supposition in the first line of this claim, it follows that $\varphi(x) \leq y_1(x)$ on $[x_0, x_1]$. In fact the uniqueness assumption on two point BVP's implies $\varphi(x_1) < y_1(x_1)$, since $\varphi(x_1) = y_0(x_1)$.

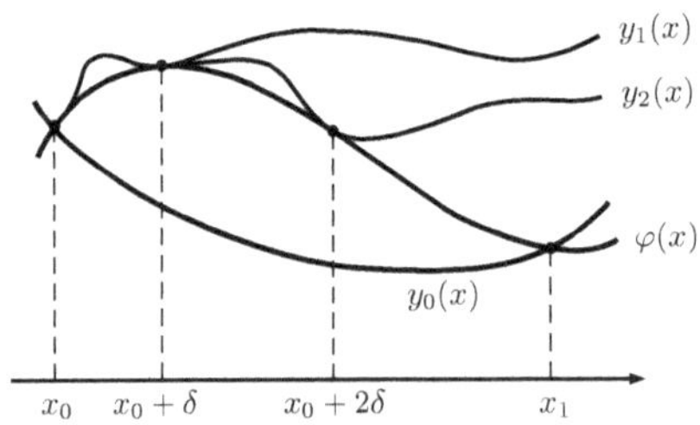

Fig. 3.10 The functions $\varphi(x)$, $y_0(x)$, $y_1(x)$, and $y_2(x)$.

Since $\varphi(x_0 + \delta) = y_1(x_0 + \delta)$ and $\varphi(x) \leq y_1(x)$ on $[x_0, x_1]$, we also have $\varphi'(x_0 + \delta) = y_1'(x_0 + \delta)$.

Now apply Corollary 3.1 again and then there exists a solution $y_2(x)$ of (3.5) satisfying $y_2(x_0 + \delta) = \varphi(x_0 + \delta) = y_1(x_0 + \delta)$, and $y_2(x_0 + 2\delta) = \varphi(x_0+2\delta)$, and $\varphi(x) \leq y_2(x)$ on $[x_0+\delta, x_0+2\delta]$. By uniqueness of two point BVP's, $y_2(x) \leq y_1(x)$ on $[x_0 + \delta, x_1]$, ($y_2(x)$ extends to all of J). Hence,

$\varphi(x) \le y_2(x) \le y_1(x)$ on $[x_0+\delta, x_0+2\delta]$, and so $\varphi(x_0+\delta) = y_1(x_0+\delta) = y_2(x_0+\delta)$, and $\varphi'(x_0+\delta) = y_1'(x_0+\delta) = y_2'(x_0+\delta)$. So $y_1(x)$ and $y_2(x)$ are solutions of the same IVP for (3.5).

We also contend that $\varphi(x) \le y_2(x)$ on $[x_0+2\delta, x_1]$. For if $y_2(x) < \varphi(x)$ for some $\hat{x} \in (x_0+2\delta, x_1]$, it must remain the case for all $x \in [\hat{x}, x_1]$ from the supposition in the first line of this claim. This implies that $y_2(x)$ intersects $y_0(x)$ at some $\tau \in [\hat{x}, x_1)$. But since $y_1(x)$ and $y_2(x)$ are solutions of the same IVP for (3.5), it follows that

$$w(x) = \begin{cases} y_1(x), & x_0 \le x \le x_0+\delta, \\ y_2(x), & x_0+\delta \le x \le x_0, \end{cases}$$

is a solution of (3.5) which intersects $y_0(x)$ at x_0 and at τ; a contradiction.

Hence $\varphi(x) \le y_2(x)$ on $[x_0+2\delta, x_1]$. So we have $\varphi(x) \le y_2(x) \le y_1(x)$ on $[x_0+\delta, x_1]$. Now let $z_1(x)$ be the solution of (3.5) defined by

$$z(x) = \begin{cases} y_1(x), & x_0 \le x \le x_0+\delta, \\ y_2(x), & x_0+\delta \le x \le x_1. \end{cases}$$

We remark here that $z_1'(x_0+2\delta) = y_2'(x_0+2\delta) = \varphi'(x_0+2\delta)$, since $\varphi(x) \le z_1(x)$ on $[x_0, x_1]$ and $z_1(x_0+2\delta) = \varphi(x_0+2\delta)$.

Again applying the Corollary 3.1, there exists a solution $y_3(x)$ of (3.5) such that $y_3(x_0+2\delta) = \varphi(x_0+2\delta) = z_1(x_0+2\delta)$ and $y_3(x_0+3\delta) = \varphi(x_0+3\delta)$, and $\varphi(x) \le y_3(x)$ on $(x_0+2\delta, x_0+3\delta)$. Arguing as above $y_3'(x_0+2\delta) = \varphi'(x_0+2\delta) = z_1'(x_0+2\delta)$ and $\varphi(x) \le y_3(x) \le z_1(x)$ on $[x_0+2\delta, x_1]$. Let $z_2(x)$ denote the solution of (3.5) defined by

$$z_2(x) = \begin{cases} z_1(x), & x_0 \le x \le x_0+2\delta, \\ y_3(x), & x_0+2\delta \le x \le x_1. \end{cases}$$

Continue this pattern until we obtain a solution $z_k(x)$ of (3.5) given by

$$z_k(x) = \begin{cases} z_{k-1}(x), & x_0 \le x \le x_0+k\delta, \\ y_{k+1}(x), & x_0+k\delta \le x \le x_1, \end{cases}$$

and $z_k(x_0+i\delta) = \varphi(x_0+i\delta)$, $0 \le i \le k$, $z_k(x_1) = \varphi(x_1)$, and $\varphi(x) \le z_k(x)$ on $[x_0, x_1]$. Hence, $z_k(x_0) = y_0(x_0)$ and $z_k(x_1) = y_0(x_1)$ and by the uniqueness assumption on two point BVP's we have $z_k(x) \equiv y_0(x)$ on $[x_0, x_1]$, a contradiction to $y_0(x) < \varphi(x) \le z_k(x)$ on (x_0, x_1).

From this contradiction we conclude that our supposition in the first line of this claim must be false. Therefore, there exists $[s_0, s_1] \subseteq [x_0, x_1]$ such that $s_1 - s_0 \le x_1 - x_0 - \delta$ and there exists a solution $z_0(x)$ of (3.5) such that $z_0(s_i) = \varphi(s_i)$ and $z_0(x) < \varphi(x)$ on (s_0, s_1). Of course, if $s_1 - s_0 \le \delta$, this contradicts the subfunction "in the small" property satisfied by φ on $[x_0, x_1]$. Hence Claim 2 is proven. $\square$

So, if $s_1 - s_0 > \delta$, we mimic the argument, proceeding through with the interval $[s_0, s_1]$ and the solution $z_0(x)$ replacing, respectively, the interval $[x_0, x_1]$ and the solution $y_0(x)$. Exactly as above, there exists $[s_{0_1}, s_{1_1}] \subseteq [s_0, s_1]$ such that $s_{1_1} - s_{0_1} \leq s_1 - s_0 - \delta$ and a solution $z_{0_1}(x)$ of (3.5) such that $z_{0_1}(s_{0_i}) = \varphi(s_{0_i})$, $i = 0, 1$, and $z_{0_1}(x) < \varphi(x)$ on (s_{0_1}, s_{1_1}). If $s_1 - s_{0_1} \leq \delta$, this contradicts the property of φ being a subfunction "in the small".

If $s_{1_1} - s_{0_1} > \delta$, repeatedly apply the argument, and after a finite number of applications, there exists a subinterval $[s_{0_j}, s_{1_j}] \subseteq [s_{0_{j-1}}, s_{1_{j-1}}]$ such that $s_{1_j} - s_{0_j} \leq s_{1_{j-1}} - s_{0_{j-1}} - \delta$, and there exists a solution $z_{0_j}(x)$ such that $z_{0_j}(s_{i_j}) = \varphi(s_{i_j})$, $i = 0, 1$, $z_{0_j}(x) < \varphi(x)$ on (s_{0_j}, s_{1_j}), and $s_{1_j} - s_{0_j} \leq \delta$, which contradicts the "in the small" property.

From this final contradiction, we conclude that no such function $y_0(x)$ exists. Therefore φ is a subfunction on I with respect to solutions of (3.5).

(Necessity). Again we will argue via contradiction. Thus now assume that φ is a subfunction on I, but that $\varphi''(x_0) < f(x_0, \varphi(x_0), \varphi'(x_0))$, for some $x_0 \in I^\circ$. By continuity, there exists $[x_1, x_2] \subseteq I^\circ$ with $x_0 \in [x_1, x_2]$ and such that $\varphi''(x) < f(x, \varphi(x), \varphi'(x))$ on $[x_1, x_2]$.

We **claim** now that there exists $\eta > 0$ such that, for any $x_1 \leq t_1 < t_2 \leq x_2$ with $t_2 - t_1 \leq \eta$, the BVP

$$\begin{cases} y'' = f(x, y, y'), \\ y(t_1) = \varphi(t_1),\ y(t_2) = \varphi(t_2), \end{cases}$$

has a solution $y(x) \leq \varphi(x)$ on $[t_1, t_2]$. To see this, define the continuous function F by

$$F(x, y, y') = \begin{cases} f(x, y, y'), & y \leq \varphi(x) \\ f(x, \varphi(x), y') - (\varphi(x) - y), & y > \varphi(x). \end{cases}$$

By Corollary 3.1, there exists $\eta > 0$ such that, for any $[t_1, t_2] \subseteq [x_1, x_2]$ with $t_2 - t_1 \leq \eta$, the BVP

$$\begin{cases} y'' = F(x, y, y'), \\ y(t_1) = \varphi(t_1),\ y(t_2) = \varphi(t_2), \end{cases}$$

has a solution $y(x)$. Now if $y(x) > \varphi(x)$ for some $x \in (t_1, t_2)$, then $y(x) - \varphi(x)$ has a positive maximum at some $t_0 \in (t_1, t_2)$. So, $y(t_0) - \varphi(t_0) > 0$,

$y'(t_0) - \varphi'(t_0) = 0$, and $y''(t_0) - \varphi''(t_0) \le 0$. This is contradictory to

$$\begin{aligned} y''(t_0) - \varphi''(t_0) &> F(t_0, y(t_0), y'(t_0)) - f(t_0, \varphi(t_0), \varphi'(t_0)) \\ &= f(t_0, \varphi(t_0), y'(t_0)) - (\varphi(t_0) - y(t_0)) - f(t_0, \varphi(t_0), \varphi'(t_0)) \\ &= f(t_0, \varphi(t_0), \varphi'(t_0)) + y(t_0) - \varphi(t_0) - f(t_0, \varphi(t_0), \varphi'(t_0)) \\ &= y(t_0) - \varphi(t_0) > 0. \end{aligned}$$

Thus, $y(x) \le \varphi(x)$ on $[t_1, t_2]$. But, if $y(x) \le \varphi(x)$, it follows that $y(x)$ is a solution of

$$\begin{cases} y'' = f(x, y, y'), \\ y(t_1) = \varphi(t_1),\ y(t_2) = \varphi(t_2), \end{cases}$$

and $y(x) \le \varphi(x)$ on $[t_1, t_2]$ as stated in the **claim**. However, it is the case that $y(x) \not\equiv \varphi(x)$ on $[t_1, t_2]$, since φ satisfies the strict inequality $\varphi''(x) < f(x, \varphi(x), \varphi'(x))$ on $[t_1, t_2]$. Hence, $y(x) < \varphi(x)$ at some points in (t_1, t_2). Yet, φ is a subfunction, hence $\varphi(x) \le y(x)$ on $[t_1, t_2]$; a contradiction. Therefore, it follows that φ satisfies

$$\varphi''(x) \ge f(x, \varphi(x), \varphi'(x)) \quad \text{on} \quad I^\circ.$$

This completes the proof. □

An application of the above theorem will be given later in this section. For now we will establish more results concerning uniqueness of solutions of BVP's via the use of inequalities.

Lemma 3.2. *Assume that $f(x, u_1, u_2)$ is continuous on $I \times \mathbb{R}^2$ and is strictly increasing in u_1 for each fixed x and u_2. Then solutions of BVP's*

$$\begin{cases} y'' = f(x, y, y'), \\ y(x_1) = y_1,\ y(x_2) = y_2, \end{cases}$$

are unique on I, when they exist.

Proof. Assume the conclusion of the lemma false. Then there exist solutions $y(x)$ and $z(x)$ and points $x_1 < x_2$ in I such that $y(x_i) = z(x_i)$, $i = 1, 2$, and $y(x) > z(x)$ on (x_1, x_2). Then $y(x) - z(x)$ has a positive maximum at some $x_0 \in (x_1, x_2)$. Hence,

$$\begin{aligned} &y(x_0) > z(x_0), \\ &y'(x_0) = z'(x_0), \\ &y''(x_0) - z''(x_0) \le 0. \end{aligned}$$

However, since f is increasing in u_1, at x_0 we have

$$y''(x_0) - z''(x_0) = f(x_0, y(x_0), y'(x_0)) - f(x_0, z(x_0), z'(x_0)) > 0;$$

a contradiction. □

Theorem 3.9. *Assume that $f(x, u_1, u_2)$ is continuous on $[a, b] \times \mathbb{R}^2$, is nondecreasing in u_1 for each fixed x and u_2, and satisfies a Lipschitz condition with respect to u_2 on each compact subset of $[a, b] \times \mathbb{R}^2$. Let $\varphi, \psi \in C[c, d] \cap C^{(2)}(c, d)$ where $[c, d] \subseteq [a, b]$, and assume that*

$$\psi'' \leq f(x, \psi, \psi'), \quad \varphi'' \geq f(x, \varphi, \varphi'),$$
$$\varphi(c) \leq \psi(c), \quad \varphi(d) \leq \psi(d).$$

Then $\varphi(x) \leq \psi(x)$ on $[c, d]$.

Proof. Assume on the contrary that at some points in (c, d), $\varphi(x) > \psi(x)$, and let $\varepsilon = \max_{c \leq x \leq d}[\varphi(x) - \psi(x)] > 0$. Choose a subinterval $[c_1, d_1] \subseteq [c, d]$ such that

$$\varphi(c_1) - \psi(c_1) = \frac{\varepsilon}{2}, \quad \varphi(d_1) - \psi(d_1) = \frac{\varepsilon}{2},$$
$$\varphi(x) - \psi(x) > \frac{\varepsilon}{2} \quad \text{on} \quad (c_1, d_1),$$
$$\varphi(x) - \psi(x) = \varepsilon \quad \text{at some points in} \quad (c_1, d_1).$$

Define the "tube" $K := \{(x, u_1, u_2) \mid c_1 \leq x \leq d_1, |\varphi(x) - u_1| \leq 1, |\varphi'(x) - u_2| \leq 1\}$. K is a compact subset of $[a, b] \times \mathbb{R}^2$, and so let k be the associated Lipschitz coefficient.

We **claim** now that there exists a function $p(x)$ satisfying the following conditions:

1) $\rho'' = (k+1)\rho'$ on $[c_1, d_1]$,
2) $0 < \rho' < 1$ on (c_1, d_1),
3) $-\frac{\varepsilon}{2} \leq \rho \leq 0$ on $[c_1, d_1]$.

To see this, solve 1) $\rho'' = (k+1)\rho'$ and we have $\rho(x) = \alpha + \beta\rho^{(k+1)x}$. Now consider condition 3), so that $\rho(d_1) = 0$. Hence, $0 = \alpha + \beta e^{(k+1)d_1}$. Now if we form $\rho(x) = -\gamma[1 - e^{(k+1)(x-d_1)}]$, then $\rho(d_1) = 0$ and so we can determine the correct γ. Note now that

$$\rho' = \gamma(k+1)e^{(k+1)(x-d_1)}$$

so that the exponent is negative for $x \in [c_1, d_1]$. Thus ρ' has a maximum of $\gamma(k+1)$ at $x = d_1$. Moreover, ρ' is positive, so ρ is increasing and has minimum value of $-\gamma[1 - e^{(k+1)(c_1-d_1)}]$ at $x = c_1$. If we choose $\gamma_0 = \min\{\frac{1}{k+1}, \frac{\varepsilon}{2[1-e^{-(k+1)(d_1-c_1)}]}\}$, then $\rho(x) = -\gamma_0[1 - e^{(k+1)(x-d_1)}]$ satisfies 1), 2) and 3).

Now define $\varphi_1(x) \equiv \varphi(x) + \rho(x)$. By 3),

$$\varphi(x) - \frac{\varepsilon}{2} \leq \varphi_1(x) \leq \varphi(x) \quad \text{on} \quad [c_1, d_1].$$

Also,

$$\varphi_1'' = \varphi'' + \rho'' \geq f(x, \varphi, \varphi') + (k+1)\rho'.$$

Note that $f(x,\varphi,\varphi'+\rho') - f(x,\varphi,\varphi') \leq |f(x,\varphi,\varphi'+\rho') - f(x,\varphi,\varphi')| \leq k|\rho'|$. Moreover, $(x,\varphi,\varphi'+\rho')$, $(x,\varphi,\varphi') \in K$. Hence

$$\begin{aligned}\varphi_1'' &\geq f(x,\varphi,\varphi') + f(x,\varphi,\varphi'+\rho') - f(x,\varphi,\varphi') + |\rho'| \\ &\geq f(x,\varphi+\rho,\varphi'+\rho') + |\rho'|, \quad \text{since } f \text{ is nondecreasing in } u_1, \\ &> f(x,\varphi_1,\varphi_1') \quad \text{on } [c_1,d_1], \text{ since } 1 \geq \rho' > 0.\end{aligned}$$

Now, $\varphi_1(c_1) - \psi(c_1) \leq \varphi(c_1) - \psi(c_1) = \frac{\varepsilon}{2}$, and $\varphi_1(d_1) - \psi(d_1) \leq \varphi(d_1) - \psi(d_1) = \frac{\varepsilon}{2}$. On the other hand, at points x in (c_1,d_1) where $\varphi(x) - \psi(x) = \varepsilon$, we have $\varphi_1(x) - \psi(x) \geq \frac{1}{2}\varepsilon$. We conclude that $\varphi_1(x) - \psi(x)$ has a positive maximum in (c_1,d_1), say at x_0. As before,

$$\begin{aligned}\varphi_1(x_0) - \psi(x_0) &> 0, \\ \varphi_1'(x_0) - \psi'(x_0) &= 0, \\ \varphi_1''(x_0) - \psi''(x_0) &\leq 0.\end{aligned}$$

If we apply $\varphi_1'' > f(x,\varphi_1,\varphi_1')$ on $[c_1,d_1]$ from above, we contradict $\varphi_1''(x_0) - \psi''(x_0) \leq 0$, for since f is nondecreasing in u_1,

$$\begin{aligned}\varphi_1''(x_0) - \psi''(x_0) &> f(x_0,\varphi_1(x_0),\varphi_1'(x_0)) - f(x_0,\psi(x_0),\psi'(x_0)) \\ &\geq f(x_0,\psi(x_0),\psi'(x_0)) - f(x_0,\psi(x_0),\psi'(x_0)), \\ &= 0.\end{aligned}$$

This completes the proof. □

Corollary 3.4. *Assume $f(x,u_1,u_2)$ satisfies the hypotheses of Theorem 3.9. Then solutions of the BVP*

$$\begin{cases} y'' = f(x,y,y'), \\ y(x_1) = y_1, \; y(x_2) = y_2, \end{cases}$$

are unique on $[a,b] \times \mathbb{R}^2$, when they exist.

Exercise 33. Prove Corollary 3.4.

We now consider an example which illustrates that the Lipschitz condition on f in the u_2 component cannot be altered substantially.

Consider the equation $y'' = |y'|^p$, $0 < p < 1$. There is no Lipschitz condition with respect to y'. Moreover if $f(x,u_1,u_2) = |u_2|^p$, f is nondecreasing in u_1. Seeking a solution of the form $y = \kappa x^r$, we have $y' = \kappa r x^{r-1}$ and

$y'' = \kappa r(r-1)x^{r-2}$. Thus, $(r-1)p = r-2$ and so $r = \frac{2-p}{1-p} = 2 + \frac{p}{1-p} > 2$. We must also have

$$\kappa r(r-1) = (\kappa r)^p.$$

Substituting for r, κ can be obtained. Note also that $y =$ constant is a solution of the differential equation. Hence, solutions of BVP's for this f satisfying $y(x_1) = y_1, y(x_2) = y_2$, are not unique, when solutions exist.

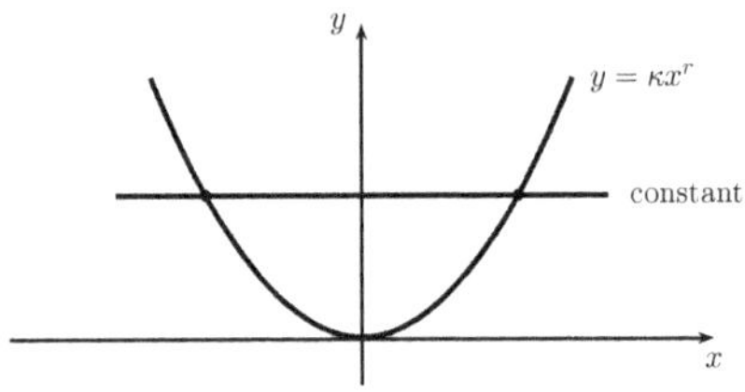

Fig. 3.11 Two solutions of a BVP.

Theorem 3.10. *Let $f(x, u_1, u_2)$ be continuous on $[a,b] \times \mathbb{R}^2$, be nondecreasing in u_1, for each fixed x and u_2, and suppose solutions of IVP's for $y'' = f(x, y, y')$ are unique. Then the conclusion of Theorem 3.9 follows.*

Proof. Suppose again that to the contrary, there exists $[c_1, d_1] \subseteq [c, d]$ such that $\varphi(c_1) = \psi(c_1)$, $\varphi(d_1) = \psi(d_1)$ and $\psi(x) < \varphi(x)$ on (c_1, d_1). Let $m = \max_{c_1 \le x \le d_1}[\varphi(x) - \psi(x)]$. Let $x_0 \in (c_1, d_1)$ be such that $\varphi(x_0) - \psi(x_0) = m$, and define

$$\varphi_1 := \varphi - m.$$

Then $\varphi_1(x_0) = \psi(x_0)$, $\varphi_1'(x_0) = \varphi'(x_0) = \psi'(x_0)$, and $\varphi_1(x_0) \le \psi(x_0)$ on

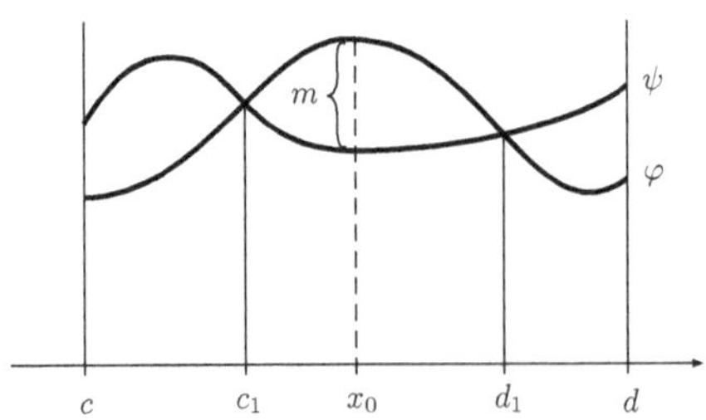

Fig. 3.12 $\varphi(c_1) = \psi(c_1)$, $\varphi(d_1) = \psi(d_1)$, and $\psi(x) < \varphi(x)$ on (c_1, d_1).

$[c_1, d_1]$, with strict inequality at some points.

Now, $\varphi_1'' = \varphi'' \geq f(x, \varphi, \varphi') \geq f(x, \varphi_1, \varphi_1')$, since f is nondecreasing in u_1. It follows from Theorem 3.7 that $\varphi_1(x) \equiv \psi(x)$, which is clearly false. Hence the conclusion holds. $\square$

Corollary 3.5. *Assume the hypotheses of Theorem* 3.10. *Then the conclusion of the Corollary* 3.4 *holds.*

3.3 Existence of Solutions of BVP's for Nonlinear ODE's

We now deal with a "uniqueness implies existence" result for nonlinear equations.

Theorem 3.11. *Concerning the differential equation,*

$$y'' = f(x, y, y'), \tag{3.6}$$

assume

i) $f(x, u_1, u_2)$ *is continuous on* $I \times \mathbb{R}^2$, *where* I *is either an open or half-open interval,*
ii) *For any interval* $[x_1, x_2] \subseteq I$ *and any solutions* $y(x)$ *and* $z(x)$ *of* (3.6), $z(x_i) = y(x_i)$, $i = 1, 2$ *implies* $z(x) \equiv y(x)$ *on* $[x_1, x_2]$, *and*
iii) *All solutions of* (3.6) *extend to* I.

Then for any $x_1 < x_2$ *belonging to* I *and any* $y_1, y_2 \in \mathbb{R}$, *the BVP*

$$\begin{cases} y'' = f(x, y, y'), \\ y(x_1) = y_1, \; y(x_2) = y_2, \end{cases}$$

has a solution, which is unique by ii).

Proof. To avoid technical complications, we will assume that solutions of IVP's for (3.6) are unique, although, the theorem is still true without this assumption. We will also assume the case $I = [a, b)$. In showing existence, we shall employ the "shooting method" for solving BVP's.

So choose x_1 and x_2 with $a \leq x_1 < x_2 < b$ and choose $y_1, y_2 \in \mathbb{R}$. Then let $y(x; m)$ denote the solution of the IVP for (3.6) which satisfies the initial conditions, $y(x_1; m) = y_1$, $y'(x_1; m) = m$. Now it follows from the fact that solutions of IVP's are unique and the Kamke Theorem, that solutions depend continuously on initial conditions. Thus, it follows that $y(x_2; m)$ is a continuous function of m, i.e., the value of y at x_2 depends continuously on m.

In other words, $y(x_2;m) : \mathbb{R} \to \mathbb{R}$ is continuous. Now $\mathbb{R}$ is a connected set and hence by the continuity of $y(x_2;m)$ on m, it follows that $J := \{(x_2, y(x_2;m)) | \, m \in \mathbb{R}\}$ is an interval on the line $x = x_2$.

(If we can show that J is neither bounded above nor below, then J will consist of the entire line $x = x_2$, and it will follow that there exists $m_0 \in \mathbb{R}$ such that $y(x_2;m_0) = y_2$. Then $y(x;m_0)$ is the desired solution of the BVP.)

We proceed to show that J is not bounded above, with there being an analogous argument for not bounded below. To this end, assume J is bounded above. This implies that there exists y_0 such that $y < y_0$, for all $(x_2, y) \in J$. Let $z(x)$ be the solution of the IVP for (3.6) with $z(x_2) = y_0$, $z'(x_2) = 0$. Note here that $y_1 \neq z(x_1)$, because $(x_2, y_0) \notin J$. There are two cases:

1) $z(x_1) > y_1$, or
2) $z(x_1) < y_1$.

Case 1: Suppose $z(x_1) > y_1$ and let $y_k(x) = y(x;k)$, $k = 1, 2, \ldots$. Then for all $k = 1, 2, \ldots$, $y_k(x_1) = y_1 < z(x_1)$, and $y_k(x_2) < y_0 = z(x_2)$. Therefore,

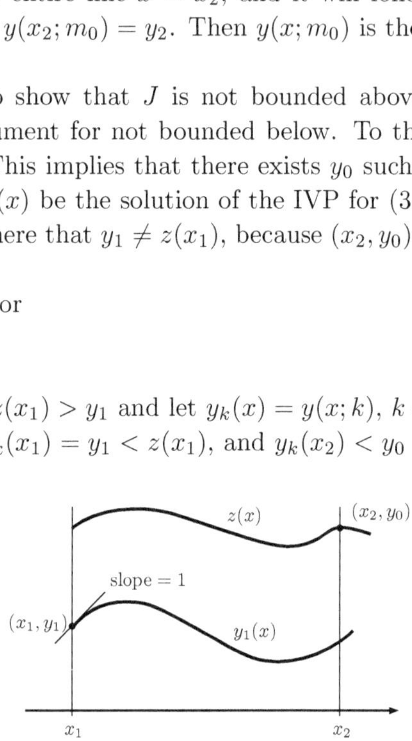

Fig. 3.13 The graphs of $z(x)$ and $y_1(x)$.

from hypothesis ii), $y_k(x) \le z(x)$ on $[x_1, x_2]$ and $y_1(x) \le y_k(x)$ on $[x_1, x_2]$, $k = 1, 2, \ldots$. So, all y_k's are trapped between $y_1(x)$ and $z(x)$ on $[x_1, x_2]$. Let $M > 0$ be such that $|z(x)| \le M$ and $|y_1(x)| \le M$ on $[x_1, x_2]$. Then for each $k \ge 1$, there exists $t_k \in (x_1, x_2)$ such that

$$|y_k'(t_k)| = \left| \frac{y_k(x_2) - y_k(x_1)}{x_2 - x_1} \right| \le \frac{2M}{x_2 - x_1},$$

and of course $|y_k(t_k)| \le M$, and $t_k \in (x_1, x_2)$. Thus, we have the three bounded sequences, $\{t_k\}$, $\{y_k(t_k)\}$, and $\{y_k'(t_k)\}$, of real numbers. First, there exists a subsequence $\{t_{k_1}\} \subseteq \{t_k\}$ and there exists $x_0 \in [x_1, x_2]$

such that $t_{k_1} \to x_0$. Second, since $\{y_{k_1}(t_{k_1})\}$ is also bounded, there exists a subsequence $\{y_{k_2}(t_{k_2})\} \subseteq \{y_{k_1}(t_{k_1})\}$ such that $y_{k_2}(t_{k_2}) \to \alpha$, for some $\alpha \in \mathbb{R}$. Finally, since $\{y'_{k_2}(t_{k_2})\}$ is bounded, there exists a subsequence $\{y'_{k_3}(t_{k_3})\}$ such that $y'_{k_3}(t_{k_3}) \to \beta$, for some $\beta \in \mathbb{R}$. In summary, we have

$$t_{k_3} \to x_0, \quad y_{k_3}(t_{k_3}) \to \alpha, \quad y'_{k_3}(t_{k_3}) \to \beta.$$

By the Kamke Convergence Theorem, there exists a further subsequence $\{y_{k_4}(x)\}$ and a solution $u(x)$ of (3.6) with $u(x_0) = \alpha$, $u'(x_0) = \beta$, such that

$$\lim_{k_4 \to \infty} y_{k_4}(x) = u(x) \text{ and } \lim_{k_4 \to \infty} y'_{k_4}(x) = u'(x) \quad \text{uniformly on } [x_1, x_2].$$

However, this is impossible, since $u'(x_1)$ is *finite*, yet $y'_{k_4}(x_1) = k_4 \longrightarrow +\infty$, as $k_4 \to +\infty$. Therefore, Case 1 is impossible.

Case 2: Suppose $z(x_1) < y_1$. Recall here that $a \le x_1 < x_2 < b$. Choose c such that $x_2 < c < b$, and let $y_k(x)$ be as in Case 1.

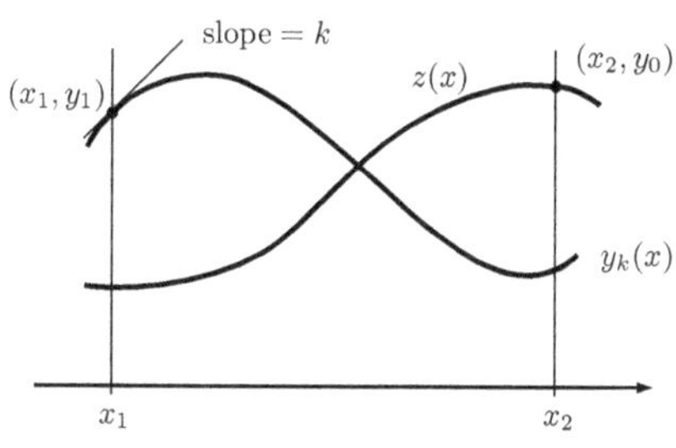

Fig. 3.14 The graphs of $z(x)$ and $y_k(x)$.

Since $y_k(x_1) > z(x_1)$ and $y_k(x_2) < z(x_2)$, for each k, it follows from hypothesis ii) that $y_k(x) < z(x)$ on $[x_2, c]$. Moreover $y_1(x) < y_k(x)$ on $[x_2, c]$, for all k. Now we can argue as in Case 1, except we will take the interval $[x_2, c]$. That is, there exist a subsequence $\{k_j\} \subseteq \{k\}$ and points $\{t_{k_j}\} \subseteq (x_2, c)$ such that

$$\begin{aligned} &t_{k_j} \to \tau, \quad \text{for some } \tau \in [x_2, c], \\ &y_{k_j}(t_{k_j}) \to \gamma, \quad \text{for some } \gamma \in \mathbb{R}, \\ &y'_{k_j}(t_{k_j}) \to \delta, \quad \text{for some } \delta \in \mathbb{R}. \end{aligned}$$

Then by the Kamke Convergence Theorem, there exists a further subsequence $\{y_{k_{j_i}}(x)\}$ and a solution of (3.6), $v(x)$, such that $v(\tau) = \gamma$, $v'(\tau) = \delta$, and $y^{(l)}_{k_j}(x) \to v^{(l)}(x)$ uniformly on compact subintervals of $[a, b)$, $l = 0, 1$;

in particular on $[x_1, c]$. Again $v'(x_1)$ is finite, yet $y'_{k_{j_i}}(x_1) = k_{j_i} \to +\infty$, as $i \to +\infty$; a contradiction. Hence, Case 2 is also impossible.

Therefore J is unbounded above. Similarly, J is unbounded below and so J consists of the entire line $x = x_2$, and as remarked above, there exists $m_0 \in \mathbb{R}$ such that $y(x_2; m_0) = y_2$, and $y(x; m_0)$ is the desired solution. □

Example 3.1. This example is an application of Theorem 3.8 and it also shows that the part of Theorem 3.11 requiring that the interval I be open or half-open is necessary.

Consider first the differential equation $y'' = -y$. We will first be concerned with hypotheses of Theorem 3.8 for this equation. BVP's for this equation with BC's of the form $y(x_1) = y_1$, $y(x_2) = y_2$ have unique solutions on any interval of length less than π, because of the disconjugacy of the equation on such intervals. Moreover, solutions of the differential equation extend to any interval. If $\varphi(x) = 1$, then $\varphi'' = 0 > -1 = -\varphi$, and so $\varphi(x) \equiv 1$ is a subfunction by Theorem 3.8 with respect to solutions of $y'' = -y$ on any interval of length less than π. (Note that $\varphi \equiv k$, $k > 0$ constant, is also a subfunction.)

We now show that in Theorem 3.11, I cannot necessarily be closed. Consider now

$$y'' = -y + \arctan y, \ -\frac{\pi}{2} < \arctan < \frac{\pi}{2}. \tag{3.7}$$

We want to show that the hypotheses of Theorem 3.11 are satisfied on $[0, \pi]$, but that there are BVP's which do not have solutions.

Claim 1. Solutions of 2 point conjugate BVP's are unique on $[0, \pi]$.

Proof of Claim 1. Assume not and let $y(x)$ and $z(x)$ be solutions of (3.7) such that $y(x_i) = z(x_i)$, $i = 1, 2$ and $y(x) > z(x)$ on (x_1, x_2), where $0 \le x_1 < x_2 \le \pi$. Let $w(x) \equiv y(x) - z(x)$. So, $w(x_i) = 0$, $i = 1, 2$ and $w(x) > 0$ on (x_1, x_2). Also

$$\begin{aligned} w'' &= y'' - z'' \\ &= -y + z + \operatorname{Arctan} y - \operatorname{Arctan} z \\ &= -w + \operatorname{Arctan} y - \operatorname{Arctan} z. \end{aligned}$$

Since $\arctan(\cdot)$ is a strictly increasing function, and since $y(x) > z(x)$, we have

$$w'' > -w \quad \text{on} \quad (x_1, x_2),$$

and so w is a subfunction with respect to solutions $y'' = -y$ on $[x_1, x_2)$ and on $(x_1, x_2]$. Now, if $x_2 - x_1 < \pi$, since the only solution of

$$\begin{cases} y'' = -y, \\ y(x_1) = 0, \ y(x_2) = 0, \end{cases}$$

is $y(x) \equiv 0$, and since $w(x_1) = w(x_2) = 0$ and w is a subfunction, it follows that $w(x) \leq 0$ on $[x_1, x_2]$. This is a contradiction, since $w(x) > 0$ on (x_1, x_2).

Thus, $[x_2,\ x_1] = [0,\ \pi]$; i.e., $x_1 = 0$, $x_2 = \pi$. So, we have $w(0) = w(\pi) = 0$, $w(x) > 0$ on $(0, \pi)$ and $w'' > -w$ on $(0, \pi)$.

Let $x_0 \in (0, \pi)$ be such that $w(x)$ has a positive maximum at $x = x_0$ and solve the IVP

$$\begin{cases} y'' = -y, \\ y(x_0) = w(x_0), \\ y'(x_0) = w'(x_0) = 0. \end{cases}$$

The solution $y(x) = w(x_0)\cos(x - x_0)$. Then

$$y''(x_0) = -y(x_0) = -w(x_0) < w''(x_0).$$

Hence, $w(x) > y(x)$ on some deleted interval $(x_0 - \delta, x_0 + \delta) \setminus \{x_0\}$. Now w

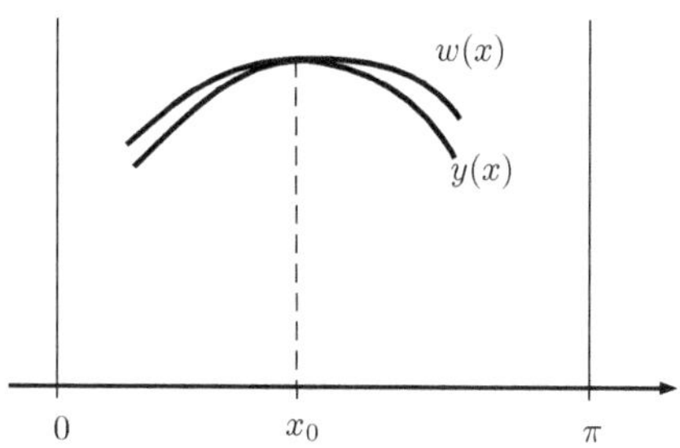

Fig. 3.15 $w(x) > y(x)$ on $(x_0 - \delta, x_0 + \delta) \setminus \{x_0\}$.

is a subfunction with respect to solutions of $y'' = -y$ on $[0, \pi)$ and on $(0, \pi]$, hence a subfunction on both $[0, x_0]$ and $[x_0, \pi]$ and both of these intervals are of length less than π. It follows that $w(x) > y(x)$ on $[0, x_0)$ and on $(x_0, \pi]$. Therefore, $w(x) > 0$ at $x = 0$ or $x = \pi$, or both.

But this contradicts $w(0) = w(\pi) = 0$. This final contradiction allows us to conclude that Claim 1 is true; i.e., solutions of 2-point conjugate BVP's for (3.7) are unique, when they exist. $\square$

Now differential equation (3.7), $y'' = -y + \text{Arctan}\, y$, satisfies a Lipschitz condition so that solutions extend (the equation is independent of x). To see that a Lipschitz condition is satisfied, notice that if $f(x, y) = -y + \text{Arctan}\, y$, then

$$\frac{\partial f}{\partial y} = -1 + \frac{1}{1 + y^2},$$

which is continuous, and also $|\frac{\partial f}{\partial y}| \leq 1$, for all (x, y, y'), and it follows that all solutions of (3.7) extend to $\mathbb{R}$.

Hence the differential equation (3.7) satisfies all of the hypotheses of Theorem 3.11 on $[0, \pi]$.

Claim 2. There exist BVP's

$$\begin{cases} y'' = -y + \text{Arctan}\, y, \\ y(0) = y_1, \; y(\pi) = y_2, \end{cases}$$

which do not have solutions.

Proof of Claim 2. Let $\varphi(x; m)$ denote the solution of the IVP

$$\begin{cases} y'' = -y + \frac{\pi}{2}, \\ y(0) = 0, \; y' = m. \end{cases}$$

So, $\varphi(x; m) = m \sin x - \frac{\pi}{2} \cos x + \frac{\pi}{2}$ and hence $\varphi(\pi; m) = \pi$, for all m.

Now let $y(x)$ be the solution of the IVP

$$\begin{cases} y'' = -y + \text{Arctan}\, y, \\ y(0) = 0, \; y' = m. \end{cases}$$

Then $y''(0) = -0 + \text{Arctan}(0) = 0 < \varphi''(0; m) = \frac{\pi}{2}$, which implies $y(x) < \varphi(x; m)$ on $(0, \delta_0)$, for some $\delta_0 > 0$. Now $\varphi'' = -\varphi + \frac{\pi}{2} > -\varphi + \text{Arctan}\, \varphi$, and so φ is a subfunction with respect to solutions of (3.7), $y'' = -y + \text{Arctan}\, y$. Since $y(x) < \varphi(x; m)$ on $(0, \delta_0)$, it follows that $y(x) < \varphi(x; m)$ on $(0, \pi]$; in particular, $y(\pi) < \varphi(\pi; m) = \pi$. We conclude if $y_2 > \pi$, then the BVP

$$\begin{cases} y'' = -y + \text{Arctan}\, y, \\ y(0) = 0, \; y(\pi) = y_2, \end{cases}$$

has no solution, for if it did, then the solution would "cross" $\varphi(x; m)$, for appropriate m, at some point in $(0, \pi]$, and we have shown this is impossible. Thus Claim 2 is also verified. □

Hence, it follows that the hypothesis of Theorem 3.11 requiring the interval to be open or half-open is necessary. And this concludes the example.

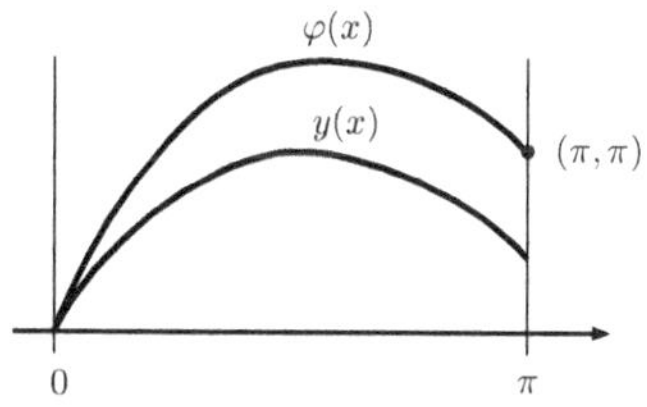

Fig. 3.16 The graphs of $\varphi(x)$ and $y(x)$.

For our next consideration, we will consider an application of Theorem 3.11. We are interested in functions which satisfy one-sided type of Lipschitz conditions: Let $f(x, y, y')$ be continuous on $I \times \mathbb{R}^2$ and assume that there are functions $k_1(x)$, $k_2(x)$, $l_1(x)$, $l_2(x)$ continuous on I, with $k_1(x) \leq k_2(x)$, $l_1(x) \leq l_2(x)$ such that

1). $k_1(x)(y - z) \leq f(x, y, y') - f(x, z, y') \leq k_2(x)(y - z)$, for all $(x, y, y'), (x, z, y') \in I \times \mathbb{R}^2$ with $y \geq z$; and
2). $l_1(x)(y' - z') \leq f(x, y, y') - f(x, y, z') \leq l_2(x)(y' - z')$, for all $(x, y, y'), (x, y, z') \in I \times \mathbb{R}^2$ with $y' \geq z'$.

If $y < z$, then from 1),

$$k_1(x)(z - y) \leq f(x, z, y') - f(x, y, y') \leq k_2(x)(z - y).$$

Now multiply by -1, and we have

3). $k_2(x)(y - z) \leq f(x, y, y') - f(x, z, y') \leq k_1(x)(y - z)$, when $y < z$.

So, $|f(x, y, y') - f(x, z, y')| \leq \max\{|k_1(x)|, |k_2(x)|\}|y - z|$. Hence on a compact subinterval, we obtain the usual Lipschitz condition. Inequality 2) can be treated in a similar manner. We can then obtain

$$\begin{aligned}|f(x, y, y') - f(x, z, z')| \leq& |f(x, y, y') - f(x, z, y') + f(x, z, y') + f(x, z, z')| \\ \leq& \max\{|k_1(x)|, |k_2(x)|\}|y - z| \\ &+ \max\{|l_1(x)|, |l_2(x)|\}|y' - z'|.\end{aligned}$$

Now, it is fairly clear that for all $(x, y, y'), (x, z, y') \in I \times \mathbb{R}^2$,

$$\begin{aligned}&\frac{1}{2}[k_2(x) + k_1(x)](y - z) - \frac{1}{2}[k_2(x) - k_1(x)]|y - z| \\ &\leq f(x, y, y') - f(x, z, y') \\ &\leq \frac{1}{2}[k_2(x) + k_1(x)](y - z) + \frac{1}{2}[k_2(x) - k_1(x)]|y - z|.\end{aligned}$$

Similarly, there is such a property satisfied by l_1 and l_2.

Define

$$g_1(x,w,w') = \frac{1}{2}(k_2+k_1)w + \frac{1}{2}(l_2+l_1)w' - \frac{1}{2}(k_2-k_1)|w| - \frac{1}{2}(l_2-l_1)|w'|,$$
$$g_2(x,w,w') = \frac{1}{2}(k_2+k_1)w + \frac{1}{2}(l_2+l_1)w' + \frac{1}{2}(k_2-k_1)|w| + \frac{1}{2}(l_2-l_1)|w'|.$$

Then from above, we have

$$g_1(x,y-z,y'-z') \le f(x,y,y') - f(x,z,z') \le g_2(x,y-z,y'-z').$$

We claim that g_1 and g_2 satisfy the Lipschitz conditions 1) and 2) on $I \times \mathbb{R}^2$; e.g., we **claim**:

$$k_1(x)(y-z) \le g_1(x,y,y') - g_1(x,z,y') \le k_2(x)(y-z).$$

There are 3 cases for this example i) $y \ge z \ge 0$, ii) $y \ge 0 > z$, iii) $0 \ge y \ge z$.

Exercise **34.** Verify the claim for cases i), ii), iii).

Thus we have $g_1(x,y,y')$ and $g_2(x,y,y')$ are both continuous on $I \times \mathbb{R}^2$ and both satisfy 1) and 2) on $I \times \mathbb{R}^2$. Therefore, solutions of IVP's for $y'' = g_1(x,y,y')$ and for $y'' = g_2(x,y,y')$ are unique on I and all solutions extend to I, respectively. (This is because 1) and 2) give the usual type of Lipschitz conditions.)

The first corollary of our next theorem exhibits our application of Theorem 3.11.

Theorem 3.12. *Let $x_0 \in I$ and let $\varphi(x)$ be the solution of the IVP*

$$\begin{cases} y'' = g_1(x,y,y'), & (3.8) \\ y(x_0) = 0,\ y'(x_0) = +1. & (3.9) \end{cases}$$

If x_1 is the first zero of $\varphi(x)$ to the right of x_0, then solutions of BVP's

$$\begin{cases} y'' = g_1(x,y,y'), \\ y(x_2) = y_2,\ y(x_3) = y_3, \end{cases}$$

are unique (when they exist) on $[x_0,x_1)$ and on $(x_0,x_1]$, and this result is best possible.

Proof. Assume on the contrary that one of these BVP's has distinct solutions $u(x)$ and $v(x)$ where say $x_0 \le x_2 < x_3 < x_1$. We can assume, without loss of generality, that $u(x) > v(x)$ on (x_2,x_3). Now set $w(x) \equiv u(x) - v(x)$; then $w(x) > 0$ on (x_2,x_3), and

$$g_1(x,u-v,u'-v') \le g_1(x,u,u') - g_1(x,v,v') = u'' - v''.$$

So, $w''(x) \geq g_1(x, w, w')$ on $[x_2, x_3]$. Moreover, it is also the case that $g_1(x, w, w')$ is positive homogeneous; (i.e., if $\lambda \geq 0$, then $g_1(x, \lambda w, \lambda w') = \lambda g(x, w, w')$). Hence, $\lambda w'' \geq g_1(x, \lambda w, \lambda w')$.

Now in what follows, we will consider 3 cases, with each case yielding a contradiction to Theorem 3.7 where λw plays the part of a lower solution and φ plays the part of an upper solution with respect to solutions of $y'' = g_1(x, y, y')$.

Case 1. $w(x) \leq \varphi(x)$ on $[x_2, x_3]$, where $x_0 < x_2 < x_3 < x_1$.

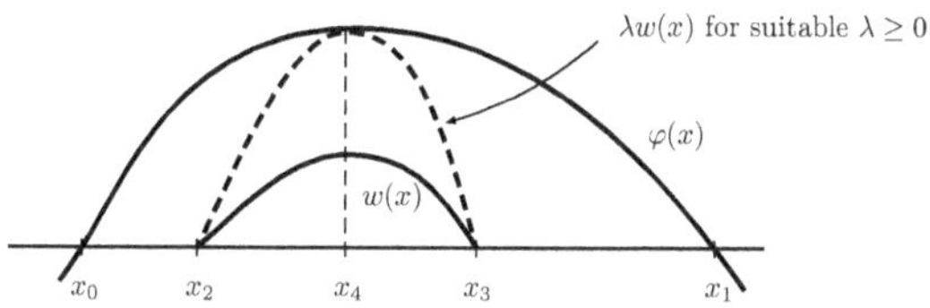

Fig. 3.17 $w(x) \leq \varphi(x)$ on $[x_2, x_3]$ with $x_0 < x_2 < x_3 < x_1$.

Then there exists $\lambda \geq 0$ such that $\lambda w \leq \varphi$ on $[x_2, x_3]$ and for some $x_4 \in (x_2, x_3)$, $\lambda w^{(i)}(x_4) = \varphi^{(i)}(x_4)$, $i = 0, 1$. By Theorem 3.7, $\lambda w \equiv \varphi$ on $[x_2, x_3]$, but this is impossible since $\varphi(x) > 0$ on (x_0, x_1) and $\lambda w(x_i) = 0$, $i = 2, 3$. Hence, Case 1 is not possible.

Case 2. $w(x) \leq \varphi(x)$ on $[x_2, x_3]$, where $x_0 = x_2 < x_3 < x_1$. Then the exists

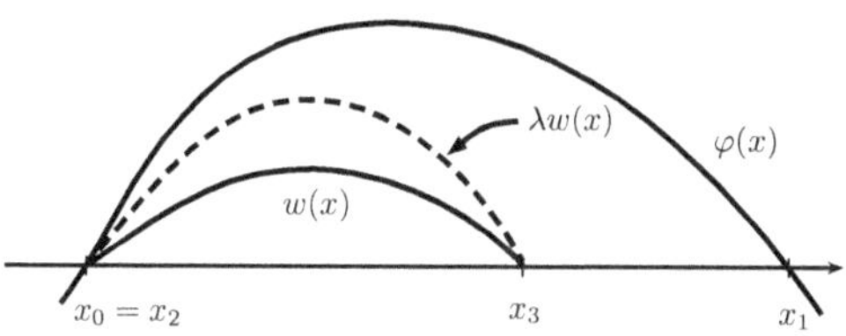

Fig. 3.18 $w(x) \leq \varphi(x)$ on $[x_2, x_3]$ with $x_0 = x_2 < x_3 < x_1$.

$\lambda \geq 0$ such that $\lambda w'(x_0) = \varphi'(x_0)$ or $\lambda w^{(i)}(x_4) = \varphi^{(i)}(x_4)$, $i = 0, 1$, some $x_4 \in (x_0, x_3)$ and $\lambda w \leq \varphi$ on $[x_0, x_3]$. Case 1 resolves the situation with x_4; otherwise $\lambda w^{(i)}(x_0) = \varphi^{(i)}(x_0)$, $i = 0, 1$, and so by Theorem 3.7, $\lambda w \equiv \varphi$ on $[x_0, x_3]$; again we have a contradiction, since $\varphi(x) > 0$ on (x_0, x_1) whereas $\lambda w(x_3) = 0$. This shows the impossibility of Case 2.

Case 3. $w(x) \geq \varphi(x)$ on a proper subinterval of $[x_0, x_3]$, where $x_0 \leq x_2 < x_3 < x_1$. Then, there exists $\lambda \geq 0$ such that $\lambda w(x) \leq \varphi(x)$ on $[x_2, x_3]$ and Cases 1 and 2 show the impossibility of Case 3.

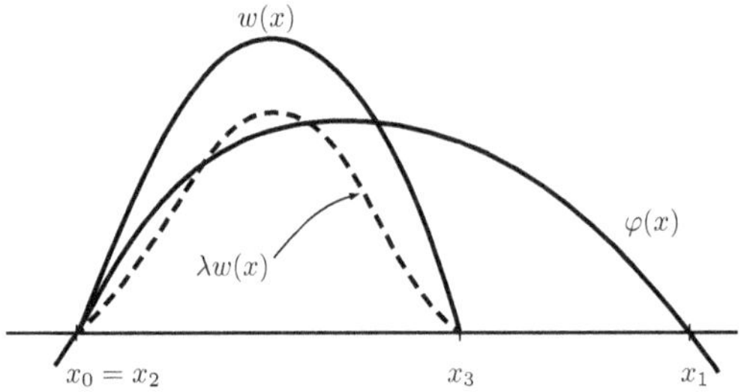

Fig. 3.19 $w(x) \geq \varphi(x)$ on a proper subinterval of $[x_0, x_3]$.

Therefore, solutions of the indicated BVP's are unique, when they exist.

In the statement of the theorem, when we say that this result is best possible, we mean that the subinterval length, $x_1 - x_0$, is the longest possible in terms of the Lipschitz coefficients, k_1, k_2, l_1, l_2 on which the indicated BVP's have at most one solution. To see that this length, $x_1 - x_0$, depends on the Lipschitz coefficients, notice that from the manner in which $g_1(x, y, y')$ is defined, it follows that the solution of (3.8), (3.9) is a solution of $y'' = k_1(x)y + \frac{1}{2}(l_2 + l_1)y' - \frac{1}{2}[l_2 - l_1]|y'|$. Solve this for interval length on $[x_0, x_1]$ since the slope is $+1$ at x_0 and the variable y remains positive then until it reaches the first zero to the right of x_0, which is $x = x_1$. Thus $x_1 - x_0$ depends on the Lipschitz coefficients. It is best possible in the sense that if $[c, d] \supsetneq [x_0, x_1)$ or $(x_0, x_1]$, then the nontrivial solution $\varphi(x)$ of (3.8), (3.9) and $y(x) \equiv 0$ are distinct solutions of

$$\begin{cases} y'' = g_1(x, y, y'), & \text{on } [c, d], \\ y(x_0) = y(x_1) = 0. \end{cases}$$

In particular, on $[c, d]$, solutions of the BVP are not unique, and so the interval length $x_1 - x_0$ is best possible in the manner stated. □

This corollary is immediate from Theorems 3.12 and 3.11.

Corollary 3.6. *Let x_0 and x_1 be as in Theorem 3.12. Then for any x_2, x_3 with $x_0 \leq x_2 < x_3 < x_1$ or $x_0 < x_2 < x_3 \leq x_1$, and for any $y_1, y_2 \in \mathbb{R}$, the BVP*

$$\begin{cases} y'' = g_1(x, y, y'), \\ y(x_2) = y_1, \ y(x_3) = y_2 \end{cases}$$

has a unique solution.

Corollary 3.7. *If x_0 and x_1 are as in Theorem 3.12, if $I \subseteq [x_0, x_1)$ or $I \subseteq (x_0, x_1]$ and if $\varphi \in C(I) \cap C^{(2)}(I^\circ)$ and $\varphi'' \geq g_1(x, \varphi, \varphi')$ on I°, then $\varphi(x)$ is a subfunction on I with respect to solutions of $y'' = g_1(x, y, y')$.*

Proof. Apply Theorems 3.8 and 3.12. □

Theorem 3.13. *Let $f(x, u_1, u_2)$ be continuous on $I \times \mathbb{R}^2$ and satisfy the Lipschitz conditions in terms of k_1, k_2, l_1, l_2 on $I \times \mathbb{R}^2$. Let x_0 and x_1 be as in Theorem 3.12. Then for any x_1, x_3 with $x_0 \leq x_2 < x_3 < x_1$ or $x_0 < x_2 < x_3 \leq x_1$ and any $y_1, y_2 \in \mathbb{R}$, the BVP*

$$\begin{cases} y'' = f(x, y, y'), \\ y(x_2) = y_1, \ y(x_3) = y_2, \end{cases}$$

has a unique solution, and this is best possible for the class of functions f satisfying the given Lipschitz conditions.

Proof. Assume $u(x)$ and $v(x)$ are distinct solutions of such a BVP. We can assume that the solutions do not intersect on (x_2, x_3). If they do, then cut the interval down to obtain an interval on which they do not cross. Assume then that $u(x) > v(x)$ on (x_2, x_3), and let $w(x) \equiv u(x) - v(x)$. Then $w(x_i) = 0$, $i = 2, 3$, and $w(x) > 0$ on (x_2, x_3). But from our Lipschitz condition, we have on $[x_2, x_3]$,

$$\begin{aligned} w''(x) &= u''(x) - v''(x) \\ &= f(x, u(x), u'(x)) - f(x, v(x), v'(x)) \\ &\geq g_1(x, u(x) - v(x), u'(x) - v'(x)) \\ &= g_1(x, w(x), w'(x)). \end{aligned}$$

From Corollary 3.7, $w(x)$ is a subfunction with respect to solutions of the differential equation (3.8), $y'' = g_1(x, y, y')$, on $[x_2, x_3]$. But the solution of (3.8) satisfying

$$y(x_2) = w(x_2) = 0, \quad y(x_3) = w(x_3) = 0$$

is $y(x) \equiv 0$. So $w(x) \leq y(x) = 0$ on $[x_2, x_3]$, which is a contradiction. Thus, solutions of all such BVP's are unique, when they exist, and consequently they exist by Theorem 3.11, and the "best possible" part is from Theorem 3.12. □

Note: $g_1(x, y, y')$ is in the class of functions discussed in Theorem 3.13.

Recall that earlier we proved in Theorem 3.3, if $f(x, y, y')$ is continuous on $I \times \mathbb{R}^2$, if the uniform Lipschitz condition is satisfied, $|f(x, y, y') - f(x, z, z')| \le K|y - z| + L|y' - z'|$, on $I \times \mathbb{R}^2$, and if $[x_1, x_2] \subseteq I$ with $\frac{K}{8}(x_2 - x_1)^2 + \frac{L}{2}(x_2 - x_1) < 1$, then the BVP

$$\begin{cases} y'' = f(x, y, y'), \\ y(x_1) = y_1, \ y(x_2) = y_2, \end{cases}$$

has a unique solution for any $y_1, y_2 \in \mathbb{R}$. The inequality concerning $x_2 - x_1$ is equivalent to saying $x_2 - x_1$ is less than the positive root of $\frac{K}{8}t^2 + \frac{L}{2}t = 1$. Now the Lipschitz condition is equivalent to our one-sided Lipschitz conditions with $k_1(x) = -K$, $k_2(x) = K$, $l_1(x) = -L$, and $l_2(x) = L$. Thus our interval lengths in terms of L and K were not best possible in Theorem 3.3 using the Contraction Mapping Principle. We now get a best possible length on the interval length in terms of K and L.

Theorem 3.14. *Let $f(x, y, y')$ be continuous on $I \times \mathbb{R}^2$ and satisfy $|f(x, y, y') - f(x, z, z')| \le K|y - z| + L|y' - z'|$ on $I \times \mathbb{R}^2$. Let $\alpha(K, L)$ be the first positive zero of $y'(x)$ where $y(x)$ is the solution of the IVP*

$$\begin{cases} y'' = -Ky - Ly', \\ y(0) = 0, \ y'(0) = 1. \end{cases}$$

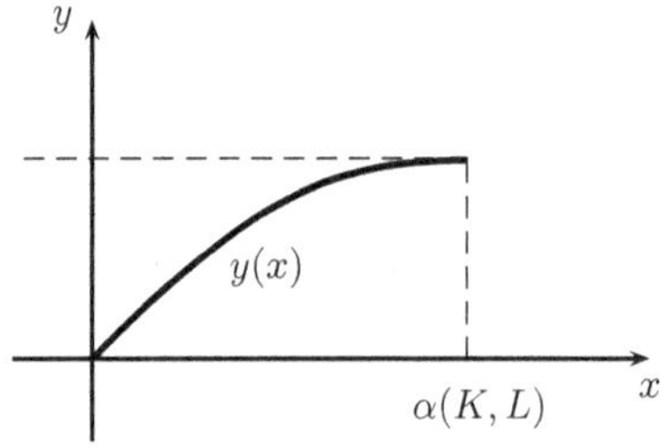

Fig. 3.20 $y'(x) = 0$ at $x = \alpha(K, L)$.

Then on any half-open subinterval of I of length $2\alpha(K, L)$, 2-point conjugate BVP's for $y'' = f(x, y, y')$ have unique solutions, and this is best possible.

Proof. We will apply Theorem 3.12 to the equation

$$y'' = g_1(x, y, y') = -Ky - L|y'|.$$

Since $g_1(x, y, y')$ is autonomous (independent of x), translations of solutions of the differential equation are also solutions, so we may as well take $x_0 = 0$ in Theorem 3.12 for our argument here. Hence, we consider the IVP

$$\begin{cases} y'' = -Ky - L|y'|, \\ y(0) = 0, \; y'(0) = +1, \end{cases}$$

and we intend to show that the interval length $x_1 - x_0 = x_1 - 0$ of Theorem 3.12 is equal to $2\alpha(k, l)$.

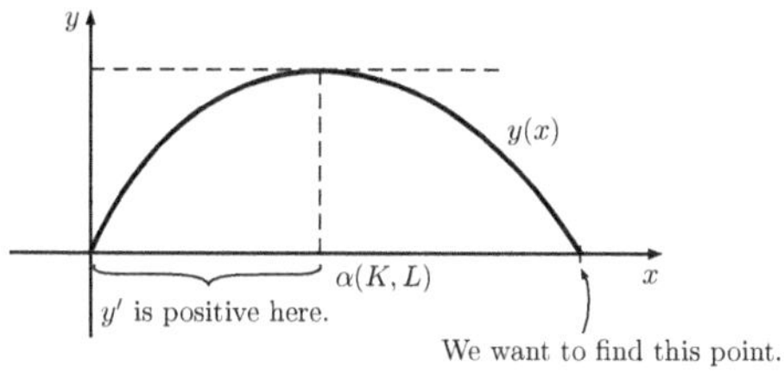

Fig. 3.21 The graph of $y(x)$.

Now from the manner in which $\alpha(K, L)$ is defined, up to $x = \alpha(K, L)$, we have $|y'| = y'$. Now at $x = \alpha(K, L)$, we have $y'' = -Ky < 0$ which implies y' changes sign at $x = \alpha(K, L)$, and so y' becomes negative to the right of $x = \alpha(K, L)$. That is, to the right of $x = \alpha(K, L)$, $y'(x) < 0$ and $y'' = -Ky - L|y'| < 0$, so it must be true that $y'(x)$ is decreasing, and hence $y'(x)$ remains negative until the first positive zero of $y(x)$ is reached, say at $x = x_1$.

We contend that $x_1 = 2\alpha(K, L)$. We argue as follows: Let $\gamma = y'(x_1) < 0$. So $y(x)$ is the solution of

$$\begin{cases} y'' = -Ky + Ly', \\ y(x_1) = 0, \; y'(x_1) = \gamma. \end{cases}$$

(Note that, for $\alpha(K, L) \le x \le x_1$, $|y'| = -y'$.) But translates of solutions are still solutions and so consider the solution $z(x)$ of

$$\begin{cases} y'' = -Ky + Ly', \quad \text{for } \; \alpha(K, L) - x_1 \le x \le 0 \\ y(0) = 0, \; y'(0) = \gamma, \end{cases}$$

i.e., $z(x)$ is a translate from the solution $y(x)$ at x_1. In particular, $z(x) = y(x + x_1)$, for $\alpha(K, L) - x_1 \le x \le 0$.

Now rotate $z(x)$ about the y-axis by replacing x by $-x$; i.e., let $w(x) = z(-x)$, for $0 \le x \le x_1 - \alpha(K, L)$.

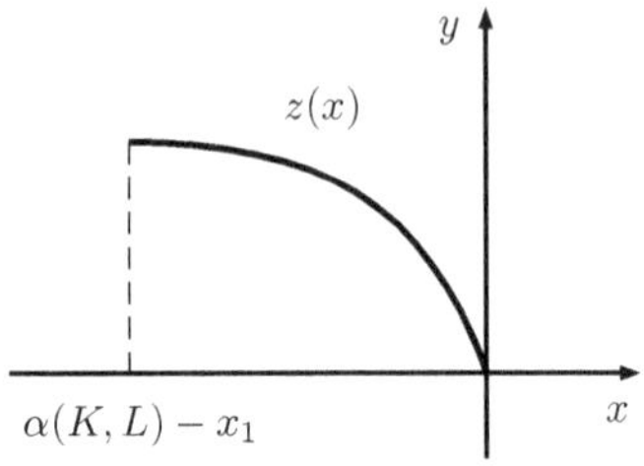

Fig. 3.22 The graph of $z(x)$.

By the Chain Rule, $w'(x) = -z'(-x)$ and $w''(x) = z''(-x)$. Recall that $z(x)$ was the solution of

$$\begin{cases} y'' = -Ky + Ly', & \text{for } \alpha(K,L) - x_1 \le x \le 0, \\ y(0) = 0, \ y'(0) = \gamma, \end{cases}$$

and by construction $x = \alpha(K,L) - x_1$ is the first zero of $z'(x)$ to the left of $x = 0$. Hence $w(x)$ is the solution of

$$\begin{cases} y'' = -Ky - Ly', & \text{for } 0 \le x \le x_1 - \alpha(K,L), \\ y(0) = 0, \ y'(0) = -\gamma > 0, \end{cases}$$

and $x = x_1 - \alpha(K,L)$ is the first zero of $w'(x)$ to the right of $x = 0$.

We recall now that the initial solution $y(x)$ discussed at the beginning of the proof satisfies

$$\begin{cases} y'' = -Ky - Ly', & \text{for } 0 \le x \le \alpha(K,L), \\ y(0) = 0, \ y'(0) = 1. \end{cases}$$

By uniqueness of solutions of IVP's,

$$w(x) = -\gamma y(x)$$

and hence the first zero to the right of $x = 0$ of the derivative $w'(x)$ is $x = \alpha(K,L)$. So, $x_1 - \alpha(K,L) = \alpha(K,L)$; i.e. $\alpha(K,L)$ is the midpoint of the segment $[0, x_1]$. Moreover, it follows immediately that $\gamma = -1$, so that the graph of $y(x)$ is symmetric about the line $x = \alpha(K,L)$. To see this, recall that $w(x)$ was constructed as a reflection about the y-axis of the translation, $z(x)$, of the solution $y(x)$. It follows that for $\alpha(K,L) \le x \le x_1 = 2\alpha(K,L)$, the graph of $y(x)$ is the mirror image of $w(x) = -\gamma y(x)$, $0 \le x \le \alpha(K,L)$. Hence

$$\lim_{x \to \alpha(K,L)^+} y(x) = \lim_{x \to \alpha(K,L)^-} -\gamma y(x),$$

$$y(\alpha(K,L)) = -\gamma y(\alpha(K,L)),$$

$$\gamma = -1,$$

and the graph of $y(x)$ is symmetric about $x = \alpha(K,L)$. □

Exercise **35.** Determine the interval length $2\alpha(K,L) = \beta(K,L)$:
1) $K = 0$, $L \neq 0$, show $\beta(0,L) = +\infty$.
2) For $L^2 = 4K$, find $\beta(K,L)$.
3) For $L^2 > 4K$, find $\beta(K,L)$.
4) For $L^2 < 4K$, find $\beta(K,L)$.
5) In each case above, compare $\beta(K,L)$ with the positive root of $\frac{Kt^2}{8} + \frac{L}{2}t = 1$.

3.4 Existence of Solutions of BVP's for Nonlinear Higher Order ODE's

In this section, we are concerned with the existence of solutions of the nth order BVP consisting of the equation

$$u^{(n)} + f(t, u, u', \dots, u^{(n-1)}) = 0, \ t \in (0,1), \tag{3.10}$$

and the general multi-point BC

$$\begin{cases} u^{(i)}(0) = g_i\left(u^{(i)}(t_1), \dots, u^{(i)}(t_m)\right), \ i = 0, \dots, n-2, \\ u^{(n-2)}(1) = g_{n-1}\left(u^{(n-2)}(t_1), \dots, u^{(n-2)}(t_m)\right), \end{cases} \tag{3.11}$$

where $n \geq 2$ and $m \geq 1$ are integers, $t_j \in [0,1]$ for $j = 1, \dots, m$ with $0 \leq t_1 < t_2 < \cdots < t_m \leq 1$, $f \in C((0,1) \times \mathbb{R}^n)$, and $g_i \in C(\mathbb{R}^m)$ for $i = 0, \dots, n-1$. By a solution of BVP (3.10), (3.11), we mean a function $u \in C^{(n-1)}[0,1] \cap C^{(n)}(0,1)$ such that $u(t)$ satisfies Eq. (3.10) on $(0,1)$, and satisfies BC (3.11).

Let $X = C^{(n-1)}[0,1]$, and for any $u \in C[0,1]$, define $||u||_\infty = \max_{t\in[0,1]} |u(t)|$. Let

$$||u|| = \max\{||u||_\infty, ||u'||_\infty, \dots, ||u^{(n-1)}||_\infty\}$$

and

$$||u||_p = \begin{cases} (\int_0^1 |u(t)|^p dt)^{1/p}, & 1 \leq p < \infty, \\ \inf\{M \ : \ \text{meas}\{t \ : \ |u(t)| > M\} = 0\}, & p = \infty, \end{cases}$$

stand for the norms in X and $L^p(0,1)$, respectively, where meas$\{\cdot\}$ denotes the Lebesgue measure of a set.

We first define the lower and upper solutions of BVP (3.10), (3.11) and a Nagumo condition.

Definition 3.6. A function $\alpha \in X \cap C^n(0,1)$ is said to be a *lower solution* of BVP (3.10), (3.11) if

$$\alpha^{(n)}(t) + f(t, \alpha(t), \alpha'(t), \dots, \alpha^{(n-1)}(t)) \geq 0 \quad \text{on } (0,1), \tag{3.12}$$

$$\begin{cases} \alpha^{(i)}(0) \le g_i\left(\alpha^{(i)}(t_1), \ldots, \alpha^{(i)}(t_m)\right), \ i = 0, \ldots, n-2, \\ \alpha^{(n-2)}(1) \le g_{n-1}\left(\alpha^{(n-2)}(t_1), \ldots, \alpha^{(n-2)}(t_m)\right). \end{cases} \tag{3.13}$$

Similarly, a function $\beta \in X \cap C^n(0,1)$ is said to be an *upper solution* of BVP (3.10), (3.11) if

$$\beta^{(n)}(t) + f(t, \beta(t), \beta'(t), \ldots, \beta^{(n-1)}(t)) \le 0 \quad \text{on } (0,1), \tag{3.14}$$

$$\begin{cases} \beta^{(i)}(0) \ge g_i\left(\beta^{(i)}(t_1), \ldots, \beta^{(i)}(t_m)\right), \ i = 0, \ldots, n-2, \\ \beta^{(n-2)}(1) \ge g_{n-1}\left(\beta^{(n-2)}(t_1), \ldots, \beta^{(n-2)}(t_m)\right). \end{cases} \tag{3.15}$$

Definition 3.7. Let $\alpha, \beta \in X$ satisfy

$$\alpha^{(i)}(t) \le \beta^{(i)}(t) \quad \text{for } t \in [0,1] \text{ and } i = 0, \ldots, n-2. \tag{3.16}$$

We say that f satisfies a *Nagumo condition* with respect to α and β if for

$$\xi = \max\{\beta^{(n-2)}(1) - \alpha^{(n-2)}(0),\ \beta^{(n-2)}(0) - \alpha^{(n-2)}(1)\}, \tag{3.17}$$

there exists a constant $C = C(\alpha, \beta)$ with

$$C > \max\{\xi,\ ||\alpha^{(n-1)}||_\infty,\ ||\beta^{(n-1)}||_\infty\} \tag{3.18}$$

and functions $\phi \in C[0, \infty)$ and $w \in L^p(0,1)$, $1 \le p \le \infty$, such that $\phi > 0$ on $[0, \infty)$,

$$|f(t, x_0, \ldots, x_{n-1})| \le w(t)\phi(|x_{n-1}|) \quad \text{on } (0,1) \times \mathbb{D}_\alpha^\beta \times \mathbb{R}, \tag{3.19}$$

and

$$\int_\xi^C \frac{v^{(p-1)/p}}{\phi(v)} dv > ||w||_p \eta^{(p-1)/p}, \tag{3.20}$$

where $(p-1)/p \equiv 1$ for $p = \infty$,

$$\mathbb{D}_\alpha^\beta = [\alpha(t), \beta(t)] \times [\alpha'(t), \beta'(t)] \times \cdots \times [\alpha^{(n-2)}(t), \beta^{(n-2)}(t)], \tag{3.21}$$

and

$$\eta = \max_{t \in [0,1]} \beta^{(n-2)}(t) - \min_{t \in [0,1]} \alpha^{(n-2)}(t). \tag{3.22}$$

Remark 3.1. Let $\alpha, \beta \in X$ satisfy (3.16). Assume that there exist $w \in L^p(0,1)$, $1 \le p \le \infty$, and $0 \le \sigma \le 1 + (p-1)/p$ such that

$$|f(t, x_0, \ldots, x_{n-1})| \le w(t)(1 + |x_{n-1}|^\sigma) \quad \text{on } (0,1) \times \mathbb{D}_\alpha^\beta \times \mathbb{R}. \tag{3.23}$$

Then f satisfies a Nagumo condition with respect to α and β with $\phi(v) = 1 + v^\sigma$.

Next, we present the main results of this section. Theorem 3.15 below is under the assumption that there exists a pair of lower and upper solutions of BVP (3.10), (3.11).

Theorem 3.15. *Assume that the following conditions hold:*

(H1) *for* $(t, x_0, \ldots, x_{n-1}) \in (0,1) \times \mathbb{R}^n$, $f(t, x_0, \ldots, x_{n-1})$ *is nondecreasing in each of* $x_0, \ldots, x_{n-3}$;

(H2) *for* $i = 1, \ldots, n-1$ *and* $(y_1, \ldots, y_m) \in \mathbb{R}^m$, $g_i(y_1, \ldots, y_m)$ *is nondecreasing in each of its arguments;*

(H3) *BVP* (3.10), (3.11) *has a lower solution* α *and an upper solution* β *satisfying* (3.16);

(H4) f *satisfies a Nagumo condition with respect to* α *and* β *with* $C = C(\alpha, \beta)$ *being the constant introduced in* (3.18).

Then BVP (3.10), (3.11) *has at least one solution* $u(t)$ *satisfying*

$$\alpha^{(i)}(t) \leq u^{(i)}(t) \leq \beta^{(i)}(t) \quad \text{for } t \in [0,1] \text{ and } i = 0, \ldots, n-2, \tag{3.24}$$

and

$$|u^{(n-1)}(t)| \leq C \quad \text{for } t \in [0,1]. \tag{3.25}$$

Remark 3.2. Notice that in (H1) we do not need the monotonicity of f in the last two variables x_{n-2} and x_{n-1}. In particular, for the case when $n = 2$, no monotonicity is required on f.

Below, we prove Theorem 3.15 via several lemmas. We assume that conditions (H1)–(H4) hold. Our first lemma introduces some useful polynomials that form a basis for the space of polynomials on $[0,1]$ with degree less than or equal to $n-1$.

Lemma 3.3. *Let*

$$p_i(t) = \frac{t^i}{i!} \quad i = 0, \ldots, n-3,$$

$$p_{n-2}(t) = \frac{t^{n-2}}{(n-2)!} - \frac{t^{n-1}}{(n-1)!},$$

and

$$p_{n-1}(t) = \frac{t^{n-1}}{(n-1)!}.$$

Then, we have

(i) $||p_i|| \le 1$ *for* $i = 0, \dots, n-1$;
(ii) $p_i(t)$, $i = 0, \dots, n-2$, *satisfies the BVP*

$$p_i^{(n)}(t) = 0, \quad t \in (0,1),$$

$$p_i^{(j)}(0) = \begin{cases} 1, & i = j, \\ 0, & i \neq j, \end{cases} \; j = 0, \dots, n-2, \quad p_i^{(n-2)}(1) = 0,$$

and p_{n-1} *satisfies the BVP*

$$p_{n-1}^{(n)}(t) = 0, \quad t \in (0,1),$$

$$p_{n-1}^{(i)}(0) = 0, \; i = 0, \dots, n-2, \quad p_{n-1}^{(n-2)}(1) = 1.$$

Proof. This can be easily verified by direct computation. □

It is well known that the Green's function for the BVP

$$-u''(t) = 0 \text{ on } (0,1), \quad u(0) = u(1) = 0,$$

is given by

$$G(t,s) = \begin{cases} t(1-s), & 0 \le t \le s \le 1, \\ s(1-t), & 0 \le s \le t \le 1. \end{cases}$$

Let $G_1(t,s) = G(t,s)$ and for $j = 2, \dots, n-1$, recursively define

$$G_j(t,s) = \int_0^t G_{j-1}(v,s)dv. \tag{3.26}$$

Then, $G_j(t,s)$ is the Green's function for the BVP

$$\begin{aligned} &-u^{(j+1)}(t) = 0, \quad t \in (0,1), \\ &u^{(i)}(0) = 0, \; i = 0, \dots, j-1, \quad u^{(j-1)}(1) = 0. \end{aligned} \tag{3.27}$$

Let α and β be given in (H3). For $u \in C^{(n-2)}[0,1]$ and $i = 0, \dots, n-2$, define

$$\tilde{u}^{[i]}(t) = \max\{\alpha^{(i)}(t), \; \min\{u^{(i)}(t), \; \beta^{(i)}(t)\}\}. \tag{3.28}$$

Then, for $i = 0, \dots, n-2$, $\tilde{u}^{[i]}(t)$ is continuous on $[0,1]$,

$$\tilde{\alpha}^{[i]}(t) = \alpha^{(i)}(t), \; \tilde{\beta}^{[i]}(t) = \beta^{(i)}(t), \text{ and } \alpha^{(i)}(t) \le \tilde{u}^{[i]}(t) \le \beta^{(i)}(t) \tag{3.29}$$

for $t \in [0,1]$ and $i = 0, \dots, n-2$. Let $C = C(\alpha, \beta)$ be given in (H4). For $u \in X$, define

$$\hat{u}^{[n-1]}(t) = \max\{-C, \; \min\{u^{(n-1)}(t), \; C\}\} \tag{3.30}$$

and a functional $F : (0,1) \times X \to \mathbb{R}$ by

$$F(t,u(\cdot)) = f(t,\tilde{u}^{[0]}(t),\tilde{u}^{[1]}(t),\ldots,\tilde{u}^{[n-2]}(t),\hat{u}^{[n-1]}(t)) + \frac{\tilde{u}^{[n-2]}(t) - u^{(n-2)}(t)}{1+(u^{(n-2)}(t))^2}. \tag{3.31}$$

Then, for $u \in X$ and $t \in (0,1)$, $F(t,u(\cdot))$ is continuous in u, and from (3.19) and (3.29), we see that

$$|F(t,u(\cdot))| \leq w(t) \max_{y\in[0,C]} \phi(y) + ||\alpha|| + ||\beta|| + 1. \tag{3.32}$$

Consider the BVP consisting of the equation

$$u^{(n)} + F(t,u(\cdot)) = 0, \quad t \in (0,1), \tag{3.33}$$

and the BC

$$\begin{cases} u^{(i)}(0) = g_i\left(\tilde{u}^{[i]}(t_1),\ldots,\tilde{u}^{[i]}(t_m)\right), \ i = 0,\ldots,n-2, \\ u^{(n-2)}(1) = g_{n-1}\left(\tilde{u}^{[n-2]}(t_1),\ldots,\tilde{u}^{[n-2]}(t_m)\right). \end{cases} \tag{3.34}$$

Lemma 3.4 below states that the polynomials introduced earlier can be used to form an equivalent integral equation to BVP (3.33), (3.34).

Lemma 3.4. *The function $u(t)$ is a solution of BVP* (3.33), (3.34) *if and only if $u(t)$ is a solution of the integral equation*

$$u(t) = \sum_{i=0}^{n-1} g_i(\tilde{u}^{[i]}(t_1),\ldots,\tilde{u}^{[i]}(t_m))p_i(t) + \int_0^1 G_{n-1}(t,s)F(s,u(\cdot))ds, \tag{3.35}$$

where p_i, $i = 0,\ldots,n-1$, is defined in Lemma 3.3 *and G_{n-1} is given by* (3.26) *with $j = n-1$.*

Proof. Suppose $u(t)$ is a solution of (3.35). Then using the BCs for p_i and G_{n-1} at $t = 0$ (see Lemma 3.3(ii) and (3.27)), we obtain

$$\begin{aligned} u^{(j)}(0) &= \sum_{i=0}^{n-1} g_i(\tilde{u}^{[i]}(t_1),\ldots,\tilde{u}^{[i]}(t_m))p_i^{(j)}(t) \\ &= g_j(\tilde{u}^{[j]}(t_1),\ldots,\tilde{u}^{[j]}(t_m)),\ j = 0,\ldots,n-2. \end{aligned}$$

A similar argument at $t = 1$ shows

$$u^{(n-2)}(1) = g_{n-1}(\tilde{u}^{[n-2]}(t_1),\ldots,\tilde{u}^{[n-2]}(t_m)).$$

This shows $u(t)$ satisfies (3.34). Equation (3.33) can be verified by differentiating (3.35) n times.

To show the converse, suppose $u(t)$ is a solutions of BVP (3.33), (3.34). Then,

$$\frac{d^n}{dt^n}\left(u(t) - \int_0^1 G_{n-1}(t,s)F(s,u(\cdot))ds\right) = 0.$$

Thus,

$$u(t) - \int_0^1 G_{n-1}(t,s)F(s,u(\cdot))ds = \nu(t),$$

where $\nu(t)$ is a polynomial of degree $n-1$. Since the functions p_i, $i = 0,\ldots,n-1$, form a basis for the space of polynomials of degree $n-1$, there exist constants a_i, $i = 0,\ldots,n-1$, such that

$$u(t) - \int_0^1 G_{n-1}(t,s)F(s,u(\cdot))ds = \sum_{i=0}^{n-1} a_i p_i(t). \tag{3.36}$$

From the properties of G_{n-1} and p_i,

$$u^{(j)}(0) = \sum_{i=0}^{n-1} a_i p_i^{(j)}(0) = a_j,\ j = 0,\ldots,n-2.$$

Since $u(t)$ satisfies (3.34), we have

$$a_j = g_j(\tilde{u}^{[j]}(t_1),\ldots,\tilde{u}^{[j]}(t_m)),\ j = 0,\ldots,n-2.$$

Similarly, we can show that

$$a_{n-1} = g_{n-1}(\tilde{u}^{[n-1]}(t_1),\ldots,\tilde{u}^{[n-1]}(t_m)).$$

Therefore, from (3.36), we see that $u(t)$ satisfies (3.35), which completes the proof of this lemma. □

The following lemma establishes the existence of a solution of BVP (3.33), (3.34).

Lemma 3.5. *BVP* (3.33), (3.34) *has at least one solution.*

Proof. For any $u \in X$, define an operator $T : X \to C[0,1]$ by

$$\begin{aligned}(Tu)(t) = &\sum_{i=0}^{n-1} g_i(\tilde{u}^{[i]}(t_1),\ldots,\tilde{u}^{[i]}(t_m))p_i(t)\\ &+ \int_0^1 G_{n-1}(t,s)F(s,u(\cdot))ds.\end{aligned} \tag{3.37}$$

Then, by Lemma 3.4, $u(t)$ is a solution of BVP (3.33), (3.34) if and only if u is a fixed point of T. From (3.32), we see that $T(X) \subseteq X$. Clearly, $T : X \to X$ is continuous. Next, we show that $T(X)$ is compact. For $u \in X$ and $i = 0, \ldots, n-1$, in view of (3.29) and the fact that g_i is continuous, there exists $d > 0$ such that

$$|g_i(\tilde{u}^{[i]}(t_1), \ldots, \tilde{u}^{[i]}(t_m))| \leq d \quad \text{on } [0,1].$$

Thus, from Lemma 3.3(i), (3.32), and (3.37), we have

$$\begin{aligned}
&(Tu)^{(j)}(t)\\
&= \sum_{i=0}^{n-1} g_i(\tilde{u}^{[i]}(t_1), \ldots, \tilde{u}^{[i]}(t_m))p_i^{(j)}(t) + \int_0^1 (\frac{\partial^j}{\partial t^j} G_{n-1}(t,s))F(s,u(\cdot))ds\\
&\leq nd + ||\alpha|| + ||\beta|| + 1 + \max_{y\in[0,C]} \phi(y) \int_0^1 w(s)ds < \infty
\end{aligned}$$

for $j = 0, \ldots, n-1$. This means that T is uniformly bounded on X and $(Tu)^{(j)}(t)$ is equicontinuous on $[0,1]$ for $j = 0, \ldots, n-2$. Now, we show that $(Tu)^{(n-1)}(t)$ is equicontinuous on $[0,1]$. From the definition of p_i and (3.37), it follows that

$$\begin{aligned}
&(Tu)^{(n-1)}(t)\\
&= -g_{n-2}(\tilde{u}^{[n-2]}(t_1), \ldots, \tilde{u}^{[n-2]}(t_m)) + g_{n-1}(\tilde{u}^{[n-2]}(t_1), \ldots, \tilde{u}^{[n-2]}(t_m))\\
&\quad - \int_0^t F(s,u(\cdot))ds.
\end{aligned}$$

Hence, the equicontinuity of $(Tu)^{(n-1)}(t)$ follows from the property of absolute continuity of integrals. Now, by the Arzelà-Ascoli theorem, $T(X)$ is compact. From the Schauder fixed point theorem, there exists a fixed point u of T in X. Thus, $u(t)$ is a solution of BVP (3.33), (3.34). This completes the proof of the lemma. □

The next two lemmas show some properties for a solution of BVP (3.33), (3.34).

Lemma 3.6. *If $u(t)$ is a solution of BVP* (3.33), (3.34), *then $u(t)$ satisfies* (3.24).

Proof. We first show that $u^{(n-2)}(t) \leq \beta^{(n-2)}(t)$ for $t \in [0,1]$. Suppose, for the sake of a contradiction, that there exists $t^* \in [0,1]$ such that

$u^{(n-2)}(t^*) > \beta^{(n-2)}(t^*)$. Without loss of generality, assume $u^{(n-2)}(t) - \beta^{(n-2)}(t)$ is maximized at t^*. If $t^* = 0$, then from (3.15), (H2), (3.29), and (3.34), we have

$$\begin{aligned} u^{(n-2)}(0) &= g_{n-2}(\tilde{u}^{[n-2]}(t_1), \ldots, \tilde{u}^{[n-2]}(t_m)) \\ &\leq g_{n-2}(\beta^{[n-2]}(t_1), \ldots, \beta^{[n-2]}(t_m)) \leq \beta^{(n-2)}(0), \end{aligned}$$

which is a contradiction. A similar contradiction occurs at $t^* = 1$. If $t^* \in (0,1)$, then $u^{(n-1)}(t^*) = \beta^{(n-1)}(t^*)$ and $u^{(n)}(t^*) \leq \beta^{(n)}(t^*)$, so from (3.18) and (3.30), $\hat{u}^{[n-1]}(t^*) = \beta^{(n-1)}(t^*)$. Now, in virtue of (3.14), (H1), (3.29), (3.31), and (3.33), it follows that

$$\begin{aligned} 0 &\geq u^{(n)}(t^*) - \beta^{(n)}(t^*) \\ &\geq -f(t^*, \tilde{u}^{[0]}(t^*), \tilde{u}^{[1]}(t^*), \ldots, \tilde{u}^{[n-2]}(t^*), \hat{u}^{[n-1]}(t^*)) \\ &\quad - \frac{\beta^{(n-2)}(t^*) - u^{(n-2)}(t^*)}{1 + (u^{(n-2)}(t^*))^2} \\ &\quad + f(t^*, \beta'(t^*), \beta'(t^*), \ldots, \beta^{(n-2)}(t^*), \beta^{(n-1)}(t^*)) \\ &\geq \frac{u^{(n-2)}(t^*) - \beta^{(n-2)}(t^*)}{1 + (u^{(n-2)}(t^*))^2} > 0. \end{aligned}$$

We again reach a contradiction. Thus, $u^{(n-2)}(t) \leq \beta^{(n-2)}(t)$ on $[0,1]$. Similarly, we can show that $u^{(n-2)}(t) \geq \alpha^{(n-2)}(t)$ on $[0,1]$. Hence,

$$\alpha^{(n-2)}(t) \leq u^{(n-2)}(t) \leq \beta^{(n-2)}(t) \quad \text{for } t \in [0,1]. \tag{3.38}$$

For $i = 0, \ldots, n-3$, from (H2) and (3.29),

$$\begin{aligned} &g_i(\alpha^i(t_1), \ldots, \alpha^{(i)}(t_m)) \\ &\leq g_i(\tilde{u}^{[i]}(t_1), \ldots, \tilde{u}^{[i]}(t_m)) \\ &\leq g_i(\beta^{(i)}(t_1), \ldots, \beta^{(i)}(t_m)), \end{aligned}$$

which together with (3.13), (3.15), and (3.34) implies that

$$\alpha^{(i)}(0) \leq u^{(i)}(0) \leq \beta^{(i)}(0). \tag{3.39}$$

Finally, integrating (3.38) and using (3.39), we see that $u(t)$ satisfies (3.24) completing the proof. □

Lemma 3.7. *If $u(t)$ is a solution of BVP (3.33), (3.34), then $u^{(n-1)}(t)$ satisfies (3.25).*

Proof. By Lemma 3.6, $u(t)$ satisfies (3.24). If (3.25) does not hold, then there exists $\tilde{t} \in [0,1]$ such that $u^{(n-1)}(\tilde{t}) > C$ or $u^{(n-1)}(\tilde{t}) < -C$. By the mean value theorem, there exists $\hat{t} \in [0,1]$ such that $u^{(n-1)}(\hat{t}) = u^{(n-2)}(1) - u^{(n-2)}(0)$. Then, from (3.17), (3.18), and (3.24), we have

$$\begin{aligned} -C < -\xi \le \alpha^{(n-2)}(1) - \beta^{(n-2)}(0) \\ \le u^{(n-1)}(\hat{t}) \le \beta^{(n-2)}(1) - \alpha^{n-2}(0) \le \xi < C. \end{aligned}$$

If $u^{(n-1)}(\tilde{t}) > C$, there exist $s_1, s_2 \in [0,1]$ such that $u^{(n-1)}(s_1) = \xi$, $u^{(n-1)}(s_2) = C$, and

$$\xi = u^{(n-1)}(s_1) \le u^{(n-1)}(t) \le u^{(n-1)}(s_2) = C \quad \text{for } t \in I, \tag{3.40}$$

where $I = [s_1, s_2]$ or $I = [s_2, s_1]$. In what follows, we only consider the case $I = [s_1, s_2]$ since the other case can be treated similarly. From (3.30) and (3.40), $\hat{u}^{[n-1]}(t) = u^{(n-1)}(t)$ on I, and in view of (3.24) and (3.28), we have $\tilde{u}^{[i]}(t) = u^{(i)}(t)$ for $t \in [0,1]$ and $i = 0, \ldots, n-2$. Thus, from (3.31), $F(t, u(\cdot)) \equiv f(t, u(t), u'(t), \ldots, u^{(n-1)}(t))$ on I. Then, by a change of variables and from (3.19) and (3.33), we obtain that

$$\begin{aligned} &\int_\xi^C \frac{v^{(p-1)/p}}{\phi(v)} dv = \int_{u^{(n-1)}(s_1)}^{u^{(n-1)}(s_2)} \frac{v^{(p-1)/p}}{\phi(v)} dv \\ &= \int_{s_1}^{s_2} \frac{u^{(n)}(s)}{\phi(u^{(n-1)}(s))} (u^{(n-1)}(s))^{(p-1)/p} ds \\ &= \int_{s_1}^{s_2} \frac{-f\left(s, u(s), u'(s), \ldots, u^{(n-1)}(s)\right)}{\phi(u^{(n-1)}(s))} (u^{(n-1)}(s))^{(p-1)/p} ds \\ &\le \int_{s_1}^{s_2} w(s)(u^{(n-1)}(s))^{(p-1)/p} ds. \end{aligned}$$

Hence, Hölder's inequality implies

$$\int_\xi^C \frac{v^{(p-1)/p}}{\phi(v)} dv \le ||w||_p \left(\int_{s_1}^{s_2} u^{(n-1)}(s) ds \right)^{(p-1)/p} \le ||w||_p \eta^{(p-1)/p},$$

where η is defined by (3.22). But this contradicts (3.20). Therefore, $u^{(n-1)}(t)$ satisfies (3.25). If $u^{(n-1)}(\tilde{t}) < -C$, by a similar argument as above, we still can show that (3.25) holds. The proof is now complete. □

We are now in a position to prove Theorem 3.15.

Proof of Theorem 3.15. Theorem 3.15 readily follows from Lemmas 3.5–3.7. □

As applications of Theorem 3.15, we give explicit conditions for the existence of solutions of BVP (3.10), (3.11). To do so, we need the following assumptions.

(A1) For any $r > 0$, there exist $\mu_r \in L^p(0,1)$ such that

$$|f(t,x_0,\ldots,x_{n-1})| \leq \mu_r(t)(1+|x_{n-1}|^\sigma) \quad \text{on } (0,1)\times\mathbb{D}_r\times\mathbb{R}, \tag{3.41}$$

where $0 \leq \sigma \leq 1+(p-1)/p$ and

$$\mathbb{D}_r = \underbrace{[-r,r]\times[-r,r]\times\ldots\times[-r,r]}_{n-1}. \tag{3.42}$$

(A2) There exist $\delta > 0$, $\theta \in L^1(0,1)$, and $\psi \in C[0,\infty)$ such that $\theta > 0$ on $(0,1)$, $\psi > 0$ on $[0,\infty)$, $\psi(\cdot)$ is locally Lipschitz on $[0,\infty)$,

$$x_0 f(t,x_0,\ldots,x_{n-1}) \leq \theta(t)|x_0|\psi(|x_{n-1}|) \quad \text{on } \mathbb{E}_\delta, \tag{3.43}$$

and

$$\int_0^\infty \frac{dv}{\psi(v)} > \int_0^1 \theta(s)ds, \tag{3.44}$$

where E_δ is defined by the set

$$\{(t,x_0,\ldots,x_{n-1}) \in (0,1)\times\mathbb{R}^n | x_i \geq \delta, i=0,\ldots,n-2,\ x_{n-1} \leq 0\}$$
$$\cup\ \{(t,x_0,\ldots,x_{n-1}) \in (0,1)\times\mathbb{R}^n | x_i \leq -\delta, i=0,\ldots,n-2,\ x_{n-1} \geq 0\}.$$

(A3) for $i=0,\ldots,n-1$, g_i is Lipschitz, i.e., there exists $c_{ij} \geq 0$, $j=1,\ldots,m$, such that

$$|g_i(x_1,\ldots,x_m) - g_i(y_1,\ldots,y_m)| \leq \sum_{j=1}^m c_{ij}|x_j - y_j| \tag{3.45}$$

for any $(x_1,\ldots,x_m), (y_1,\ldots,y_m) \in \mathbb{R}^m$.

(A4) For the constants c_{ij} given in (A3), there exists $0 < \Lambda < 1$ such that

$$\sum_{i=0}^{n-2}\sum_{j=1}^m \frac{c_{ij}}{i!} + \sum_{j=1}^m \frac{c_{(n-1)j}}{(n-2)!} \leq \Lambda. \tag{3.46}$$

The following theorem gives explicit conditions for the existence of solutions of BVP (3.10), (3.11).

Theorem 3.16. *Assume that* (H1), (H2), *and* (A1)–(A4) *hold. Then BVP* (3.10), (3.11) *has at least one solution $u(t)$. Moreover, $u(t)$ is nontrivial if one of the following conditions holds:*

(B1) *there exists a subset* S *of* $(0,1)$ *with positive measure such that* $f(t,0,\ldots,0) \not\equiv 0$ *for* $t \in S$;

(B2) *there exists* $i_0 \in \{0,\ldots,n-1\}$ *such that* $g_{i_0}(0,\ldots,0) \neq 0$.

As applications of our results, we also obtain existence results for the BVP consisting of Eq. (3.10) and the nonhomogeneous BC

$$\begin{cases} u^{(i)}(0) = \sum_{j=1}^{m} a_{ij} u^{(i)}(t_j) + \lambda_i, \ i = 0,\ldots,n-2, \\ u^{(n-2)}(1) = \sum_{j=1}^{m} a_{(n-1)j} u^{(n-2)}(t_j) + \lambda_{n-1}, \end{cases} \tag{3.47}$$

where $a_{ij}, \lambda_i \in \mathbb{R}$ for $i = 0,\ldots,n-1$ and $j = 1,\ldots,m$.

Corollary 3.8. *Let* a_{ij} *and* λ_i *be given as in* (3.47). *Assume that* (H1), (A1), *and* (A2) *hold together with the following conditions:*

(C1) $a_{ij} \geq 0$ *for* $i = 0,\ldots,n-1$ *and* $j = 1,\ldots,m$;

(C2) *there exists* $0 < \Lambda < 1$ *such that*

$$\sum_{i=0}^{n-2} \sum_{j=1}^{m} \frac{a_{ij}}{i!} + \sum_{j=1}^{m} \frac{a_{(n-1)j}}{(n-2)!} \leq \Lambda. \tag{3.48}$$

Then, for each $(\lambda_0,\ldots,\lambda_{n-1}) \in \mathbb{R}^n$, *BVP* (3.10), (3.47) *has at least one solution* $u(t)$. *Moreover, if either (B1) holds or* $\sum_{i=0}^{n-1} \lambda_i^2 \neq 0$, *then* $u(t)$ *is nontrivial.*

In the following, we assume (H1), (H2), and (A1)–(A4) hold. The following lemma is a special case of [Graef and Kong (2007), Lemma 3.7].

Lemma 3.8. *Let* θ *and* ψ *be given in* (A2). *Then the initial value problem* (IVP)

$$z'(t) = -\theta(t)\psi(|z(t)|), \quad z(0) = 0, \tag{3.49}$$

has a unique solution $z(t)$ *satisfying* $z(t) \leq 0$ *on* $[0,1]$.

Now, we prove Theorem 3.16.

Proof of Theorem 3.16. Let $z(t)$ be the unique solution of IVP (3.49). For any $k > 0$, we show that the BVP consisting of the equation

$$u^{(n-1)}(t) = z(t), \quad t \in [0,1], \tag{3.50}$$

and the BC

$$\begin{cases} u^{(i)}(0) = k + \left|g_i\left(u^{(i)}(t_1),\ldots,u^{(i)}(t_m)\right)\right|, \ i = 0,\ldots,n-3, \\ u^{(n-2)}(0) = k + \left|g_{n-2}\left(u^{(n-2)}(t_1),\ldots,u^{(n-2)}(t_m)\right)\right| \\ \qquad\qquad + \left|g_{n-1}\left(u^{(n-2)}(t_1),\ldots,u^{(n-2)}(t_m)\right)\right| \end{cases} \tag{3.51}$$

has a unique solution $\beta_k(t)$.

For any $k > 0$ and $u \in C^{(n-2)}[0,1]$, define an operator $A_k : C^{(n-2)}[0,1] \to C^{(n-2)}[0,1]$ by

$$
\begin{aligned}
(A_k u)(t) = {} & \sum_{i=0}^{n-2} (k + |g_i(u^{(i)}(t_1), \ldots, u^{(i)}(t_m))|) \frac{t^i}{i!} \\
& + |g_{n-1}(u^{(n-2)}(t_1), \ldots, u^{(n-2)}(t_m))| \frac{t^{n-2}}{(n-2)!} \\
& + \frac{1}{(n-2)!} \int_0^t (t-s)^{n-2} z(s) ds. \qquad (3.52)
\end{aligned}
$$

In the above definition, we take $0^0 = 1$. Clearly, a solution of BVP (3.50), (3.51) is a fixed point of A_k. For any $u, v \in C^{(n-2)}[0,1]$, $t \in [0,1]$, and $j = 0, 1, \ldots, n-2$, from (3.45), (3.46), and (3.52), we have

$$
\begin{aligned}
& |(A_k u)^{(j)}(t) - (A_k v)^{(j)}(t)| \\
= & \left| \sum_{i=j}^{n-2} |g_i(u^{(i)}(t_1), \ldots, u^{(i)}(t_m))| \frac{t^i}{i!} \right. \\
& + |g_{n-1}(u^{(n-2)}(t_1), \ldots, u^{(n-2)}(t_m))| \frac{t^{n-2}}{(n-2)!} \\
& - \sum_{i=0}^{n-2} |g_i(v^{(i)}(t_1), \ldots, v^{(i)}(t_m))| \frac{t^i}{i!} \\
& \left. - |g_{n-1}(v^{(n-2)}(t_1), \ldots, v^{(n-2)}(t_m))| \frac{t^{n-2}}{(n-2)!} \right| \\
\leq & \sum_{i=0}^{n-2} |g_i(u^{(i)}(t_1), \ldots, u^{(i)}(t_m)) - g_i(v^{(i)}(t_1), \ldots, v^{(i)}(t_m))| \frac{t^i}{i!} \\
& + |g_{n-1}(u^{(n-2)}(t_1), \ldots, u^{(n-2)}(t_m)) \\
& - g_{n-1}(v^{(n-2)}(t_1), \ldots, v^{(n-2)}(t_m))| \frac{t^{n-2}}{(n-2)!} \\
\leq & \sum_{i=0}^{n-2} \sum_{j=1}^{m} c_{ij} |u^{(i)}(t_j) - v^{(i)}(t_j)| \frac{t^i}{i!} \\
& + \sum_{j=1}^{m} c_{(n-1)j} |u^{(n-2)}(t_j) - v^{(n-2)}(t_j)| \frac{t^{n-2}}{(n-2)!} \\
\leq & \left(\sum_{i=0}^{n-2} \sum_{j=1}^{m} \frac{c_{ij}}{i!} + \sum_{j=1}^{m} \frac{c_{(n-1)j}}{(n-2)!} \right) ||u - v|| \leq \Lambda ||u - v||.
\end{aligned}
$$

Thus, A_k is a contraction mapping. Hence, for any $k > 0$, A_k has a unique fixed point β_k in X, and consequently, BVP (3.50), (3.51) has a unique solution $\beta_k(t)$. Choose k_1 large enough so that

$$k_1 + \int_0^1 z(s)ds \geq 0 \tag{3.53}$$

and

$$\beta^{(i)}(t) := \beta_{k_1}^{(i)}(t) \geq \delta \quad \text{for } t \in [0,1] \text{ and } i = 0, \ldots, n-2, \tag{3.54}$$

where δ is given in (A2). From (3.49) and (3.50), it is clear that

$$\beta^{(n)}(t) = z'(t) = -\theta(t)\psi(|z(t)|) = -\theta(t)\psi(|\beta^{(n)}(t)|). \tag{3.55}$$

In view of (3.54) and the fact that $\beta^{(n-1)}(t) = z(t) \leq 0$ on $[0,1]$, then, from (3.43) and (3.55), it follows that

$$\beta^{(n)}(t) + f(t, \beta(t), \beta'(t), \ldots, \beta^{(n-1)}(t)) \leq 0 \quad \text{for } t \in (0,1),$$

i.e., $\beta(t)$ satisfies (3.14). Moreover, in view of (3.52), the fact that β_{k_1} is a fixed point of A_{k_1}, and (3.53), it is easy to see that (3.15) holds. Thus, $\beta(t)$ is an upper solution of BVP (3.10), (3.11).

Now, for any $k > 0$, consider the BVP consisting of the equation

$$u^{(n-1)}(t) = -z(t), \quad t \in [0,1], \tag{3.56}$$

and the BC

$$\begin{cases} u^{(i)}(0) = -k - \left|g_i\left(u^{(i)}(t_1), \ldots, u^{(i)}(t_m)\right)\right|, \ i = 0, \ldots, n-3, \\ u^{(n-2)}(0) = -k - \left|g_{n-2}\left(u^{(n-2)}(t_1), \ldots, u^{(n-2)}(t_m)\right)\right| \\ \qquad\qquad - \left|g_{n-1}\left(u^{(n-2)}(t_1), \ldots, u^{(n-2)}(t_m)\right)\right|. \end{cases} \tag{3.57}$$

Then, a solution of BVP (3.56), (3.57) is a fixed point of the operator B_k defined by

$$\begin{aligned}(B_k u)(t) = &-\sum_{i=0}^{n-2}(k + |g_i(u^{(i)}(t_1), \ldots, u^{(i)}(t_m))|)\frac{t^i}{i!} \\ &-|g_{n-1}(u^{(n-2)}(t_1), \ldots, u^{(n-2)}(t_m))|\frac{t^{n-2}}{(n-2)!} \\ &-\frac{1}{(n-2)!}\int_0^t (t-s)^{n-2} z(s)ds.\end{aligned}$$

Using an argument similar to the one above, we can show that there exists k_2 large enough so that BVP (3.56), (3.57) with $k = k_2$ has a unique solution $\alpha(t)$ satisfying $\alpha^{(i)} \leq -\delta$ on $[0,1]$, $i = 0, \ldots, n-2$. Then, arguing as above, we can see that α is a lower solution of BVP (3.10), (3.11). Clearly, $\alpha(t)$ and $\beta(t)$ satisfy (3.16).

Let $r = \max\{||\alpha||, ||\beta||\}$. Then (A1) implies that (3.23) holds with $\mu(t) = \mu_r(t)$. Thus, by Remark 3.1, f satisfies a Nagumo condition with respect to α and β. Therefore, by Theorem 3.15, BVP (3.10), (3.11) has at least one solution. Finally, it is obvious that if either (B1) or (B2) holds, then $u(t)$ is nontrivial. This completes the proof of the theorem. □

Proof of Corollary 3.8. For $(y_1, \ldots, y_m) \in \mathbb{R}^m$ and $i = 0, \ldots, n-1$, let

$$g_i(y_1, \ldots, y_m) = \sum_{j=1}^{m} a_{ij} y_j + \lambda_i.$$

Then, (C1) implies that (H2) holds. Moreover, noting (C2), we see that (A3) and (A4) hold with $c_{ij} = a_{ij}$ for $i = 0, \ldots, n-1$ and $j = 1, \ldots, m$. From the assumption that $\sum_{i=0}^{n-1} \lambda_i^2 \neq 0$, (B2) holds. Corollary 3.8 then follows from Theorem 3.16. □

Finally, in this section, we give several examples to illustrate the results.

Example 3.2. Consider the BVP consisting of the equation

$$u'' - u + (u')^2 - 1/2 = 0, \ t \in (0, 1), \tag{3.58}$$

and the BC

$$u(0) = u(1) = u^3(1/2). \tag{3.59}$$

We claim that BVP (3.58), (3.59) has at least one nontrivial solution $u(t)$ satisfying

$$-t/2 - 1/4 \leq u(t) \leq t/2 + 1/4 \quad \text{for } t \in [0, 1]. \tag{3.60}$$

In fact, with $n = 2$, $m = 1$, $t_1 = 1/2$, $f(t, x_0, x_1) = -x_0 + x_1^2 - 1/2$, $g_0(x) = g_1(x) = x^3$, we see that BVP (3.58), (3.59) is of the form of BVP (3.10), (3.11). By Remark 3.2, it is clear that (H1) and (H2) hold.

Let $\alpha(t) = -t/2 - 1/4$ and $\beta(t) = t/2 + 1/4$ for $t \in [0, 1]$. We can easily check that $\alpha(t)$ and $\beta(t)$ are lower and upper solutions of BVP (3.58), (3.59), respectively, and satisfy (3.16). Moreover, it is easy to see that

$$|f(t, x_0, x_1)| \leq 2 + x_1^2 \leq 2(1 + x_1^2) \quad \text{on } (0, 1) \times \mathbb{D}_\alpha^\beta \times \mathbb{R},$$

where $\mathbb{D}_\alpha^\beta$ is defined by (3.21) with the above $\alpha(t)$ and $\beta(t)$. Thus, (3.23) holds with $p = \infty$, $w(t) = 2$, and $\sigma = 2$. Then, from Remark 3.1, f satisfies a Nagumo condition with respect to α and β. By Theorem 3.15, BVP (3.58), (3.59) has one solution $u(t)$ satisfying (3.60). Finally, in view of Eq. (3.58), $u(t)$ is nontrivial.

Example 3.3. Consider the BVP consisting of the equation

$$u''' + tu + t^2(u')^{1/3} - 10t^{-1/2}u'' = 0, \ t \in (0,1), \tag{3.61}$$

and the BC

$$u(0) = u^{1/3}(1/2) + 2, \quad u'(0) = u'(1) = u^{1/3}(1/2). \tag{3.62}$$

We claim that BVP (3.61), (3.62) has at least one nontrivial solution $u(t)$ satisfying

$$-t^2 - 7t \le u(t) \le t^2 + 7t + 4 \tag{3.63}$$

and

$$-2t - 7 \le u'(t) \le 2t + 7 \tag{3.64}$$

for $t \in [0,1]$.

In fact, with $n = 3$, $m = 1$, $t_1 = 1/2$, $f(t, x_0, x_1, x_2) = tx_0 + t^2 x_1^{1/3} - 10t^{-1/2}x_2$, $g_0(x) = x^{1/3} + 2$, and $g_1(x) = g_2(x) = x^{1/3}$, we see that BVP (3.61), (3.62) is of the form of BVP (3.10), (3.11). Clearly, (H1) and (H2) hold.

Let $\alpha(t) = -t^2 - 7t$ and $\beta(t) = t^2 + 7t + 4$ for $t \in [0,1]$. Then we can check that $\alpha(t)$ and $\beta(t)$ are lower and upper solutions of BVP (3.61), (3.62), respectively, and satisfy (3.16). Moreover, it is easy to see that

$$\begin{aligned} &|f(t, x_0, x_1, x_2)| \\ &\le 15 + 10t^{-1/2}|x_2| \le 15t^{-1/2}(1 + |x_2|) \quad \text{on } (0,1) \times \mathbb{D}_\alpha^\beta \times \mathbb{R}, \end{aligned}$$

where $\mathbb{D}_\alpha^\beta$ is defined by (3.21) with the above $\alpha(t)$ and $\beta(t)$. Thus, (3.23) holds with $p = 1$, $w(t) = 15t^{-1/2}$, and $\sigma = 1$. Then, from Remark 3.1, f satisfies a Nagumo condition with respect to α and β. By Theorem 3.15, BVP (3.61), (3.62) has one solution $u(t)$ satisfying (3.63) and (3.64). Finally, in view of BC (3.62), $u(t)$ is nontrivial.

Example 3.4. Consider the BVP consisting of the equation

$$u^{(4)} - \exp(u'')\left(\pi - \arctan u + (u''')^2\right) = 0, \ t \in (0,1), \tag{3.65}$$

and the BC

$$\begin{cases} u(0) = (1/4)u(1/3) + \lambda_0, \ u'(0) = (1/8)u'(1/2) + \lambda_1, \\ u''(0) = (1/8)u''(2/3) + \lambda_2, \ u''(1) = u''(3/4) + \lambda_3. \end{cases} \tag{3.66}$$

We claim that, for each $(\lambda_0, \lambda_1, \lambda_2, \lambda_3) \in \mathbb{R}^4$, BVP (3.65), (3.66) has at least one nontrivial solution.

In fact, with $n = m = 4$, $t_1 = 1/3$, $t_2 = 1/2$, $t_3 = 2/3$, $t_4 = 3/4$, $f(t, x_0, x_1, x_2, x_3) = -e^{x_2}\left(\pi - \arctan x_0 + x_3^2\right)$, $a_{01} = 1/4$, $a_{02} = a_{03} = a_{04} = 0$, $a_{11} = 0$, $a_{12} = 1/8$, $a_{13} = a_{14} = 0$, $a_{21} = a_{22} = 0$, $a_{23} = 1/8$, $a_{24} = 0$, $a_{31} = a_{32} = a_{33} = 0$, and $a_{34} = 1$, it is easy to see that BVP (3.65), (3.66) is of the form of BVP (3.10), (3.47). Clearly, (H1), (C1), (C2) with $\Lambda = 15/16$, and (B1) hold.

For any $r > 0$, since

$$\begin{aligned} &|f(t, x_0, x_1, x_2, x_3)| \\ &\leq e^r\left(3\pi/2 + x_3^2\right) \leq (3\pi e^r/2)\left(1 + x_3^2\right) \quad \text{on } (0,1) \times \mathbb{D}_r \times \mathbb{R}, \end{aligned}$$

where $\mathbb{D}_r$ is defined by (3.42), we see that (A1) holds with $\mu_r(t) = 3\pi e^r/2$, $p = \infty$, and $\sigma = 2$.

For any $\delta > 0$, if $x_i \geq \delta$, $i = 0, 1, 2$, then we have

$$f(t, x_0, x_1, x_2, x_3) \leq -e^{\delta}\left(\pi/2 + x_3^2\right) < 0,$$

and if $x_i \leq -\delta$, $i = 0, 1, 2$, then we have

$$f(t, x_0, x_1, x_2, x_3) \geq -e^{-\delta}\left(3\pi/2 + x_3^2\right).$$

Let $\theta(t) = 1$, $\psi(x_3) = e^{-\delta}\left(3\pi/2 + x_3^2\right)$. Clearly, $\theta \in L^1(0,1)$ and $\psi \in C[0,\infty)$ satisfy $\theta > 0$ on $(0,1)$, $\psi > 0$ on $[0,\infty)$, $\psi(\cdot)$ is locally Lipschitz on $[0,\infty)$, and (3.43) holds. Note that

$$\int_0^1 \theta(s)ds = 1$$

and

$$\int_0^\infty \frac{dv}{\psi(v)} = e^{\delta}\int_0^\infty \frac{dv}{3\pi/2 + v^2} = e^{\delta}\sqrt{\pi/6} > 1 \quad \text{if } \delta \text{ is large.}$$

Thus, (3.44) holds for a large δ. Thus, (A2) holds. The conclusion now readily follows from Corollary 3.8.

Bibliography

Agarwal R. P. (1986). *Boundary Value Problems for Higher Order Differential Equations* (World Scientific Publishing Co., Inc., Teaneck, NJ).

Bailey P. B., Shampine L. F. and Waltman P. E. (1968). *Nonlinear Two Point Boundary Value Problems* (Academic Press, New York).

Bernfeld S. R. and Lakshmikantham V. (1974). *An Introduction to Nonlinear Boundary Value Problems* (Academic Press, New York).

Boas R. P. (1996). *A Primer of Real Functions*, Fourth Edition, Caras Mathematical Monographs, Vol. 13 (Math. Association of America, Washington, D. C.).

Eloe P. W. and Henderson J. (2016). *Nonlinear Interpolation and Boundary Value Problems*, Trends in Abstract and Applied Analysis, Vol. 2 (World Scientific Publishing Co. Pte. Ltd., Hackensack, NJ).

Erbe L. (1970). Nonlinear boundary value problems for second order differential equations, *J. Differential Equations* **7**, pp. 459–472.

Graef J. R. and Kong L. (2007), Existence of solutions for nonlinear boundary value problems, *Comm. Appl. Nonlinear Anal.* **14**, pp. 39–60.

Graef J. R., Kong L. and Kong Q. (2011), Higher order multi-point boundary value problems, *Math. Nachr.* **284** (2011), pp. 39–52.

Hartman P. (1964). *Ordinary Differential Equations* (Wiley, New York).

Jackson L. K. (1968). Subfunctions and second order ordinary differential equations, *Adv. Math.* **2**, pp. 307–363.

Jackson L. K. (1976). A Nagumo condition for ordinary differential equations, *Proc. Amer. Math. Soc.* **57**, pp. 93–96.

Jackson L. K. (1977). Boundary value problems for ordinary differential equations, *Studies in Ordinary Differential Equations*, Stud. in Math., Vol. 14 (Editor: J. K. Hale) (Math. Association of America, Washington, D. C.), pp. 93–127.

Klaasen G. (1971). Differential inequalities and existence theorems for second and third order boundary value problems, *J. Differential Equations* **10**, pp. 529–537.

Lasota A. and Opial Z. (1967). On the existence and uniqueness of solutions of a boundary value problem for an ordinary second order differential equation, *Colloq. Math.* **18**, pp. 1–5.

Schrader K. W. (1969). Existence theorems for second order boundary value problems, *J. Differential Equations* **5**, pp. 572–584.

Waltman P. E. (1968). A nonlinear boundary value problem, *J. Differential Equations* **4**, pp. 597–603.

Chapter 4

Existence of Solutions II

A "uniqueness implies existence" result is established for conjugate BVP's for a third order ODE. A compactness condition on bounded sequences of solutions plays an important role.

4.1 Existence of Solutions of BVP's Associated with $y''' = f(x, y, y', y'')$

This section will be devoted to proving the following theorem.

Theorem 4.1. *Assume that*

$$y''' = f(x, y, y', y'') \tag{4.1}$$

satisfies the conditions

(A) $f : (a, b) \times \mathbb{R}^3 \to \mathbb{R}$ *is continuous,*
(B) *All solutions of* (4.1) *extend to* (a, b), *and*
(C) *For any* $a < x_1 < x_2 < x_3 < b$ *and any solutions* $y(x)$ *and* $z(x)$ *of* (4.1), *it follows that* $y(x_i) = z(x_i)$, $i = 1, 2, 3$, *implies* $y(x) = z(x)$ *on* $[x_1, x_3]$.

Then for any $a < x_1 < x_2 < x_3 < b$ *and* $y_1, y_2, y_3 \in \mathbb{R}$, *the BVP*

$$\begin{cases} y''' = f(x, y, y', y''), \\ y(x_1) = y_1, \ y(x_2) = y_2, \ y(x_3) = y_3 \end{cases} \tag{4.2}$$

has a solution.

Note: This is a "uniqueness implies existence theorem," for (C) guarantees uniqueness.

Before attempting a proof of Theorem 4.1, it is necessary that we prove "a number" of lemmas and theorems.

Remark 4.1. In the proof of Theorem 3.11, we made use of the following: If $y'' = f(x, y, y')$ satisfies (A) and (B), if $[c, d] \subseteq (a, b)$, and if $\{y_k(x)\}_{k=1}^{\infty}$ is a sequence of solutions of $y'' = f(x, y, y')$, with $|y_k(x)| \leq M$ on $[c, d]$ for some $M > 0$ and for all k, then there exists a subsequence $\{y_{k_j}(x)\}$ such that $\lim_{j\to\infty} y_{k_j}(x)$ and $\lim_{j\to\infty} y'_{k_j}(x)$ exist uniformly on each compact subinterval of (a, b).

The argument used there might lead one to make a conjecture such as:

Conjecture: If (4.1), $y''' = f(x, y, y', y'')$ satisfies (A) and (B), if $[c, d] \subseteq (a, b)$, and if $\{y_k(x)\}$ is a sequence of solutions of (4.1) with $|y_k(x)| \leq M$ on $[c, d]$, for some $M > 0$ and for all k, then there is a subsequence $\{y_{k_j}(x)\}$ such that $\lim_{j\to\infty} y_{k_j}^{(i)}(x)$ exists uniformly on each compact subinterval of (a, b), for $i = 0, 1, 2$.

However the conjecture is false. The differential equation $y''' = -(y')^3$ satisfies the hypotheses (A) and (B), yet the conclusion of the conjecture does not hold. The details of this example will be discussed after the proof of Theorem 4.1 is completed. It is the case though that if hypothesis (C) is also taken into account, we will prove the following:

Compactness Condition: Assume that Eq. (4.1), $y''' = f(x, y, y', y'')$ satisfies (A), (B), and (C) of Theorem 4.1. Then if $[c, d] \subset (a, b)$ and if $\{y_k(x)\}$ is a sequence of solutions of (4.1) such that $|y_k(x)| \leq M$ on $[c, d]$ for some $M > 0$ and for all k, then there exists a subsequence $\{y_{k_j}(x)\}$ such that $\{y_{k_j}^{(i)}(x)\}$ converges uniformly on each compact subinterval of (a, b), for $i = 0, 1, 2$.

Before we prove the "Compactness Condition", we must prove a number of lemmas.

Lemma 4.1. *Let $[a, b]$ be a compact interval and let $M > 0$ be given. Then there exists $N > 0$ depending only on M and $b - a$, such that, if $y(x) \in C^{(2)}[a, b]$, $|y(x)| \leq M$ on $[a, b]$, and $|y'(x)| + |y''(x)| \geq N$ on $[a, b]$, then there exists $x_0 \in (a, b)$ such that $y'(x_0) = 0$.*

Proof. Assume the conclusion is false. Let $[a, b]$ and $M > 0$ be given.

Assume $y \in C^{(2)}[a,b]$ with $|y(x)| \le M$ on $[a,b]$, and assume

$$|y'(x)| + |y''(x)| \ge \mathcal{K} + \frac{2M}{b-a} + 1 \quad \text{on} \quad [a,b],$$

where we will determine $\mathcal{K}$.

By the Mean Value Theorem, there exists $x_1 \in (a,b)$ such that

$$|y'(x_1)| = \left|\frac{y(b)-y(a)}{b-a}\right| \le \frac{2M}{b-a}.$$

There are 2 cases here and both are similar. We consider only the case

$$0 < y'(x_1) \le \frac{2M}{b-a} \quad \text{and} \quad a < x_1 \le \frac{a+b}{2}.$$

Now, if $y(x_1) = M$, then $y'(x_1) = 0$ and we are done. So assume $y(x_1) \neq M$. (Note that we are assuming $y'(x_1) \neq 0$, for if $y'(x_1) = 0$, we are done.) Define $\eta = \frac{b-a}{8}$.

Case 1: Assume $y''(x_1) \le 0$. Then $y''(x_1) \le -\mathcal{K}$ in order for the lower bound on $|y'(x)| + |y''(x)|$ above to hold. Thus, $y'(x)$ is decreasing on a right neighborhood of $x = x_1$. If, in fact $0 \le y'(x) \le \frac{2M}{b-a}$, then it will follow that $y''(x) \le -\mathcal{K}$, on $[x_1, b]$. So, by Taylor's Theorem,

$$\begin{aligned} y(b) &= y(x_1) + y'(x_1)(x_1 - b) + \frac{y''(\xi)(x_1-b)^2}{2}, \quad \text{some} \ \ x_1 < \xi < b, \\ &< M + \frac{2M(b-a)}{b-a} - \frac{\mathcal{K}(b-a)^2}{4} \\ &\le -M, \end{aligned}$$

if $\mathcal{K} \ge \frac{32M}{(b-a)^2} = \frac{M}{2\eta^2}$. Hence, $|y(b)| > M$, a contradiction. We conclude there exists $x_1 \le x_0 < b$ such that $y'(x_0) = 0$.

Case 2: Assume $y''(x_1) > 0$, [we will work our way across $[a,b]$ on subintervals of length η]. Then $y''(x_1) > \mathcal{K}$. Assume $y''(x) \ge \frac{1}{2}\mathcal{K}$ on $[x_1, x_1+\eta]$. Again, by Taylor's Theorem,

$$y(x_1+\eta) = y(x_1) + y'(x_1)\eta + \frac{y''(\xi)}{2}\eta^2, \quad \text{some} \ \ \xi \in (x_1, x_1+\eta).$$

As in Case 1, $y(x_1+\eta) > -M + \frac{1}{4}\mathcal{K}\eta^2 \ge M$, provided $\mathcal{K} \ge \frac{8M}{\eta^2}$. So, $y(x_1+\eta) > M$, a contradiction.

From this contradiction, we conclude there exists $x_1 < x_2 < x_1+\eta$ such that $y''(x_2) = \frac{1}{2}\mathcal{K}$. It follows that, since y'' is positive up to x_2, we can assume x_2 is the first point such that $y''(x_2) = \frac{1}{2}\mathcal{K}$. Thus,

$$y'(x_2) > \frac{1}{2}\mathcal{K} + \frac{2M}{b-a} \quad \text{on} \quad [x_1, x_2).$$

Now assume $y'(x) \geq \frac{1}{2}\mathcal{K} + \frac{2M}{b-a}$ on $[x_2, x_2+\eta)$. Then

$$\begin{aligned} y(x_2+\eta) &= y(x_2) + y'(\xi)\eta, \quad \text{some} \quad \xi \in (x_2, x_2+\eta) \\ &> -M + \frac{1}{2}\mathcal{K}\eta + \eta\frac{2M}{b-a} \\ &= -M + \frac{1}{2}\mathcal{K}\eta + \frac{M}{4} \\ &\geq M, \end{aligned}$$

provided $\mathcal{K} \geq \frac{7M}{2\eta}$. Hence, $y(x_2+\eta) > M$, a contradiction.

From this contradiction, there exists $x_2 < x_3 < x_2+\eta$ such that

$$y'(x_3) = \frac{1}{2}\mathcal{K} + \frac{2M}{b-a},$$

and we take x_3 to be the first such point; i.e., $y'(x) > \frac{1}{2}\mathcal{K} + \frac{2M}{b-a}$on $[x_2, x_3)$. So y' is decreasing on $[x_2, x_3)$. Thus, $y''(x_3) < -\frac{1}{2}\mathcal{K}$. Then in a right neighborhood of x_3, $y''(x) < -\frac{1}{2}\mathcal{K}$ and $y'(x)$ is decreasing.

Assume $0 < y'(x) \leq \frac{1}{2}\mathcal{K} + \frac{2M}{b-a}$ on $[x_3, b)$. Then, by Taylor's Theorem,

$$y(x_3) = y(b) + y'(b)(x_3 - b) + \frac{y''(\xi)(x_3-b)^2}{2}, \quad \text{some} \quad \xi \in (x_3, b).$$

Now the second term on the right side is nonpositive. Then

$$y(x_3) < M - \frac{\mathcal{K}}{4}(b-a)^2 \leq -M,$$

provided $\mathcal{K} \geq \frac{M}{\eta^2}$. (Recall by definition, $(b - x_3) \geq \frac{b-a}{4}$). Hence, $|y(x_3)| > M$, a contradiction. From this contradiction, we conclude $0 < y'(x) \leq \frac{1}{2}\mathcal{K} + \frac{2M}{b-a}$ on $[x_3, b)$ is false.

We conclude $y'(x_0) = 0$, for some $x_3 < x_0 < b$. □

We now apply the Schauder-Tychonoff Theorem (Theorem 3.4) to show the existence of solutions for 2-point conjugate BVP's for (4.1) provided the interval is small enough.

Lemma 4.2 (Local existence of 2-point problems). *Let* $f(x, u_1, u_2, u_3)$ *be continuous on* $[a, b] \times \mathbb{R}^3$. *Then for* $M, N, P > 0$, *there exists* $\delta(M, N, P) > 0$ *such that each of the* 2*-point BVP's*

$$\begin{cases} y''' = f(x, y, y', y''), \\ y(x_1) = y_1, \; y'(x_1) = y_2, \; y(x_2) = y_3, \end{cases}$$

$$\begin{cases} y''' = f(x, y, y', y''), \\ y(x_1) = y_3, \; y(x_2) = y_1, \; y'(x_2) = y_2, \end{cases}$$

has a solution provided $a \le x_1 < x_2 \le b$, $x_2 - x_1 \le \delta$,

$$|y_1| + |y_2|\tfrac{(x_2-x_1)}{4} + |y_3| \le M,$$
$$\max\left\{|y_2|, \left|\tfrac{2(y_3-y_1)}{x_2-x_1} - y_2\right|\right\} \le N,$$
$$2\left|\tfrac{y_3-y_1-y_2(x_2-x_1)}{(x_2-x_1)^2}\right| \le P.$$

Proof. The details of the proof will be for the first 2-point BVP. Let $M, N, P > 0$ be given. Let

$$X := \{(x, u_1, u_2, u_3) | \, a \le x \le b, |u_1| \le 2M, |u_2| \le 2N, |u_3| \le 2P\}.$$

Then X is a compact subset of $[a, b] \times \mathbb{R}^3$. Let $Q = \max\{|f(x, u_1, u_2, u_3)| : (x, u_1, u_2, u_3) \in X\}$, and let $\mathcal{B} = C^{(2)}[x_1, x_2]$ where $[x_1, x_2] \subseteq [a, b]$. Then $\mathcal{B}$ is a Banach space with

$$\|h\| := \max|h(x)| + \max|h'(x)| + \max|h''(x)|, \quad \text{for} \quad h \in X, \ x_1 \le x \le x_2.$$

Now define $K \subseteq \mathcal{B}$ by

$$K := \{h \in \mathcal{B} \,|\, |h(x)| \le 2M, |h'(x)| \le 2N, |h''(x)| \le 2P, \ \text{on} \ [x_1, x_2]\},$$

and define $T : \mathcal{B} \to \mathcal{B}$ by

$$(Th)(x) := w(x) + \int_{x_1}^{x_2} G(x, s) f(s, h(s), h'(s), h''(s))\, ds,$$

where $w(x)$ is the solution of the BVP

$$\begin{cases} y''' = 0, \\ y(x_1) = y_1, \ y'(x_1) = y_2, \ y(x_2) = y_3, \end{cases}$$

and $G(x, s)$ is the Green's function for the BVP

$$\begin{cases} y''' = 0, \\ y(x_1) = y'(x_1) = y(x_2) = 0. \end{cases}$$

(The fact that $(Th)(\cdot) \in \mathcal{B}$ follows from properties of the Green's function. Moreover the fact that fixed points of T are solutions for the first 2-point BVP follows from arguments analogous to those made in Theorem 3.1 of Chapter 3.)

Observe that $w(x) = y_1 u_1(x) + y_2 u_2(x) + y_3 u_3(x)$ where u_1, u_2, u_3 are solutions of $y''' = 0$ satisfying, respectively, the conditions

$$\begin{cases} u_1(x_1) = 1, \\ u_1'(x_1) = 0, \\ u_1(x_2) = 0; \end{cases} \qquad \begin{cases} u_2(x_1) = 0, \\ u_2'(x_1) = 1, \\ u_2(x_2) = 0; \end{cases} \qquad \begin{cases} u_3(x_1) = 0, \\ u_3'(x_1) = 0, \\ u_3(x_2) = 1. \end{cases}$$

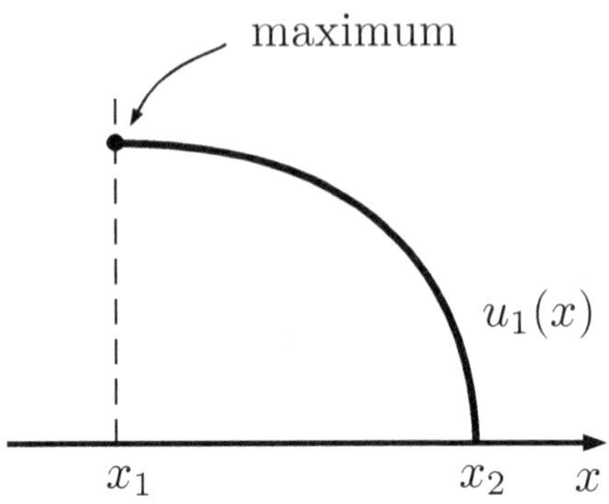

Fig. 4.1 The graph of $u_1(x)$.

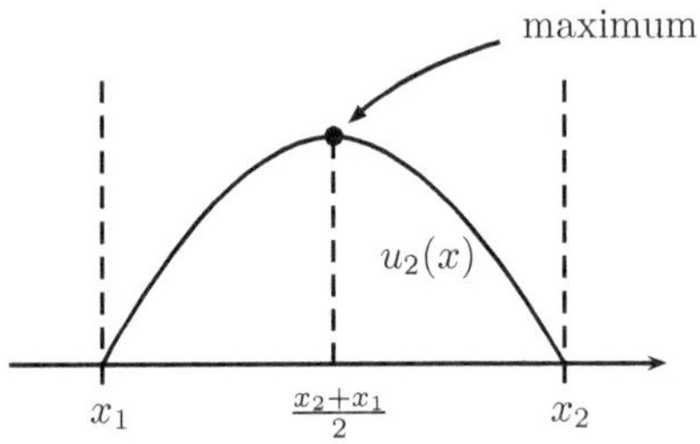

Fig. 4.2 The graph of $u_2(x)$.

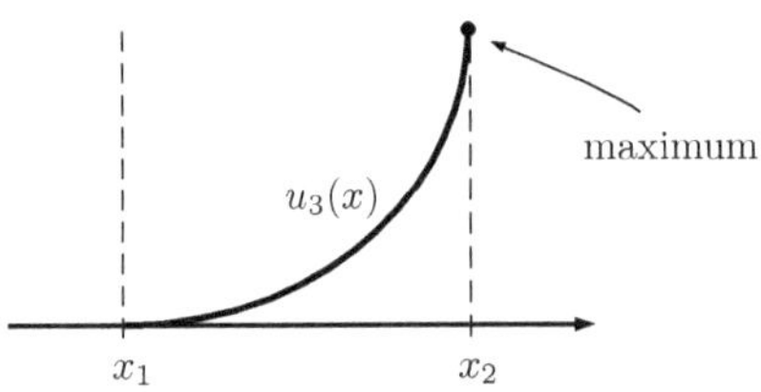

Fig. 4.3 The graph of $u_3(x)$.

The solutions are quadratic in $(x - x_1)$, and in particular, $u_1(x) = 1 - \frac{(x-x_1)^2}{(x_2-x_1)^2}$, with maximum at $x = x_1$, $u_2(x) = (x - x_1) - \frac{(x-x_1)^2}{(x_2-x_1)}$, with maximum at $x = \frac{x_2+x_1}{2}$, and $u_3(x) = \frac{(x-x_1)^2}{(x_2-x_1)^2}$, with maximum at $x = x_2$.

It then follows that $|w(x)| \leq |y_1| + |y_2|\frac{(x_2-x_1)}{4} + |y_3| \leq M$, on $[x_1, x_2]$. Also $w'(x) = y_1u_1'(x) + y_2u_2'(x) + y_3u_3'(x)$ is linear and hence will have its maximum at an end point. Hence,

$$|w'(x)| \le \max\left\{|y_2|, \frac{2(y_3-y_2)}{x_2-x_1} - y_2\right\} \le N, \quad \text{on} \quad [x_1, x_2].$$

Now $w''(x)$ is constant. So,

$$|w''(x)| = 2\left|\frac{y_3 - y_1 - y_2(x_2-x_1)}{(x_2-x_1)^2}\right| \le P, \quad \text{on} \quad [x_1, x_2].$$

For the second 2-point BVP in the statement of the lemma, to obtain u_1, u_2, u_3, we get the same corresponding bounds by interchanging x_1 and x_2.

Thus, for $h \in K$ and with the above bounds on $|w|$, $|w'|$, and $|w''|$, we have

$$\begin{aligned}
|(Th)(x)| &\le |w(x)| + \int_{x_1}^{x_2} |G(x,s)||f(s, h(s), h'(s), h''(s))|\, ds \\
&\le M + Q\int_{x_1}^{x_2} |G(x,s)|\, ds \\
&\le M + Q\frac{2(x_2-x_1)^3}{81}, \quad \text{on} \quad [x_1, x_2].
\end{aligned}$$

Similarly,

$$\begin{aligned}
|(Th)'(x)| &\le |w'(x)| + Q\int_{x_1}^{x_2} |G(x,s)| \left|\frac{\partial G(x,s)}{\partial x}\right| ds \\
&\le N + Q\frac{(x_2-x_1)^2}{6}, \quad \text{on} \quad [x_1, x_2],
\end{aligned}$$

and

$$|(Th)''(x)| \le P + Q\frac{2(x_2-x_1)}{3}, \quad \text{on} \quad [x_1, x_2].$$

So T maps K into K, provided $x_2 - x_1 \le \delta$, where

$$\delta = \min\left\{\sqrt[3]{\frac{81M}{2Q}}, \sqrt{\frac{6N}{Q}}, \frac{3P}{2Q}\right\}.$$

Furthermore, since K is a closed, bounded convex subset of $\mathcal{B}$, and since T is continuous on K and maps K onto a set with compact closure (see the proof of Theorem 3.5 of Chapter 3 for the analogous argument), it follows by the Schauder-Tychonoff Fixed Point Theorem (Theorem 3.3) that T has a fixed point in K. This fixed point is a solution of the first 2-point BVP in the statement of the lemma. □

Of much importance eventually will be the following corollary.

Corollary 4.1. *Let $f(x, u_1, u_2, u_3)$ be continuous on $[a, b] \times \mathbb{R}^3$ and let $M > 0$ be given. Then there exists $\delta(M) > 0$ such that each of the BVP's*

$$\begin{cases} y''' = f(x, y, y', y''), \\ y(x_1) = y(x_2) = y_1, \ y'(x_1) = 0, \end{cases} \qquad \begin{cases} y''' = f(x, y, y', y''), \\ y(x_1) = y(x_2) = y_1, \ y'(x_2) = 0, \end{cases}$$

has a solution provided $a \le x_1 < x_2 \le b$, $x_2 - x_1 \le \delta$, and $|y_1| \le M$. Furthermore, in this case $\delta(M)$ can be chosen such that for the solution, $y(x)$, of each of the problems, we have $|y'| \le 1$ and $|y''| \le 1$ on $[x_1, x_2]$.

Proof. We need only consider the modification of Lemma 4.2 required for the proof of the corollary. Define

$$X := \{(x, u_1, u_2, u_3) | \, a \le x \le b, |u_1| \le 2M, |u_2| \le 1, |u_3| \le 1\},$$

$\mathcal{B} := C^{(2)}[x_1, x_2]$ and

$$K := \{h \in \mathcal{B} | \, |h(x)| \le 2M, |h'(x)| \le 1, |h''(x)| \le 1 \ \text{ on } \ [x_1, x_2]\}.$$

Now in each problem above, the corresponding

$$w(x) \equiv y_1$$

satisfies each BVP. Then take

$$\delta(M) = \min \left\{ \sqrt[3]{\frac{81M}{2Q}}, \sqrt{\frac{3}{Q}}, \frac{3}{4Q} \right\}$$

and apply Lemma 4.2. □

The next lemma says that uniqueness of solutions of 3-point BVP's for Eq. (4.1) implies uniqueness of solutions of 2-point BVP's for Eq. (4.1), when such solutions exist.

Lemma 4.3. *Assume that* (4.1) *satisfies conditions* (A), (B), *and* (C) *of Theorem* 4.1. *Then for any $a < x_1 < x_2 < b$ and any solutions $y(x)$ and $z(x)$ of* (4.1), *$y(x_i) = z(x_i)$ for $i = 1, 2$ and either $y'(x_1) = z'(x_1)$ or $y'(x_2) = z'(x_i)$ imply $y(x) \equiv z(x)$ on $[x_1, x_2]$.*

Proof. The lemma is true as stated, but again to avoid technical problems, we will assume that solutions of IVP's on $(a, b) \times \mathbb{R}^3$ are unique.

Assume that the lemma is false. Then there exist $a < x_1 < x_2 < b$ and solutions $y(x)$ and $z(x)$ with $y(x_i) = z(x_i)$, $i = 1, 2$, $y'(x_1) = z'(x_1)$ or $y'(x_2) = z'(x_2)$ and $y(x) \not\equiv z(x)$ on $[x_1, x_2]$. To consider a particular case,

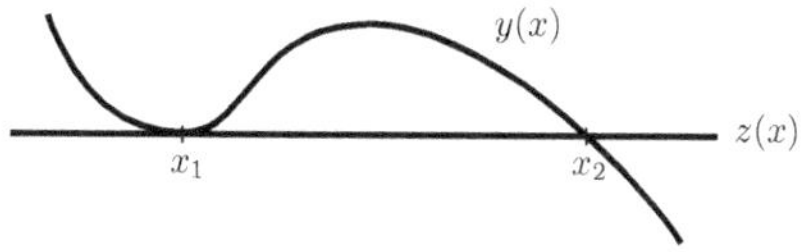

Fig. 4.4 The solutions $y(x)$ and $z(x)$.

assume that $y'(x_1) = z'(x_1)$ and $y(x) > z(x)$ on (x_1, x_2). By uniqueness of solutions of IVP's it must be the case that $y''(x_1) > z''(x_1)$.

Now choose $z''(x_1) < m < y''(x_1)$, and let $u(x)$ be the solution of the IVP for (4.1) satisfying

$$\begin{aligned} u(x_1) &= y(x_1) = z(x_1), \\ u'(x_1) &= y'(x_1) = z'(x_1), \\ u''(x_1) &= m. \end{aligned}$$

It follows that there exist $a < c < x_1 < d < x_2$ such that $z(x) < u(x) < y(x)$ on $[c, d] \setminus \{x_1\}$.

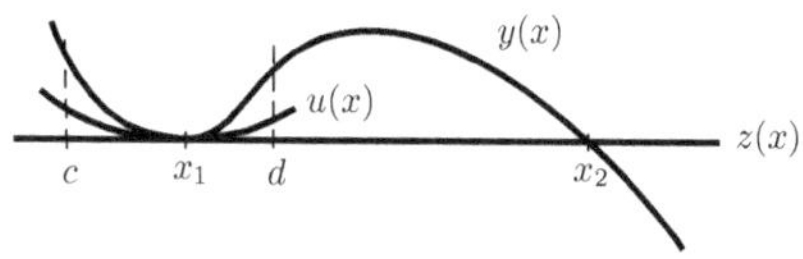

Fig. 4.5 $z(x) < u(x) < y(x)$ on $[c, d] \setminus \{x_1\}$.

Then let $\varepsilon > 0$ and let $u_\varepsilon(x)$ be the solution of (4.1) with

$$\begin{aligned} u_\varepsilon(x_1) &= y(x_1) = u(x_1) = z(x_1), \\ u_\varepsilon'(x_1) &= u'(x_1) + \varepsilon, \\ u_\varepsilon''(x_1) &= u''(x_1). \end{aligned}$$

Since solutions of IVP's for (4.1) are unique, solutions depend continuously on initial conditions. Hence there exists $\varepsilon_0 > 0$ such that $z(c) < u_{\varepsilon_0}(c) < y(c)$ and $z(d) < u_{\varepsilon_0}(d) < y(d)$. So, $u_{\varepsilon_0}(x_1) = y(x_1) = z(x_1)$, $u_{\varepsilon_0}(x) - z(x)$ has a zero on (c, x_1), and $u_{\varepsilon_0} - y(x)$ has a zero on (x_1, d). Hypothesis (B) implies $u_{\varepsilon_0}(x)$ extends to (a, b) and since $y(d) > u_{\varepsilon_0}(d) > z(d)$, it follows that there exists $\tau_\varepsilon(d, x_2]$ such that either $u_{\varepsilon_0}(\tau) - z(\tau) = 0$ or $u_{\varepsilon_0}(\tau) - y(\tau) = 0$. From property (C), it follows that $u_{\varepsilon_0}(x) \equiv z(x)$ or $u_{\varepsilon_0}(x) \equiv y(x)$ on whatever interval contains the three zeros of the appropriate difference. This is a contradiction for $u_{\varepsilon_0}'(x_1) > y'(x_1) = z'(x_1)$.

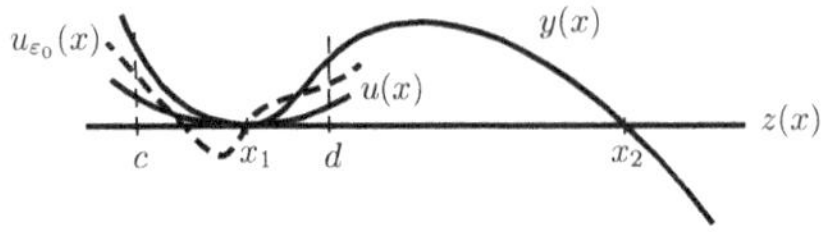

Fig. 4.6 $z(c) < u_{\varepsilon_0}(c) < y(c)$ and $z(d) < u_{\varepsilon_0}(d) < y(d)$.

Thus, it follows that each of the 2-point BVP's in the statement of the lemma has at most one solution. □

We now prove the "Compactness Condition".

Theorem 4.2. *Assume that Eq.* (4.1) *satisfies* (A), (B), *and* (C) *of Theorem* 4.1. *Then given any compact* $[c,d] \subset (a,b)$, *any* $M > 0$, *and any sequence* $\{y_k(x)\}$ *of solutions of* (4.1) *with* $|y_k(x)| \le M$ *on* $[c,d]$, *for all* $k \ge 1$, *then there exists a subsequence* $\{y_{k_j}(x)\}$ *such that* $\{y_{k_j}^{(i)}(x)\}$ *converges uniformly on each compact subinterval of* (a,b), *for* $i = 0,1,2$.

Proof. Assume the theorem is false; i.e., assume there exist $[c,d] \subset (a,b)$ and a sequence of solutions $\{y_k(x)\}$ of (4.1) such that $|y_k(x)| \le M$ on $[c,d]$, for some $M > 0$ and for all $k \ge 1$, but no subsequence satisfies the conclusion of the theorem. Then $\lim_{k\to\infty}\{|y_k'(x)| + |y_k''(x)|\} = +\infty$ uniformly on $[c,d]$; i.e., given $h > 0$, there exists $K > 0$ such that $|y_k'(x)| + |y_k''(x)| > h$, for all $x \in [c,d]$ and for all $k \ge K$. (Otherwise, there would exist some $\bar{h} > 0$ for which no such K exists. This in turn would imply the existence of a subsequence $\{k_j\} \subseteq \{k\}$ and a sequence $\{x_j\} \subseteq [c,d]$ such that $|y_{k_j}'(x_j)| + |y_{k_j}''(x_j)| \le \bar{h}$, for all j. But if this were the case, the Kamke Theorem could be applied to obtain a further subsequence of solutions satisfying the requirement of the theorem).

Now let $\delta = \delta(M)$ be the δ of the Corollary 4.1 for Eq. (4.1), $M > 0$ and the interval $[c,d]$. Partition $[c,d]$ into p equal subintervals of length less than $\frac{\delta}{3}$. For example, let $p = \lfloor \frac{d-c}{\frac{\delta}{3}} \rfloor + 1$.

Now let N be the number in Lemma 4.1 corresponding to $M > 0$ and our subinterval length, $\frac{d-c}{p}$. Finally, let k_0 be an integer such that $|y_{k_0}'(x)| + |y_{k_0}''(x)| > \max\{N, 2\}$ on $[c,d]$. (We can do this since the limit $\to +\infty$ as $k \to +\infty$). It follows from Lemma 4.1 that $y_{k_0}'(x) = 0$ for at least one point in each of the p open subintervals of our subdivision of $[c,d]$.

Take $t_1 < t_2 < t_3 < t_4$ as four successive division points in the partition of $[c,d]$, and let $t_1 < x_1 < t_2 < x_2 < t_3 < x_3 < t_4$ be such that $y_{k_0}'(x_i) = 0$, $i = 1,2,3$.

Now $y''_{k_0}(x_2) \neq 0$, since $|y'_{k_0}(x)| + |y''_{k_0}(x)| > \max\{N, 2\}$, for all $x \in [c, d]$. Assume the case $y''_{k_0}(x_2) < 0$. So $y_{k_0}(x)$ has a relative maximum at $x = x_2$. There are now various cases, and in each case, a contradiction will arise.

Case 1: Either $y_{k_0}(x_1) \geq y_{k_0}(x_2)$ or $y_{k_0}(x_3) \geq y_{k_0}(x_2)$. Assume without loss of generality the latter case. Suppose $y_{k_0}(x_2) = y_1$. Then there exists $\tau \in (x_2, x_3]$ such that $y_{k_0}(\tau) = y_1$. Thus $y_{k_0}(x)$ is a solution of (4.1) and satisfies the 2-point BVP $y_{k_0}(x_2) = y_{k_0}(\tau) = y_1$ and $y'_{k_0}(x_2) = 0$ and $\tau - x_2 < \delta(M)$. This is the type of 2-point BVP dealt with in the Corollary 4.1.

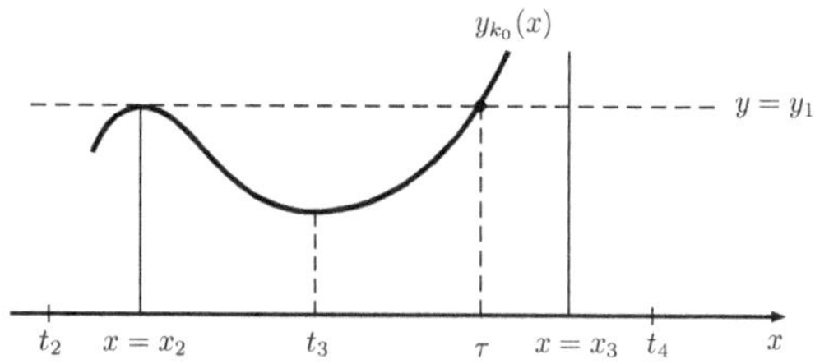

Fig. 4.7 $y_{k_0}(x_3) \geq y_{k_0}(x_2)$ and $y_{k_0}(\tau) = y_1$.

By Lemma 4.3, solutions of such 2-point BVP's are unique, hence $y_{k_0}(x)$ must be the solution satisfying the conditions of Corollary 4.1. But by Corollary 4.1, $|y'_{k_0}| \leq 1$, and $|y''_{k_0}(x)| \leq 1$ on the interval on which it is a solution of the 2-point BVP. So, $|y'_{k_0}(x)| + |y''_{k_0}(x)| \leq 2$ on $[x_2, \tau]$, which is a contradiction.

Case 2: Suppose $y_{k_0}(x_1) < y_{k_0}(x_2)$ and $y_{k_0}(x_3) < y_{k_0}(x_2)$. There are two subcases:

i) $y_{k_0}(x_1) = y_{k_0}(x_3)$. Assume that $y_{k_0}(x_1) = y_1 = y_{k_0}(x_3)$, and note that $x_3 - x_1 < \delta(M)$. Then $y_{k_0}(x)$ is a solution of the type of BVP dealt with in the Corollary 4.1, since $y_{k_0}(x_1) = y_{k_0}(x_3) = y_1$ and $y'_{k_0}(x_1) = 0$, and $x_3 - x_1 < \delta(M)$. The same contradiction as in Case 1 arises.

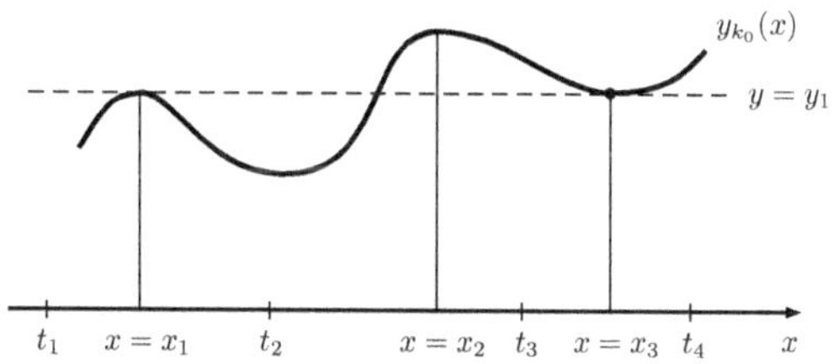

Fig. 4.8 $y_{k_0}(x_1) = y_{k_0}(x_3)$.

ii) $y_{k_0}(x_1) \neq y_{k_0}(x_3)$. Without loss of generality, assume $y_{k_0}(x_1) < y_{k_0}(x_3) < y_{k_0}(x_2)$. This time suppose $y_{k_0}(x_3) = y_1$. Then there exists $\tau \in (x_1, x_3)$ such that $y_{k_0}(\tau) = y_1$. Thus again, we have the situation where $y_{k_0}(\tau) = y_{k_0}(x_3) = y_1$, $y'_{k_0}(x_3) = 0$ and the interval length $x_3 - \tau \leq \delta(M)$. The same contradiction as in Case 1 arises again.

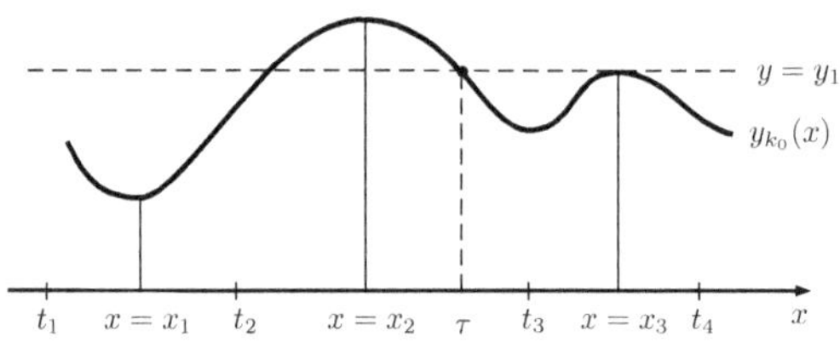

Fig. 4.9 $y_{k_0}(x_1) < y_{k_0}(x_3) < y_{k_0}(x_2)$ and $y''_{k_0}(x_3) < 0$.

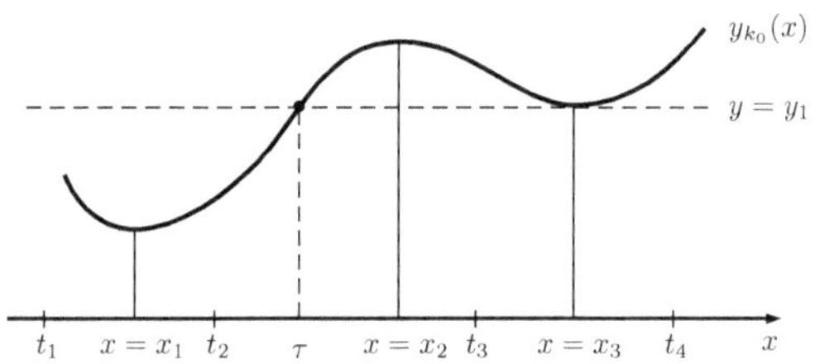

Fig. 4.10 $y_{k_0}(x_1) < y_{k_0}(x_3) < y_{k_0}(x_2)$ and $y''_{k_0}(x_3) > 0$.

Thus, both cases are impossible and it follows that the theorem is true as stated. □

Before finally presenting the proof of Theorem 4.1, we establish a "local existence" result for solutions of 3-point BVP's for (4.1). For that local existence result, we make use of the following theorem from algebraic topology.

Theorem 4.3 (Brouwer Invariance of Domain Theorem). *Let G be an open subset of $\mathbb{R}^n$. If $\varphi : G \to \mathbb{R}^n$ is continuous and one-to-one, then $\varphi(G)$ is an open subset of $\mathbb{R}^n$, and φ is a homeomorphism.*

The following result says that we can solve any 3-point conjugate BVP if we choose boundary conditions close enough to those of a known solution, as well as provides a continuous dependence on boundary conditions for solutions of the 3-point BVP (4.2).

Lemma 4.4. *Assume that* $y''' = f(x, y, y', y'')$, (4.1) *satisfies* (A), (B), *and* (C) *of Theorem* 4.1. *Let* $z(x)$ *be an arbitrary but fixed solution of* (4.1). *Then for any* $a < x_1 < x_2 < x_3 < b$, *any* c *and* d *with* $a < c < x_1$, $x_3 < d < b$, *and any* $\varepsilon > 0$, *there exists* $\delta > 0$, *such that* $|x_i - t_i| < \delta$ *and* $|z(x_i) - y_i| < \delta$, *for* $i = 1, 2, 3$, *imply that* (4.1) *has a solution* $y(x)$ *with* $y(t_i) = y_i$, $i = 1, 2, 3$, *and* $|z(x) - y(x)| < \epsilon$ *on* $[c, d]$.

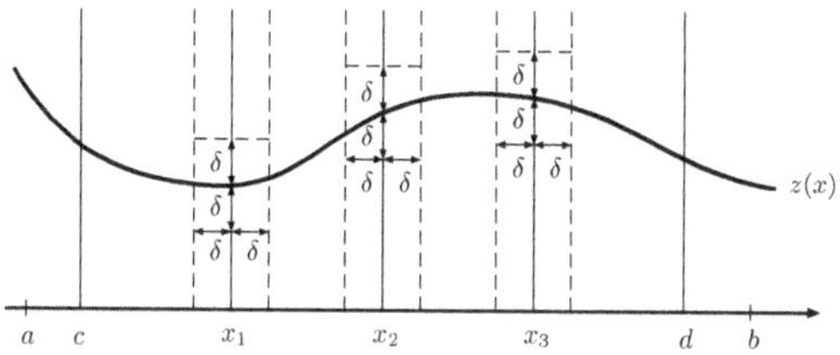

Fig. 4.11 Continuous dependence of solutions on the initial data.

Proof. The lemma is true as stated, but to avoid certain technical arguments, we assume that solutions of IVP's for (4.1) are unique. Define the open subset G of $\mathbb{R}^6$ by

$$G := \{(t_1, t_2, t_3, c_1, c_2, c_3) \in \mathbb{R}^6 | \, a < t_1 < t_2 < t_3 < b \text{ and } (c_1, c_2, c_3) \in \mathbb{R}^3\}.$$

Then define a mapping $\varphi : G \to \mathbb{R}^6$ by

$$\varphi(t_1, t_2, t_3, c_1, c_2, c_3) = (t_1, t_2, t_3, y(t_1), y(t_2), y(t_3)),$$

where $y(x)$ is the solution of (4.1) with $y^{(i-1)}(x_0) = c_i$, $i = 1, 2, 3$ where $x_0 \in (a, b)$ is a fixed selected point.

The uniqueness of solutions of IVP's implies that solutions of IVP's for (4.1) depend continuously on initial conditions. It follows that φ is a continuous function. It is also the case that φ is one-one. To see this, assume that $\varphi(\hat{t}_1, \hat{t}_2, \hat{t}_3, \hat{c}_1, \hat{c}_2, \hat{c}_3) = \varphi(t_1, t_2, t_3, c_1, c_2, c_3)$. So, $\hat{t}_i = t_i$, $i = 1, 2, 3$ and $y(\hat{t}_i) = y(t_i) = w(t_i)$, $i = 1, 2, 3$, where $y(x)$ and $w(x)$ are solutions of (4.1) with $y^{(i-1)}(x_0) = \hat{c}_i$, $w^{(i-1)}(x_) = c_i$, $i = 1, 2, 3$. By property (C) and uniqueness of solutions of IVP's, we have $y(x) \equiv w(x)$ on (a, b). So, $\hat{c}_i = c_i$, $i = 1, 2, 3$. Therefore, φ is one-to-one. Thus, by the Brouwer Invariance of Domain Theorem, $\varphi(G)$ is open and φ^{-1} is continuous on $\varphi(G)$.

We shall show that the lemma is true through the use of the continuity of φ^{-1} and the openness of $\varphi(G)$. So let $a < x_1 < x_2 < x_3 < b$ be chosen, and let $a < c < x_1$, $x_3 < d < b$ and $\varepsilon > 0$ be chosen also. By continuity with respect to initial conditions, there exists $\eta > 0$ such that $|z^{(i-1)}(x_0) - c_i| < \eta$ for $i = 1, 2, 3$, with $z(x)$ as the fixed solution, implies $|y(x) - z(x)| < \varepsilon$ on $[c, d]$ where $y(x)$ is the solution of (4.1) with $y^{(i-1)}(x_0) = c_i$, $i = 1, 2, 3$.

Now $(x_1, x_2, x_3, z(x_1), z(x_2), z(x_3)) \in \varphi(G)$ and $\varphi(G)$ is open and $\varphi^{-1} : \varphi(G) \to G$ is continuous. Hence there exists $\delta > 0$ such that $|t_i - x_i| < \delta$, $i = 1, 2, 3$, and $|y_i - z(x_i)| < \delta$, $i = 1, 2, 3$. Hence, $(t_1, t_2, t_3, y_1, y_2, y_3) \in \varphi(G)$.

By the continuity of φ^{-1}, $\varphi^{-1}(t_1, t_2, t_3, y_1, y_2, y_3)$ is in the open cube of half-edge η centered at $\varphi^{-1}(x_1, x_2, x_3, z(x_1), z(x_2), z(x_3)) = (x_1, x_2, x_3, z(x_0), z'(x_0), z''(x_0))$. From the continuity with respect to initial conditions above, this implies (4.1) has a solution $y(x)$ with $y(t_i) = y_i$, $i = 1, 2, 3$ and $|y(x) - z(x)| < \varepsilon$ on $[c, d]$. □

We now present the proof of Theorem 4.1 concerning uniqueness implies existence for solutions of 3-point BVP's for (4.1). For the convenience of the reader we restate the theorem here.

Theorem 4.1. *Assume that*

$$y''' = f(x, y, y', y''), \tag{4.1}$$

satisfies the conditions:

(A) *f is continuous on $(a, b) \times \mathbb{R}^3$,*
(B) *Solutions of* (4.1) *extend to (a, b), and*
(C) *For any $a < x_1 < x_2 < x_3 < b$ and any solutions $y(x)$ and $z(x)$ of* (4.1), *$z(x_i) = y(x_i)$, $i = 1, 2, 3$, implies $y(x) \equiv z(x)$ on $[x_1, x_2]$.*

Then for any $a < x_1 < x_2 < x_3 < b$ and any $y_1, y_2, y_3 \in \mathbb{R}$, the BVP

$$\begin{cases} y''' = f(x, y, y', y''), \\ y(x_i) = y_i, \ i = 1, 2, 3, \end{cases}$$

has a solution.

Proof. The proof is based on the following property of the reals: If $S \subseteq \mathbb{R}$ is both an open and a closed set, then $S = \emptyset$ or $S = \mathbb{R}$; i.e., $\mathbb{R}$ is connected. Moreover, we shall again assume that solutions of IVP's for (4.1) are unique, however the theorem is true as stated.

Now let $z(x)$ be an arbitrary but fixed solution of Eq. (4.1). Let x_1, x_2, x_3 be such that $a < x_1 < x_2 < x_3 < b$ and let $y_1, y_2, y_3 \in \mathbb{R}$. Now define

$$S := \{r \in \mathbb{R} \mid \text{there exists a solution } y(x) \text{ of (4.1) with } y(x_1) = z(x_1), \\ y(x_2) = z(x_2), \text{ and } y(x_3) = r\}.$$

First, $S \neq \emptyset$, since $z(x_3) \in S$. It is our goal to establish that S is both an open and closed subset of $\mathbb{R}$.

S is open: Let $s \in S$. Hence from the definition of S, there exists a solution $y(x)$ such that $y(x_1) = z(x_1)$, $y(x_2) = z(x_2)$, and $y(x_3) = s$. By Lemma 4.4, there exists $\varepsilon > 0$ such that for any $u_1, u_2, u_3 \in \mathbb{R}$ with

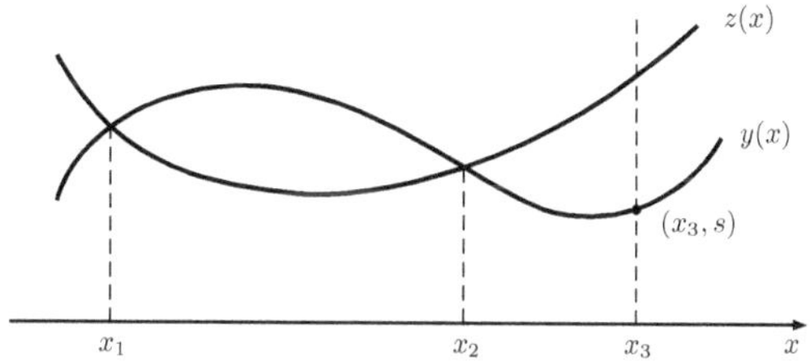

Fig. 4.12 $y(x_1) = z(x_1)$, $y(x_2) = z(x_2)$, and $y(x_3) = s$.

$|y(x_i) - u_i| < \varepsilon$, (4.1) has a solution $u(x)$ with $u(x_i) = u_i$, $i = 1, 2, 3$. In particular for any $u_1, u_2, u_3 \in \mathbb{R}$ with $|y(x_1) - u_1| < \varepsilon$, $|y(x_2) - u_2| < \varepsilon$, $|s - u_3| < \varepsilon$, (4.1) has a solution $u(x)$ with $u(x_i) = u_i$, $i = 1, 2, 3$. So, $(s - \varepsilon, s + \varepsilon) \subset S$ and hence S is open.

S is closed: Assume S is not closed. Then there exists r_0 which is a limit point (or boundary point) of S, but $r_0 \notin S$. Then there exists a sequence of distinct points $\{r_k\} \subset S$ such that $r_k \to r_0$. Without loss of generality, we can assume $\{r_k\}$ is strictly monotone, say strictly monotone increasing.

For each $k \geq 1$, let $y_k(x)$ be the solution of (4.1) with $y_k(x_1) = z(x_1)$, $y_k(x_2) = z(x_2)$, and $y_k(x_3) = r_k$. Since $r_{k+1} > r_k$ and since $y_k(x_1) = y_{k+1}(x_1)$ and $y_k(x_2) = y_{k+1}(x_2)$, it follows from Lemma 4.3 and condition (C) that $y_k(x) < y_{k+1}(x)$ on $(a, x_1) \cup (x_2, b)$ and $y_k(x) > y_{k+1}(x)$ on (x_1, x_2), for all $k = 1, 2, \ldots$, and $y_k(x)$ and $y_{k+1}(x)$ are not tangent at x_1 and x_2 by

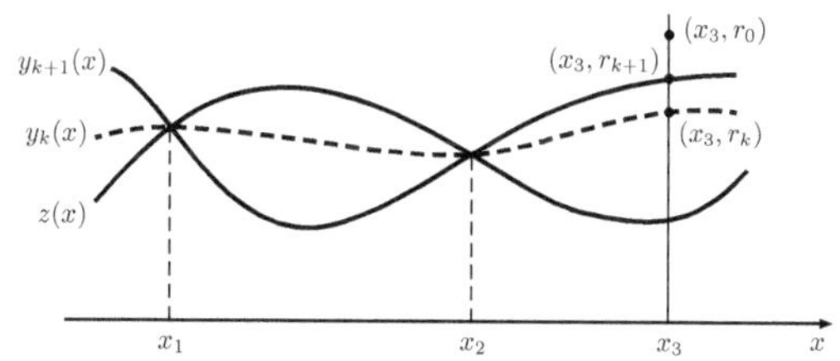

Fig. 4.13 $y_k(x) < y_{k+1}(x)$ on $(a, x_1) \cup (x_2, b)$ and $y_k(x) > y_{k+1}(x)$ on (x_1, x_2).

Lemma 4.3. Since $r_0 \notin S$, it follows from Theorem 4.2 that, $\{y_k(x)\}$ cannot be a uniformly bounded sequence of solutions on any compact subinterval of (a, b).

In any interval where the y_k's are increasing, the y_k's take on arbitrarily large values.

Designate $w(x)$ as the solution of the IVP for (4.1) with $w(x_3) = r_0$, $w'(x_3) = 0$, $w''(x_3) = 0$. Now choose $\delta > 0$ such that $a < x_1 - \delta < x_1 + \delta < x_2 < x_3 - \delta < x_3 + \delta < b$. Then, for sufficiently large K, there are

i) points in $(x_1 - \delta, x_1)$, where $y_K(x) > w(x)$,
ii) points in $(x_1, x_1 + \delta)$, where $y_K(x) < w(x)$,
iii) points in $(x_3 - \delta, x_3)$ and in $(x_3, x_3 + \delta)$, where $y_K(x) > w(x)$.

Since $y_k(x_3) < w(x_3) = r_0$, for all k, it follows that, for K sufficiently large, $y_K(x) - w(x)$ has a zero in $(x_1 - \delta, x_1 + \delta)$, $[x_1 + \delta, x_3)$, $(x_3, x_3 + \delta)$. This is a contradiction to condition (C). Therefore S is also closed.

Therefore $S \equiv \mathbb{R}$; thus choose $r = y_3 \in S$ and we have a solution to

$$\begin{cases} y''' = f(x, y, y', y''), \\ y(x_1) = z(x_1), \ \ y(x_2) = z(x_2), \ \ y(x_3) = y_3. \end{cases}$$

Now define

$$S' := \{r \in \mathbb{R} \mid \text{there exists a solution } y(x) \text{ of (4.1) with } y(x_1) = z(x_1), \\ y(x_2) = r, \text{ and } y(x_3) = y_3\}.$$

Then $S' \notin \emptyset$ since the solution produced above is such that $r = z(x_2) \in S'$. The same argument above shows $S' \equiv \mathbb{R}$. So choosing $r = y_2 \in S'$, we have a solution to

$$\begin{cases} y''' = f(x, y, y', y''), \\ y(x_1) = z(x_1), \ \ y(x_2) = y_2, \ \ y(x_3) = y_3. \end{cases}$$

Finally, define

$$S'' := \{r \in \mathbb{R} \mid \text{there exists a solution } y(x) \text{ of (4.1) with } y(x_1) = r, \\ y(x_2) = y_2, \text{ and } y(x_3) = y_3\}.$$

Again, it will follow that $S'' \equiv \mathbb{R}$ and so taking $r = y_1 \in S''$ we have a solution to

$$\begin{cases} y''' = f(x, y, y', y''), \\ y(x_1) = y_1, \ \ y(x_2) = y_2, \ \ y(x_3) = y_3. \end{cases}$$

The proof is complete. □

The following example shows that the "Conjecture" in Remark 4.1 is false. Consider the IVP

$$\begin{cases} y''' = -(y')^3, \\ y(0) = y'(0) = 0, \ y''(0) = m > 0. \end{cases}$$

Since $f(x, y_1, y_2, y_3) = -(y_2)^3$ has continuous partial derivatives, this IVP has a unique solution. Call the solution $y(x)$. On its interval of existence, we have

$$y'''(x) + (y'(x))^3 = 0.$$

Thus,

$$y'''(x)y''(x) + (y'(x))^3 y''(x) = 0,$$

and integrating over $[0, x]$, we have

$$\frac{1}{2}(y''(x))^2 + \frac{1}{4}(y'(x))^4 = C = \frac{1}{2}m^2. \tag{4.3}$$

Now, from the ordinary differential equation and the initial conditions, $y''' < 0$ on a right neighborhood of $x = 0$; thus $y''' < 0$ until y' vanishes. So, $y''(x)$ decreases until y' vanishes (say $y'(x_1) = 0$). Then $y'' > 0$ on $(0, x_0)$ and $y'' < 0$ on (x_0, x_1), for some $0 < x_0 < x_1$.

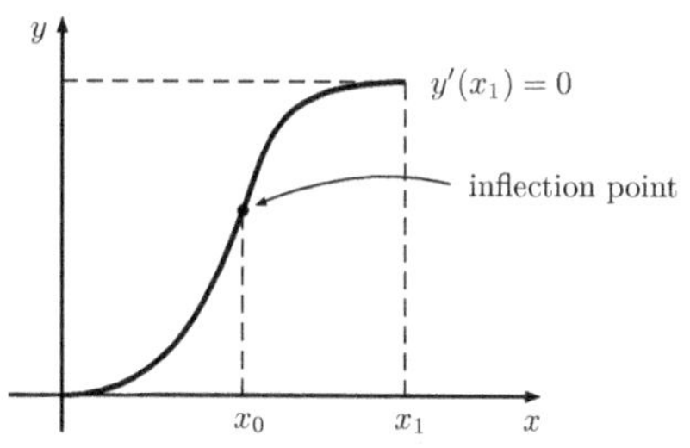

Fig. 4.14 $y'' > 0$ on $(0, x_0)$, $y'' < 0$ on (x_0, x_1), and $y'(x_1) = 0$.

Solving (4.3) for $y''(x)$, for $0 \le x \le x_0$, we have

$$y''(x) = \sqrt{m^2 - \frac{1}{2}(y'(x))^4}, \quad \text{or} \quad \frac{y''(x)y'(x)}{\sqrt{m^2 - \frac{1}{2}(y'(x))^4}} = y'(x).$$

Hence,

$$\int_0^{x_0} \frac{y''(x)y'(x)}{\sqrt{m^2 - \frac{1}{2}(y'(x))^4}} dx = \int_0^{x_0} y'(x)dx = y(x_0).$$

(Note: $y''(x_0) = 0$ and so (4.3) yields $y'(x_0) = \sqrt[4]{2m^2}$.)

Setting $s = y'(x)$,

$$\int_0^{\sqrt[4]{2m^2}} \frac{s}{\sqrt{m^2 - \frac{1}{2}s^4}} ds = y(x_0).$$

Setting $s = \sqrt[4]{2m^2 r}$,

$$\int_0^{\sqrt[4]{2m^2}} \frac{s}{\sqrt{m^2 - \frac{1}{2}s^4}} ds = \frac{\sqrt{2}}{2} \int_0^1 \frac{dr}{\sqrt{1-r^2}} = \frac{\sqrt{2}}{2} \operatorname{Arcsin} r \Big|_0^1 = \frac{\pi\sqrt{2}}{4}.$$

Hence, $y(x_0) = \frac{\pi\sqrt{2}}{4}$.

Integrating over $[x_0, x_1]$ and using $y''(x) < 0$ and making the same changes of variables,

$$\int_{\sqrt[4]{2m^2}}^{y'(x_1)=0} \frac{sds}{\sqrt{m^2 - \frac{1}{2}s^4}} = \frac{\pi\sqrt{2}}{4}.$$

It follows as in the previous steps that $y(x_1) = \frac{\pi\sqrt{2}}{2}$ and that the graph of $y(x)$ from x_0 to x_1 is symmetric to the graph from 0 to x_0, (i.e. symmetric about the point $(x_0, y(x_0))$. Well, extending the solution $y(x)$ on past x_1, we would solve

$$\begin{cases} y''' = -(y')^3, \\ y(x_1) = \frac{\pi\sqrt{2}}{2}, \ y'(x_1) = 0, \ y''(x_1) = -m, \end{cases}$$

and find that $y(x)$ vanishes at some point $x_2 = 2x_1$, that $y'(x_2) = 0$, $y''(x_2) = m$. Also $y(x)$ is symmetric about the line $x = x_1$.

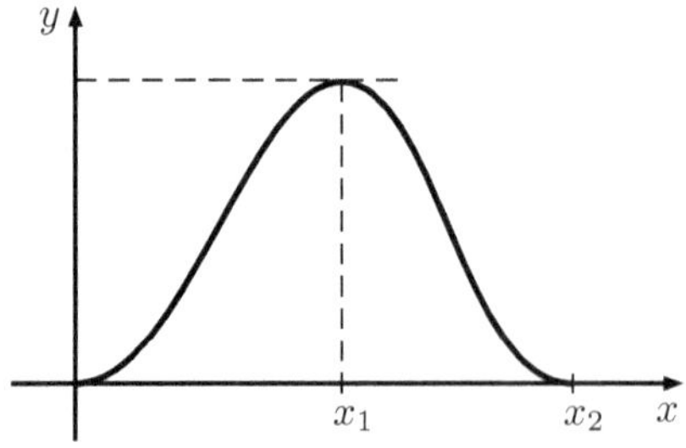

Fig. 4.15 The graph of $y(x)$.

Now, at x_2, $y(x)$ satisfies

$$\begin{cases} y''' = -(y')^3, \\ y(x_2) = y'(x_2) = 0, \ y''(x_2) = m. \end{cases}$$

The ordinary differential equation is independent of x, so translations of solutions are still solutions of the ordinary differential equation. Thus, we can translate this IVP back to $x = 0$, we obtain the same graph as before which we then translate back to the interval $[x_2, 2x_2]$, We continue this construction and obtain that $y(x)$ exists on all of $\mathbb{R}$, $0 \leq y(x) \leq \frac{\pi\sqrt{2}}{2}$, all $x \in \mathbb{R}$, $y(x)$ is periodic, and $\text{Per}(y(x)) = 4x_0$.

Now, for all $k \in \mathbb{N}$, let $y_k(x)$ be the solution of

$$\begin{cases} y''' = -(y')^3, \\ y(0) = y'(0) = 0, \ y''(0) = k. \end{cases}$$

Then, each $y_k(x)$ extends to all of $\mathbb{R}$, $|y_k(x)| \leq \frac{\pi\sqrt{2}}{2}$ on all of $\mathbb{R}$, and $f(x, y_1, y_2, y_3) = -y_2^3$ is of course continuous.

From above $y_k''(x) = \sqrt{k^2 - \frac{1}{2}(y_k'(x))^4}$ on $[0, x_{0k}]$, where $4x_{0k} = \text{Per}(y_k(x))$. Then,

$$\begin{aligned} \frac{\text{Per}(y_k(x))}{4} = x_{0k} = \int_0^{x_{0k}} dx &= \int_0^{x_{0k}} \frac{y_k''(x)}{\sqrt{k^2 - \frac{1}{2}(y_k'(x))^4}}\, dx \\ &= \frac{1}{\sqrt[4]{2}\sqrt{2k}} \int_0^1 u^{-\frac{3}{4}}(1-u^2)^{-\frac{1}{2}}\, du \\ &= \frac{1}{\sqrt[4]{2}\sqrt{2k}} \left[\frac{1}{2} B\left(\frac{1}{4}, \frac{1}{2}\right)\right] \\ &= \frac{1}{2^{\frac{7}{4}}\sqrt{k\pi}} \left[\Gamma\left(\frac{1}{4}\right)\right]^2. \end{aligned}$$

Thus, as k increases, the oscillations are packed together tighter, however, each $y_k(x)$ still attains its maximum value $\frac{\pi\sqrt{2}}{2}$ on each subinterval of length $4x_{0k}$. It follows that, regardless of the compact subinterval $[c, d]$, $\{y_k''(x)\}$ has no convergent subsequence on $[c, d]$.

Thus, the "Conjecture" of Remark 4.1 is false.

The hypotheses of Theorem 4.1 actually yield existence of a larger family of solutions of BVP's for Eq. (4.1). Recall that the hypotheses of Theorem 4.1 imply that all 2-point conjugate BVP's for (4.1) have at most one solution as was proven in Lemma 4.3. We now show that those hypotheses imply the existence of solutions of all such 2-point BVP's for (4.1).

Theorem 4.4. *If Eq.* (4.1) *satisfies* (A), (B), *and* (C), *then all 2-point conjugate BVP's for* (4.1) *have solutions; that is, for any* $a < x_1 < x_2 < b$ *and any* $y_1, y_2, y_3 \in \mathbb{R}$, *there is a solution* $y(x)$ *of* (4.1) *satisfying,*

$$y(x_1) = y_1, \; y'(x_1) = y_2, \; y(x_2) = y_3, \tag{4.4}$$

as well as a solution $z(x)$ *of* (4.1) *satisfying,*

$$z(x_1) = y_1, \; z(x_2) = y_2, \; z'(x_2) = y_3, \tag{4.5}$$

where both $y(x)$ *and* $z(x)$ *are unique by Lemma* 4.3.

Proof. We employ again the shooting method as well as Theorem 4.1. We will give only the proof for conditions (4.4), with those for (4.5) being similar.

So, let $y(x; m)$ denote a solution of (4.1) with

$$y(x_1; m) = y_1, \quad y'(x_1; m) = y_2, \quad y''(x_1; m) = m.$$

Define $S := \{r \in \mathbb{R} \mid y(x_2; m) = r, \text{ for some } m \in \mathbb{R}\}$. Since solutions depend continuously upon initial conditions, $S \subseteq \mathbb{R}$ is an interval. In order that, there exists $m_0 \in \mathbb{R}$ such that (4.4) has a solution, it suffices to show that S is neither bounded above nor below; hence $S \equiv \mathbb{R}$. We look at one case.

Assume S is bounded above. So there exists $y_0 \in \mathbb{R}$ such that $r < y_0$, for all $r \in S$. Let $x_2 < x_3 < b$ be fixed and let $w(x)$ be the solution guaranteed

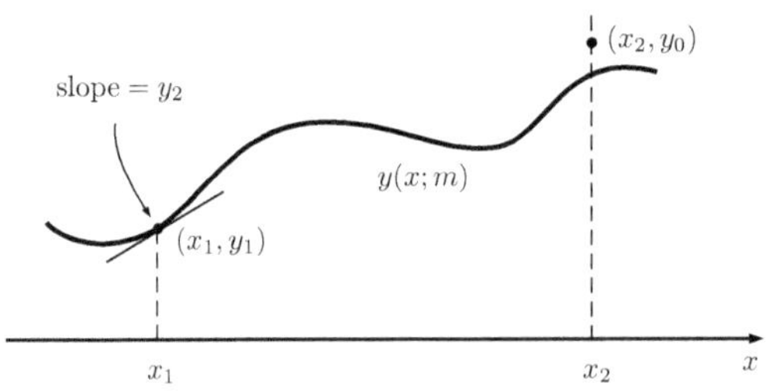

Fig. 4.16 The graph of $y(x; m)$.

by Theorem 4.1 satisfying

$$w(x_1) = y_1, \; w(x_2) = y_0, \; w(x_3) = 0.$$

Then $w'(x_1) \neq y_2$, since $y_0 \notin S$. Now let $y_k(x) = y(x; k)$, $k = 1, 2, 3, \ldots$. Then $y_k(x) < y_{k+1}(x)$ on $(a, b) \setminus \{x_1\}$ by Lemma 4.3. We have two cases:

i) $w'(x_1) > y_2$. Since $y_k(x_2) < y_0 = w(x_2)$, for all k, and since 3-point

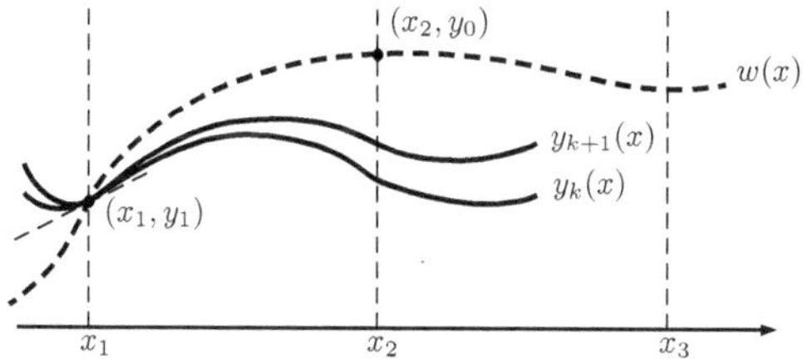

Fig. 4.17 $w'(x_1) > y_2$ and $y_k(x) \le w(x)$ on $[x_1, x_2]$.

BVP's have unique solutions, it follows that $y_1(x) \le y_k(x) \le w(x)$ on $[x_1, x_2]$, for all k. Then by Theorem 4.2, there exists a subsequence $\{y_{k_j}(x)\}$ such that $\{y_{k_j}^{(i)}(x)\}$ converges uniformly on compact subintervals of (a, b), $i = 0, 1, 2$. But this is impossible, since $y''_{k_j}(x_1) = k_j \to +\infty$, as $j \to +\infty$.

ii) $w'(x_1) < y_2$. Since $y_k(x_2) < y_0 = w(x_2)$, and since 3-point BVP's

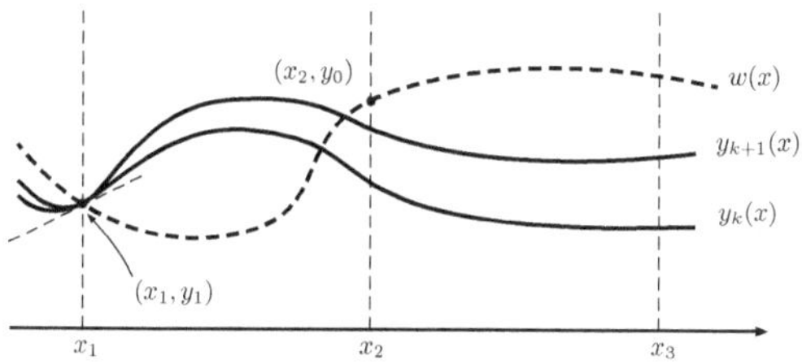

Fig. 4.18 $w'(x_1) < y_2$ and $y_k(x) < w(x)$ on $[x_2, x_3]$.

have unique solutions, it follows that $y_1(x) < y_k(x) < w(x)$ on $[x_2, x_3]$, for all k. The same contradiction as in i) occurs, and so ii) is also impossible.

Thus S is not bounded above. Similarly, S is not bounded below and thus $S \equiv \mathbb{R}$. Hence there exists $m_0 \in \mathbb{R}$, such that $y(x_2; m_0) = y_3$, and $y(x; m_0)$ is the desired solution of BVP (4.1), (4.4).

For BVP (4.1), (4.5), the shooting method would be utilized by shooting from right to left. □

4.2 A Converse of Lemma 4.3

In this section, we are concerned with a converse result for Lemma 4.3. In considering this converse, we will state the conclusion of Lemma 4.3 as the following condition:

(D) For any $a < x_1 < x_2 < b$ and any solutions $y(x)$ and $z(x)$ of (4.1), it follows that $y(x_i) = z(x_i)$, $i = 1, 2$, and either $y'(x_1) = z'(x_1)$ or $y'(x_2) = z'(x_2)$ imply $y(x) = z(x)$ on $[x_1, x_2]$.

In particular, we proved in Lemma 4.3 that, for (4.1), conditions (A), (B) and (C) imply condition (D). In the statement of Lemma 4.3, hypotheses (A) and (B) are perhaps reasonable, hence we were primarily interested in the implication of (C) to (D). Lemma 4.3 says that (C) is sufficient for (D). We now show that it is also necessary in the presence of (A) and (B).

Theorem 4.5. *Assume that* (4.1) *satisfies conditions* (A), (B), *and* (D). *Then* (4.1) *satisfies condition* (C).

Proof. Assume the conclusion of the theorem is false. Then there exist distinct solutions $y(x)$ and $z(x)$ of (4.1) and points $a < x_1 < x_2 < x_3 < b$ such that $y(x_i) = z(x_i)$, $i = 1, 2, 3$. In view of hypothesis (D), $y'(x_i) \neq z'(x_i)$, for each $i = 1, 2, 3$, hence without loss of generality we can assume $y(x) > z(x)$ on (x_1, x_2) and $y(x) < z(x)$ on (x_2, x_3).

For each $n \geq 1$, let $y_n(x)$ be a solution on (a, b) of (4.1) satisfying the initial conditions,

$$y_n^{(i)}(x_1) = y^{(i)}(x_1), \quad i = 0, 1,$$
$$y_n''(x_1) = y''(x_1) + n.$$

Then it follows from (D) that, for each $n \geq 1$,

$$y(x) < y_n(x) < y_{n+1}(x),$$

on $(a, b) \setminus \{x_1\}$. For each $n \geq 1$, let

$$E_n = \{x \in [x_2, x_3] |\, y_n(x) \leq z(x)\}.$$

Employing a tedious, but not difficult argument, it can be shown from our hypotheses that the sets $E_n \neq \emptyset$, for each $n \geq 1$. Thus, $E_{n+1} \subset E_n \subset$

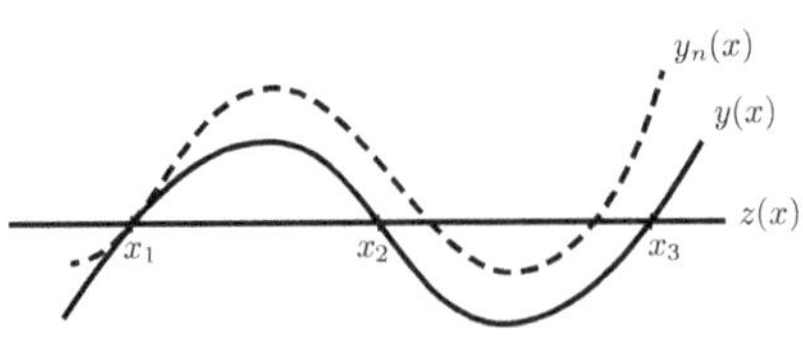

Fig. 4.19 The graphs of $y(x)$, $y_n(x)$, and $z(x)$.

(x_2, x_3), for each $n \geq 1$, and each E_n is nonnull and compact. It follows that

$$\bigcap_n E_n \equiv E \neq \emptyset.$$

Next, we observe that the set E consists of a single point x_0 with $x_2 < x_0 < x_3$. In fact, if $t_1, t_2 \in E$ with $x_2 < t_1 < t_2 < x_3$, then the same type of argument one uses to show that the foregoing sets E_n are nonull leads to the conclusion that the entire interval $[t_1, t_2] \subseteq E$. However, by the monotonicity condition on $\{y_n(x)\}$, it follows that

$$y(x) < y_n(x) \leq z(x)$$

on $[t_1, t_2]$, for all $n \geq 1$. Thus there exists $M > 0$ such that $|y_n(x)| \leq M$, on $[t_1, t_2]$, for all $n \geq 1$.

However, if this is the case we can then invoke Theorem 4.2 to obtain a subsequence $\{y_{n_j}(x)\}$ such that $\{y_{n_j}^{(i)}(x)\}$ converges uniformly on compact subintervals of (a, b), for each $i = 0, 1, 2$. (The proof of Theorem 4.2 did not make direct use of hypotheses (C), but rather used the fact that (D) was satisfied.) Yet this is impossible, since $y''_{n_j}(x_1) = y'' + n_j \to +\infty$, as $j \to +\infty$. Thus, we conclude that $E = \{x_0\}$ with $x_2 < x_0 < x_3$, and

$$\lim_{n\to\infty} y_n(x_0) := y_0 \leq z(x_0).$$

Now we claim this is not possible. There are two cases:

(i) First, assume $y_0 = z(x_0)$. Given $\varepsilon > 0$, let $z(x; \varepsilon)$ be the solution of

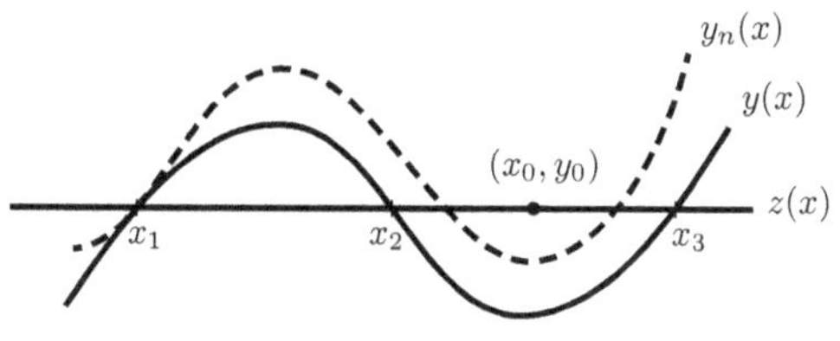

Fig. 4.20 The case $y_0 = z(x_0)$.

(4.1) satisfying the initial conditions,

$$z^{(i)}(x_1; \varepsilon) = z^{(i)}(x_1), \ i = 0, 1$$
$$z''(x_1; \varepsilon) = z''(x_1) - \varepsilon.$$

Then $z(x; \varepsilon) < z(x)$ on (x_1, b) by (D), and for ε sufficiently small, there exists $[t_1, t_2] \subsetneq [x_2, x_3]$ such that $z(t_i; \varepsilon) = y(t_i)$, $i = 1, 2$, $y(x) < z(x; \varepsilon)$ on (t_1, t_2), and in particular $z(x_0; \varepsilon) > y(x_0)$. The relations enjoyed by $y(x)$ and $z(x)$ are now shared by $y(x)$ and $z(x; \varepsilon)$. One could then define a new sequence of sets $\{F_n\}$ by

$$F_n = \{x \in [t_1, t_2] \mid y_n(x) \leq z(x; \varepsilon)\}.$$

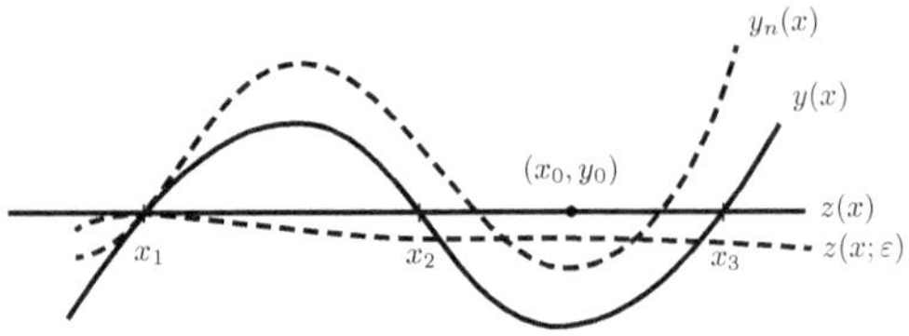

Fig. 4.21 $z(x;\varepsilon) < z(x)$ on (x_1, b).

As before, each $F_n \neq \emptyset$ and $\bigcap_n F_n = \{x_0\}$, (notice $F_n \subset E_n$), and

$$\lim_{n\to\infty} y_n(x_0) \equiv y_0 \leq z(x_0;\varepsilon) < z(x_0) = y_0,$$

which is a contradiction. Thus case (i) is not possible.

(ii) Now assume $y(x_0) < y_0 < z(x_0)$. In this case, for $0 \leq \lambda \leq 1$, let $z(x;\lambda)$ be the solution of the IVP for (4.1) such that

$$z^{(i)}(x_1;\lambda) = \lambda_y^{(i)}(x_1) + (1-\lambda)z^{(i)}(x_1),\ i = 0, 1, 2.$$

Then the set of ordered triples

$$L \equiv \{(z(x_1;\lambda), z'(x_1;\lambda), z''(x_1;\lambda)\,|\, 0 \leq \lambda \leq 1\}$$

is a line segment in $\mathbb{R}^3$ and is the continuous image of the mapping $h : [0,1] \to \mathbb{R}^3$ given by

$$h(\lambda) = (z(x_1;\lambda), z'(x_1;\lambda), z''(x_1;\lambda)).$$

Solutions of (4.1) depend continuously upon initial conditions, and hence the mapping $g : L \to \mathbb{R}^3$ defined by

$$g(z(x_1;\lambda), z'(x_1;\lambda), z''(x_1;\lambda)) = (z(x_0;\lambda), z'(x_0;\lambda), z''(x_0;\lambda))$$

is continuous. If we let $p_1 : \mathbb{R}^3 \to \mathbb{R}$ denote the projection from $\mathbb{R}^3$ onto the first component; that is $p_1(t_1, t_2, t_3) = t_1$, then p_1 is continuous. Consequently $p_1 \circ g \circ h : [0,1] \to \mathbb{R}$ is continuous. Now $p_1 \circ g \circ h(0) = z(x_0) > y_0 > y(x_0) = p_1 \circ g \circ h(1)$, and it follows that there exists $0 < \lambda_0 < 1$ such that $p_1 \circ g \circ h(\lambda_0) = y_0$. Hence there is a solution $z(x;\lambda_0)$ of (4.1) such that

$$z^{(i)}(x_1;\lambda_0) = \lambda_0 y^{(i)}(x_1) + (1-\lambda_0)z^{(i)}(x_1),\ i = 0, 1, 2,$$

and such that

$$z(x_0;\lambda_0) = y_0.$$

We also have

$$z(x_1;\lambda_0) = y(x_1) = z(x_1).$$

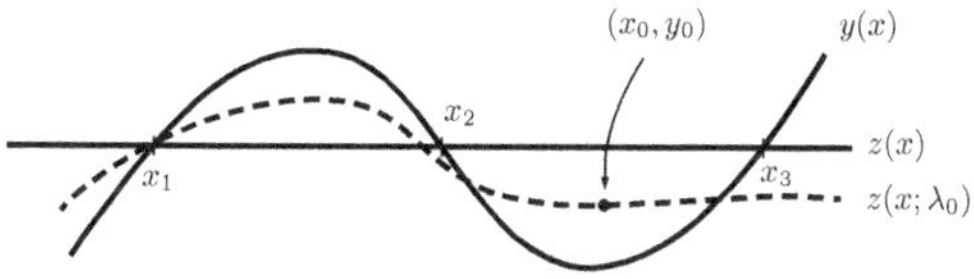

Fig. 4.22 The case $y(x_0) < y_0 < z(x_0)$.

By continuity there exists $\eta > 0$ such that $[x_0 - \eta, x_0 + \eta] \subset (x_2, x_3)$ and such that

$$z(x;\lambda_0) < z(x)$$

on $[x_0 - \eta, x_0 + \eta]$. Then with the given sequence of solutions $\{y_n(x)\}$, we have

$$\lim_{n\to\infty} y_n(x) > z(x) > z(x;\lambda_0),$$

for all $x \in [x_0 - \eta, x_0 + \eta] \setminus \{x_0\}$, and

$$\lim_{n\to\infty} y_n(x_0) = y_0 = z(x_0;\lambda_0).$$

This is the same situation as in case (i) where $y_0 = z(x_0)$, except now $y_0 = z(x_0;\lambda_0)$, and hence a contradiction will again arise.

Since both cases yield a contradiction, $y_0 \le z(x_0)$ is impossible and hence Eq. (4.1) must satisfy (C). □

As a consequence of Theorem 4.1 and Theorem 4.5, we have immediately:

Corollary 4.2. *Assume that Eq.* (4.1) *satisfies* (A), (B), *and* (D). *Then all* 3*-point conjugate BVP's for* (4.1) *have unique solutions on* (a, b).

As a consequence of Theorem 4.4 and Theorem 4.5, we have:

Corollary 4.3. *Assume that Eq.* (4.1) *satisfies* (A), (B), *and* (D). *Then all* 2*-point conjugate BVP's for* (4.1) *have unique solutions on* (a, b).

Bibliography

Agarwal R. P. (1986). *Boundary Value Problems for Higher Order Differential Equations* (World Scientific Publishing Co., Inc., Teaneck, NJ).

Agarwal R. P. (1997). Compactness condition for boundary value problems, *Equadiff (Brno)* **9**, pp. 1–23.

Bernfeld S. R. and Lakshmikantham V. (1974). *An Introduction to Nonlinear Boundary Value Problems* (Academic Press, New York).

Eloe P. W. and Henderson J. (2016). *Nonlinear Interpolation and Boundary Value Problems*, Trends in Abstract and Applied Analysis, Vol. 2 (World Scientific Publishing Co. Pte. Ltd., Hackensack, NJ).

Hartman P. (1958). Unrestricted n-parameter families, *Rend. Circ. Mat. Palermo* **7**, No. 2, pp. 123–142.

Hartman P. (1971). On n-parameter families and interpolation problems for nonlinear ordinary differential equations, *Trans. Amer. Math. Soc.* **154**, pp. 201–226.

Henderson J. and Jackson L. K. (1983). Existence and uniqueness of solutions of k-point boundary value problems for ordinary differential equations, *J. Differential Equations* **48**, No. 3, pp. 373–386.

Jackson L. K. (1973). Existence and uniqueness of solutions of boundary value problems for third order differential equations, *J. Differential Equations* **13**, pp. 432–437.

Jackson L. K. (1973). Uniqueness of solutions of boundary value problems for ordinary differential equations, *SIAM J. Appl. Math.* **24**, No. 4, pp. 535–538.

Jackson L. K. (1976). A compactness condition for solutions of ordinary differential equations, *Proc. Amer. Math. Soc.* **57**, No. 1, 89–92.

Jackson L. K. (1977). Boundary value problems for ordinary differential equations, *Studies in Ordinary Differential Equations*, Stud. in Math. **14** (Editor: J. K. Hale) (Math. Association of America, Washington, D. C.), pp. 93–127.

Jackson L. K. and Klaasen G. (1970). Uniqueness of solutions of boundary value problems for ordinary differential equations, *SIAM J. Appl. Math.* **19**, No. 3, 542–546.

Jackson L. K. and Schrader K. W. (1970). Subfunctions and third order differential equations, *J. Differential Equations* **8**, 180–194.

Jackson L. K. and Schrader K. W. (1971). Existence and uniqueness of solutions of boundary value problems for third order differential equations, *J. Differential Equations* **9** , 46–54.

Klaasen G. (1973). Existence theorem for boundary value problems for nth order ordinary differential equations, *Rocky Mountain J. Math.* **3**, 457–473.

Schrader K. W. (1985). Uniqueness implies existence for solutions of nonlinear boundary value problems, *Abstracts Amer. Math. Soc.* **6**, 235.

Chapter 5

Solution Matching

In this chapter, we are first concerned with solutions of three-point BVP's which are obtained by "solution matching" of two-point BVP's. Then, solutions of five-point nonlocal BVP's are obtained by matching solutions of three-point nonlocal BVP's.

5.1 Introduction

We first will be concerned with solutions of three-point BVP's on an interval $[x_1, x_3]$ for the third order ODE,

$$y''' = f(x, y, y', y''), \tag{5.1}$$

which are obtained by "solution matching" of two-point BVP's for (5.1). In particular, we will give sufficient conditions such that if $y_1(x)$ is a solution of a two-point BVP for (5.1) on $[x_1, x_2]$ and $y_2(x)$ is a solution of a two-point BVP for (5.1) on $[x_2, x_3]$, then

$$y(x) = \begin{cases} y_1(x), & x_1 \le x \le x_2, \\ y_2(x), & x_2 \le x \le x_3, \end{cases}$$

is a solution of a three-point BVP for (5.1) on $[x_1, x_3]$.

We will assume that $f : [x_1, x_3] \times \mathbb{R}^3 \to \mathbb{R}$ is continuous and that solutions of IVP's for (5.1) exist and are unique on $[x_1, x_3]$. Moreover, $x_2 \in (x_1, x_3)$ will be fixed throughout. In addition to these hypotheses we assume there exists a function $g : [x_1, x_3] \times \mathbb{R}^3 \to \mathbb{R}$ such that:

(A) For each $v_3, u_3 \in \mathbb{R}$

$$f(x, v_1, v_2, v_3) - f(x, u_1, u_2, u_3) > g(x, v_1 - u_1, v_2 - u_2, v_3 - u_3)$$

when $x \in (x_1, x_2]$, $u_1 - v_1 \ge 0$, and $u_2 - v_2 < 0$, or when $x \in [x_2, x_3)$, $u_1 - v_1 \le 0$, and $u_2 - v_2 < 0$;

(B) There exists $\delta_1 > 0$ such that, for each $0 < \delta < \delta_1$, the IVP

$$\begin{cases} y''' = g(x, y, y', y''), \\ y(x_2) = 0, \ y''(x_2) = 0, \ y'(x_2) = \delta, \end{cases}$$

has a solution z such that z' does not change sign on $[x_1, x_3]$;

(C) There exists $\delta_2 > 0$ such that, for each $0 < \delta < \delta_2$, the IVP

$$\begin{cases} y''' = g(x, y, y', y''), \\ y(x_2) = y'(x_2) = 0, \ y''(x_2) = \delta, \ (-\delta), \end{cases}$$

has a solution z on $[x_2, x_3]$, $([x_1, x_2])$, such that z'' does not change sign on $[x_2, x_3]$, $([x_1, x_2])$; and

(D) For each $w \in \mathbb{R}$,

$$g(x, v_1, v_2, w) \geq g(x, u_1, u_2, w),$$

when $x \in (x_1, x_2]$, $u_1 - v_1 \geq 0$, and $v_2 > u_2 \geq 0$, or when $x \in [x_2, x_3)$, $u_1 - v_1 \leq 0$, and $v_2 > u_2 \geq 0$.

5.1.1 *Uniqueness and existence of solutions by matching*

In this section, solutions of (5.1) satisfying

$$y(x_1) = y_1, \ y(x_2) = y_2, \ y^{(j)}(x_2) = m, \tag{4_j}$$

and

$$y(x_2) = y_2, \ y^{(j)}(x_2) = m, \ y(x_3) = y_3, \tag{5_j}$$

where $j = 1, 2$, $y_1, y_2, y_3 \in \mathbb{R}$ and $m \in \mathbb{R}$, are matched to obtain a unique solution of the BVP for (5.1) satisfying

$$y(x_1) = y_1, \ y(x_2) = y_2, \ y(x_3) = y_3. \tag{5.2}$$

Theorem 5.1. *Let $y_1, y_2, y_3 \in \mathbb{R}$ be as above and assume that* (A)–(D) *are satisfied. Then given $m \in \mathbb{R}$, each of the BVP's* (5.1), (4_j) *and* (5.1), (5_j), $j = 1, 2$, *has at most one solution.*

Proof. We will consider only the proof of the theorem for (5.1), (4_1) with arguments for the other cases being similar.

Thus, let's assume there are distinct solutions α and β of (5.1), (4_1) and let $w := \alpha - \beta$. Then $w(x_1) = w(x_2) = w'(x_2) = 0$. By uniqueness of solutions of IVP's for (5.1), we may assume that $w''(x_2) < 0$. It follows that there exists $x_1 < r_1 < x_2$ such that $w''(r_1) = 0$ and $w''(x) < 0$ on $(r_1, x_2]$. Then, $w'(x) > 0$ and $w(x) < 0$

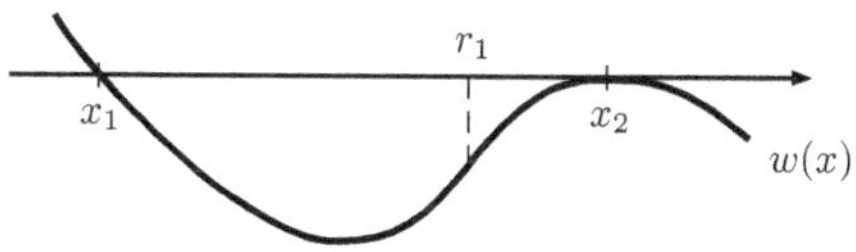

Fig. 5.1 The graph of $w(x)$.

on $[r_1, x_2)$. Now let $0 < \delta < \min\{\delta_2, -w''(x_2)\}$ and let w_δ satisfy the criteria of hypothesis (C) relative to the interval $[x_1, x_2]$; i.e.,

$$\begin{cases} w_\delta''' = g(x, w_\delta, w_\delta', w_\delta''), \\ w_\delta^{(i)}(x_2) = 0, \ i = 0, 1, \quad w_\delta''(x_2) = -\delta, \end{cases}$$

and w_δ'' does not change sign in $[x_1, x_2]$.

Now set $z \equiv w - w_\delta$. Then $z(x_2) = z'(x_2) = 0$, and $z''(x_2) < 0$. Moreover, by (C), $z''(r_1) \geq 0$, and hence there exists $r_1 \leq r_2 \leq x_2$ such that $z''(r_2) = 0$ and $z''(x) < 0$ on $(r_2, x_2]$. Consequently, $z'(x) > 0$ and $z(x) < 0$ on $[r_2, x_2)$.

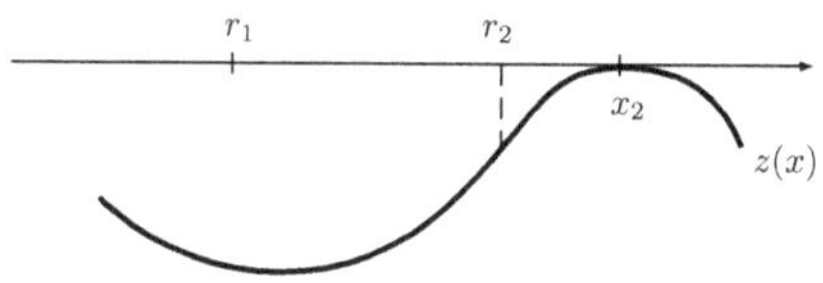

Fig. 5.2 The graph of $z(x)$.

The following contradiction arises; first

$$z'''(r_2) = \lim_{x \to r_2^+} \frac{z''(x)}{x - r_2} \leq 0,$$

whereas, from (A) and (D)

$$\begin{aligned} z'''(r_2) &= w'''(r_2) - w_\delta'''(r_2) \\ &> g(r_2, w(r_2), w'(r_2), w''(r_2)) - g(r_2, w_\delta(r_2), w_\delta'(r_2), w_\delta''(r_2)) \\ &\geq 0. \end{aligned}$$

Thus (5.1), (4_1) has at most one solution. The remainder of the theorem follows by similar arguments. □

We now prove that under the same conditions of Theorem 5.1, solutions of (5.1), (5.2) are unique when they exist.

Theorem 5.2. *Assume that hypotheses* (A)–(D) *are satisfied. Then* (5.1), (5.2) *has at most one solution.*

Proof. Again we argue by contradiction. Assume that, for some relations $y_1, y_2, y_3 \in \mathbb{R}$, there are distinct solutions α and β of (5.1), (5.2) and let $w = \alpha - \beta$. Then

$$w(x_1) = w(x_2) = w(x_3) = 0.$$

From Theorem 5.1, $w'(x_2), w''(x_2) \neq 0$. Assume without loss of generality that $w'(x_2) > 0$. Then there exists points $x_1 < r_1 < x_2 < r_2 < x_3$ such that $w'(r_i) = 0$, $i = 1, 2$, and $w'(x) > 0$ on (r_1, r_2).

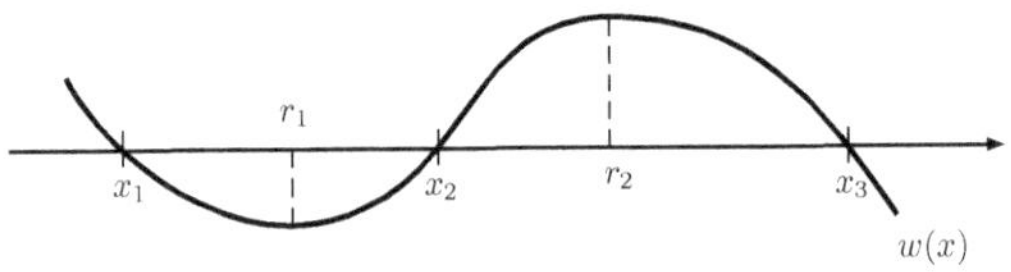

Fig. 5.3 The graph of $w(x)$.

Now let $0 < \delta < \min\{\delta_1, w'(x_2)\}$, let z_δ be a solution satisfying the conditions of (B), and set $z \equiv w - z_\delta$. Then $z(x_2) = 0$, $z'(x_2) > 0$, $z''(x_2) = w''(x_2) \neq 0$, and $z'(r_2) \leq 0$.

There are two cases in which the argument for each is similar:

(i) Suppose $z''(x_2) > 0$. In view of the fact that $z''(x_2) > 0$ and $z'(r_2) \leq 0$, there exists $x_2 < r_3 < r_2$ such that $z''(r_3) = 0$ and $z''(x) > 0$ on $[x_2, r_3)$. Then $z(x) > 0$ and $z'(x) > 0$ on $(x_2, r_3]$. Now

$$z'''(r_3) = \lim_{x \to r_3^-} \frac{z''(x)}{x - r_3} \leq 0,$$

yet, by virtue of (A) and (D) again,

$$\begin{aligned} z'''(r_3) &= w'''(r_3) - z_\delta'''(r_3) \\ &> g(r_3, w(r_3), w'(r_3), w''(r_3)) - g(r_3, z_\delta(r_3), z_\delta'(r_3), z_\delta''(r_3)) \\ &\geq 0. \end{aligned}$$

a contradiction.

(ii) Suppose $z''(x_2) < 0$. This case is resolved by making the analogous argument on $[x_1, x_2]$.

Thus the assumption concerning distinct solutions α and β is false and the proof is complete.

□

Now given $m \in \mathbb{R}$, let $\alpha(x,m)$, $\beta(x,m)$, $u(x,m)$ and $v(x,m)$ denote the solutions, when they exist, of the BVP's for (5.1) given, respectively, by (4_1), (5_1), (4_2), and (5_2).

Theorem 5.3. *Suppose that* (A)–(D) *are satisfied and that, for each* $m \in \mathbb{R}$, *there exist solutions of* (5.1), (4_j) *and* (5.1), (5_j), $j = 1, 2$; *that is, for each* $m \in \mathbb{R}$, α, β, u *and* v *all exist. Then* $u'(x_2, m)$ *and* $\alpha''(x_2, m)$ *are strictly increasing functions of* m *with ranges all of* $\mathbb{R}$, *and* $v'(x_2, m)$ *and* $\beta''(x_2, m)$ *are strictly decreasing functions of* m *with ranges all of* $\mathbb{R}$.

Proof. The "strictness" of the conclusion arises from Theorem 5.1. We will prove the theorem with respect to the solution $\alpha(x,m)$. Let $m_1 > m_2$ and let

$$w(x) \equiv \alpha(x, m_1) - \alpha(x, m_2).$$

Then $w(x_1) = w(x_2) = 0$, $w'(x_2) > 0$, and $w''(x_2) \neq 0$. Contrary to the conclusion of the theorem, assume that $w''(x_2) < 0$. Since there exists $x_1 < r_1 < x_2$ such that $w'(r_1) = 0$ and $w'(x) > 0$ on $(r_1, x_2]$, it follows by continuity that there exists $r_1 < r_2 < x_2$ such that $w''(r_2) = 0$ and $w''(x) < 0$ on $(r_2, x_2]$. We also have $w(x) < 0$ on $[r_2, x_2)$.

Now let $0 < \delta < \min\{\delta_2, -w''(x_2)\}$ and let w_δ be a solution of the IVP satisfying the conditions of (C). Set $z \equiv w - w_\delta$. Then

$$z(x_2) = 0, \quad z'(x_2) = w'(x_2) > 0, \quad z''(x_2) < 0.$$

Furthermore, $z''(r_2) \geq 0$; thus there exists $r_2 \leq r_3 < x_2$ such that $z''(r_3) = 0$ and $z''(x) < 0$ on $(r_3, x_2]$. Then $z'(x) > 0$ and $z(x) < 0$ on $[r_3, x_2)$. As in the other proofs above, we can then argue that $z'''(r_3) \leq 0$ and $z'''(r_3) > 0$, again a contradiction. Thus $w''(x_2) > 0$ and consequently, $\alpha''(x_2, m)$ is a strictly increasing function of m.

In order to show that $\{\alpha''(x_2, m) | m \in \mathbb{R}\} = \mathbb{R}$, let $k \in \mathbb{R}$ and consider the solution $u(x,k)$ of the BVP (5.1), (4_2) with u as specified above. Consider also the solution $\alpha(x, u'(x_2, k))$ of (5.1), (4_1). Then $\alpha(x, u'(x_2, k))$ and $u(x,k)$ are solutions of the same type problem (5.1), (4_1), and hence by Theorem 5.1, the two functions are identical. Therefore

$$\alpha''(x_2, u'(x_2, k)) = u''(x_2, k) = k,$$

and the statement concerning the range of $\alpha''(x_2, m)$ is verified. The other three parts are established in a similar manner. □

Theorem 5.4. *Assume the hypotheses of Theorem* 5.3 *hold. Then* (5.1), (5.2) *has a unique solution.*

Proof. We have a choice here in that we can make use of the pair of solutions α and β or the pair u and v. We shall choose the latter pair. By Theorem 5.3, there exists a unique $m_0 \in \mathbb{R}$ such that $u'(x_2, m_0) = v'(x_2, m_0)$. Of course it is true that $u''(x_2, m_0) = m_0 = v''(x_2, m_0)$. Then

$$y(x) = \begin{cases} u(x, m_0), & x_1 \le x \le x_2, \\ v(x, m_0), & x_2 \le x \le x_3, \end{cases}$$

is a solution of (5.1), (5.2), and by Theorem 5.2, $y(x)$ is the unique solution. □

5.1.2 *Uniqueness and existence of solutions for a larger class of BVP's by matching*

In this section results analogous to those of the preceding section are given for a larger class of two-point and three-point BVP's for (5.1).

More precisely, let $\mu, \nu \in \{0, 1\}$ be arbitrary, but fixed, in this section; then solutions of (5.1) satisfying,

$$y^{(\mu)}(x_1) = y_1, \; y(x_2) = y_2, \; y^{(j)}(x_2) = m, \tag{6_j}$$

and

$$y(x_2) = y_2, \; y^{(j)}(x_2) = m, \; y^{(\nu)}(x_3) = y_3, \tag{7_j}$$

where $j = 1, 2$, $y_1, y_2, y_3 \in \mathbb{R}$ and $m \in \mathbb{R}$, are matched to obtain a unique solution of the BVP for (5.1) satisfying

$$y^{(\mu)}(x_1) = y_1, \; y(x_2) = y_2, \; y^{(\nu)}(x_3) = y_3. \tag{5.3}$$

The proofs of the following analogues are essentially the same as those presented in the previous section. Since only the slightest modifications are required when μ or ν equals one, we will omit the proofs.

Theorem 5.5. *Let $y_1, y_2, y_3 \in \mathbb{R}$ be as above and assume that* (A)–(D) *are satisfied. Then given $m \in \mathbb{R}$, each of the BVP's* (5.1), (6_j) *and* (5.1), (7_j), *$j = 1, 2$, has at most one solution.*

Theorem 5.6. *Assume that hypotheses* (A)–(D) *are satisfied. Then* (5.1), (5.3) *has at most one solution.*

Now given $m \in \mathbb{R}$, let $\sigma(x, m)$, $\rho(x, m)$, $y(x, m)$, and $z(x, m)$ denote the solutions, when they exist, of the BVP's for (5.1) given, respectively, by (6_1), (7_1), (6_2), and (7_2).

Theorem 5.7. *Suppose that* (A)–(D) *are satisfied and that, for each $m \in \mathbb{R}$, there exist solutions of* (5.1), (6_j) *and* (5.1), (7_j), *$j = 1, 2$. Then*

$\sigma''(x_2, m)$ and $y'(x_2, m)$, ($\rho''(x_2, m)$ and $z'(x_2, m)$), are strictly increasing, (decreasing), functions of m with ranges all of $\mathbb{R}$.

Theorem 5.8. *Assume the hypotheses of Theorem 5.7 hold. Then (5.1), (5.3) has a unique solution.*

5.2 Five-Point Nonlocal BVP's for nth Order ODE's by Solution Matching

We are concerned with the existence and uniqueness of solutions of BVP's on an interval $[a, c]$ for the nth order ordinary differential equation,

$$y^{(n)} = f(x, y, y', \ldots, y^{(n-1)}), \tag{5.4}$$

satisfying the 5-point nonlocal boundary conditions,

$$y(a) - y(x_1) = y_1,\ y^{(i-1)}(b) = y_{i+1},\ 1 \le i \le n-2,\ y(x_2) - y(c) = y_n, \tag{5.5}$$

where $a < x_1 < b < x_2 < c$ and $y_1, \ldots, y_n \in \mathbb{R}$.

It is assumed throughout that $f : [a, c] \times \mathbb{R}^n \to \mathbb{R}$ is continuous and that solutions of IVP's for (5.4) are unique and exist on all of $[a, c]$. Moreover, the points $a < x_1 < b < x_2 < c$ are fixed throughout.

Monotonicity conditions will be imposed in f, and sufficient condition will be given such that, if $y_1(x)$ is a solution of a 3-point nonlocal BVP on $[a, b]$, and if $y_2(x)$ is a solution of another 3-point nonlocal BVP on $[b, c]$, then $y(x)$ defined by

$$y(x) = \begin{cases} y_1(x), & a \le x \le b \\ y_2(x), & b \le x \le c, \end{cases}$$

will be a desired solution of (5.4), (5.5). In particular, a monotonicity condition will be imposed on $f(x, r_1, \ldots, r_n)$ insuring that each 3-point BVP for (5.4) satisfying any one of the following conditions has at most one solution:

$$y(a) - y(x_1) = y_1,\ y^{(i-1)}(b) = y_{i+1},\ 1 \le i \le n-2,\ y^{(n-2)}(b) = m, \tag{5.6}$$

$$y(a) - y(x_1) = y_1,\ y^{(i-1)}(b) = y_{i+1},\ 1 \le i \le n-2,\ y^{(n-1)}(b) = m, \tag{5.7}$$

$$y^{(i-1)}(b) = y_{i+1},\ 1 \le i \le n-2,\ y^{(n-2)}(b) = m,\ y(x_2) - y(c) = y_n, \tag{5.8}$$

$$y^{(i-1)}(b) = y_{i+1},\ 1 \le i \le n-2,\ y^{(n-1)}(b) = m,\ y(x_2) - y(c) = y_n, \tag{5.9}$$

where $m \in \mathbb{R}$.

The monotonicity hypothesis on f which will play a fundamental role in uniqueness of solutions is given by:

(A) For all $w \in \mathbb{R}$,

$$f(x, v_1, \ldots, v_{n-1}, v_{n-1}, w) > f(x, u_1, \ldots, un-2, u_{n-1}, w),$$

(a) when $x \in (a, b]$, $(-1)^{n-i} u_i \geq (-1)^{n-i} v_i$, $1 \leq i \leq n-2$, and $v_{n-1} > u_{n-1}$, or

(b) when $x \in [b, c)$, $v_i \geq u_i$, $1 \leq i \leq n-2$, and $v_{n-1} > u_{n-1}$.

5.2.1 *Uniqueness of solutions*

We establish that under (A), solutions of 3-point BVP's, as well as 5-point BVP's, are unique when they exist.

Theorem 5.9. *Let $y_1 \ldots, y_n \in \mathbb{R}$ be given and assume condition (A) is satisfied. Then, given $m \in \mathbb{R}$, each of the BVP's for* (5.4) *satisfying any of condition* (5.6), (5.7), (5.8) *or* (5.9) *has at most one solution.*

Proof. We will establish the result only for (5.4), (5.6). Arguments for the other BVP's are very similar.

In order to reach a contradiction, we assume that for some $m \in \mathbb{R}$, there are distinct solutions, α and β, of (5.4), (5.6), and set $w = \alpha - \beta$. Then,

$$w(a) - w(x_1) = w^{(i-1)}(b) = 0, \quad 1 \leq i \leq n-1.$$

By uniqueness of solutions of IVP's for (5.4), we may assume with no loss of generality that $w^{(n-1)}(b) < 0$. It follows from the boundary conditions satisfied by w that, there exists $a < r < b$ such that

$$w^{(n-1)}(r) = 0 \text{ and } w^{(n-1)}(x) < 0 \text{ on } (r, b].$$

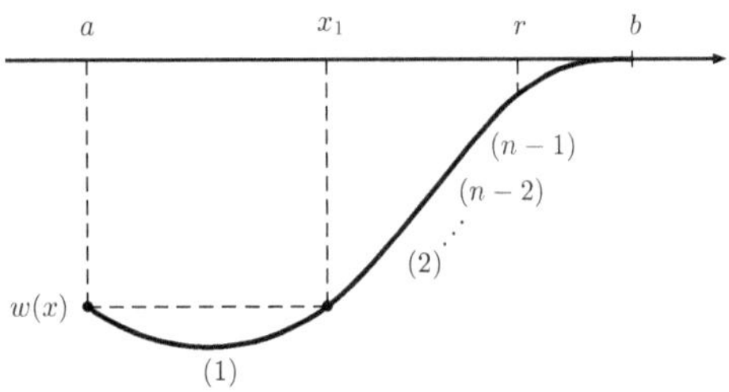

Fig. 5.4 The graph of $w(x)$.

Since $w^{(i-1)}(b) = 0$, $1 \leq i \leq n-1$, it follows in turn that

$$(-1)^{(n-j)} w^{(j)}(x) > 0, \; 0 \leq j \leq n-2, \text{ on } [r, b).$$

This leads to

$$\begin{aligned} w^{(n)}(r) &= \lim_{x\to r^+} \frac{w^{(n-1)(x)-w^{(n-1)}(r)}}{x-r} \\ &= \lim_{x\to r^+} \frac{w^{(n-1)(x)}}{x-r} \\ &\le 0. \end{aligned}$$

However, from condition (A), we have

$$\begin{aligned} w^{(n)}(r) &= \alpha^{(n)}(r) - \beta^{(n)}(r) \\ &= f(r, \alpha(r), \alpha'(r), \ldots, \alpha^{(n-2)}(r), \alpha^{(n-1)}(r)) \\ &\quad - f(r, \beta(r), \beta'(r), \ldots, \beta^{(n-2)}(r), \beta^{(n-1)}(r)) \\ &= f(r, \alpha(r), \alpha'(r), \ldots, \alpha^{(n-2)}(r), \alpha^{(n-1)}(r)) \\ &\quad - f(r, \beta(r), \beta'(r), \ldots, \beta^{(n-2)}(r), \alpha^{(n-1)}(r)) \\ &> 0, \end{aligned}$$

which is a contradiction. Thus, (5.4), (5.6) has at most one solution. □

Theorem 5.10. *Let $y_1, \ldots, y_n \in \mathbb{R}$ be given. Assume condition* (A) *is satisfied. Then, the BVP* (5.4), (5.5) *has at most one solution.*

Proof. Again, we argue by contradiction. Assume for some values $y_1, \ldots, y_n \in \mathbb{R}$, there exist distinct solutions α and β of (5.4), (5.5). Also, let $w = \alpha - \beta$. Then

$$w(a) - w(x_1) = w^{(i-1)}(b) = w(x_2) - w(c) = 0, \quad 1 \le i \le n-1.$$

By Theorem 5.9, $w^{(n-2)}(b) \neq 0$ and $w^{(n-1)}(b) \neq 0$. We assume with no loss of generality that $w^{(n-2)}(b) > 0$. Then, from the boundary conditions satisfied by w, there are points $a < r_1 < b < r_2 < c$ such that

$$w^{(n-2)}(r_1) = w^{(n-2)}(r_2) = 0, \text{ and } w^{(n-2)}(x) > 0 \text{ on } (r_1, r_2).$$

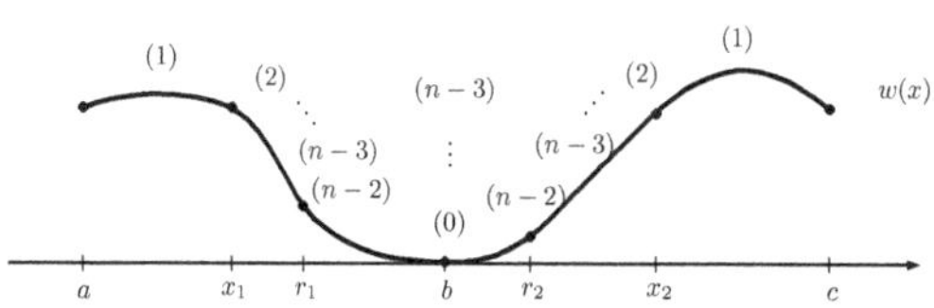

Fig. 5.5 The graph of $w(x)$.

There are two cases to analyze; that is, $w^{(n-1)}(b) > 0$ and $w^{(n-1)}(b) < 0$.

The arguments for the two cases are analogous, and therefore we will treat only the first case $w^{(n-1)}(b) > 0$.

In view of the fact that $w^{(n-1)}(b) > 0$ and $w^{(n-2)}(r_2) = 0$, there exists $b < r < r_2$ so that $w^{(n-1)}(r) = 0$ and $w^{(n-1)}(x) > 0$ on $[b, r)$. Then, $w^{(j)}(x) > 0$, on $(b, r]$ for $0 \le j \le n-2$. This leads to

$$\begin{aligned} w^{(n)}(r) &= \lim_{x\to r^-} \frac{w^{(n-1)}(x) - w^{(n-1)}(r)}{x-r} \\ &= \lim_{x\to r^-} \frac{w^{(n-1)}(x)}{x-r} \\ &\le 0. \end{aligned}$$

However, again from condition (A), we have

$$\begin{aligned} w^{(n)}(r) &= \alpha^{(n)}(r) - \beta^{(n)}(r) \\ &= f(r, \alpha(r), \alpha'(r), \ldots, \alpha^{(n-2)}(r), \alpha^{(n-1)}(r)) \\ &\quad - f(r, \beta(r), \beta'(r), \ldots, \beta^{(n-2)}(r), \beta^{(n-1)}(r)) \\ &= f(r, \alpha(r), \alpha'(r), \ldots, \alpha^{(n-2)}(r), \alpha^{(n-1)}(r)) \\ &\quad - f(r, \beta(r), \beta'(r), \ldots, \beta^{(n-2)}(r), \alpha^{(n-1)}(r)) \\ &> 0, \end{aligned}$$

which contradicts the initial assumption. Thus, (5.4), (5.5) has at most one solution. □

5.2.2 *Existence of solutions*

In this section, we show that solutions of (5.4) satisfying each of (5.6), (5.7), (5.8), and (5.9) are monotone functions of m. Then, we will use these monotonicity properties to obtain solutions of (5.4), (5.5). For notation purposes, given $m \in \mathbb{R}$, let $\alpha(x,m), u(x,m), \beta(x,m)$, and $v(x,m)$ denote the solutions, when they exist, of the BVP's for (5.4) satisfying, respectively, (5.6), (5.7), (5.8), and (5.9).

Theorem 5.11. *Suppose that the monotonicity hypothesis (A) is satisfied and that, for each $m \in \mathbb{R}$, there exist solutions of (5.4) satisfying each of the conditions (5.6), (5.7), (5.8), and (5.9). Then, $\alpha^{(n-1)}(b,m)$ and $\beta^{(n-1)}(b,m)$ are, respectively, strictly increasing and decreasing functions of m with ranges all of $\mathbb{R}$.*

Proof. The "strictness" of the conclusion arises from Theorem 5.10. Let $m_1 > m_2$ and let $w(x) = \alpha(x, m_1) - \alpha(x, m_2)$. Then,

$$w(x_1) - w(a) = w^{(i-1)}(b) = 0, \quad 1 \le i \le n-2,$$
$$w^{(n-2)}(b) = m_1 - m_2 > 0, \quad w^{(n-1)}(b) \ne 0.$$

Contrary to the conclusion, assume $w^{(n-1)}(b) < 0$. Since there exists $a < r_1 < b$ so that $w^{(n-2)}(r_1) = 0$ and $w^{(n-2)}(x) > 0$ on $(r_1, b]$, it follows that there exists $r_1 < r_2 < b$ such that $w^{(n-1)}(r_2) = 0$ and $w^{(n-1)}(x) < 0$ on $(r_2, b]$. We also have $(-1)^{n-j} w^{(j)}(x) > 0$, on $[r_2, b)$ for $0 \le j \le n-2$.

As in the previous proofs, we arrive at the contradiction, $0 < w^{(n)}(r_2) \le 0$. Thus, $w^{(n-1)}(b) > 0$, and as a consequence, $\alpha^{(n-1)}(b, m)$ is a strictly increasing function of m.

We next argue that $\{\alpha^{(n-1)}(b, m) \mid m \in \mathbb{R}\} = \mathbb{R}$. Let $k \in \mathbb{R}$ and consider the solution $u(x, k)$ of (5.4), (5.7) with u as defined above. Consider also the solution $\alpha(x, u^{(n-2)}(b, k))$ of (5.4), (5.6). Then, $\alpha(x, u^{(n-2)}(b, k))$ and $u(x, k)$ are solutions of the same type BVP (5.4), (5.6). Hence, by Theorem 5.10, the functions are identical. Therefore,

$$\alpha^{(n-1)}(b, u^{(n-2)}(b, k)) = u^{(n-1)}(b, k) = k,$$

and the range of $\alpha^{(n-1)}(b, m)$, as a function of m, is all of $\mathbb{R}$.

The argument for $\beta^{(n-1)}(b, m)$ is quite similar. □

In a very similar way, we also have a monotonicity result on the $(n-2)$nd derivatives of $u(x, m)$ and $v(x, m)$.

Theorem 5.12. *Suppose the hypotheses of Theorem* 5.11 *are satisfied. Then,* $u^{(n-2)}(b, m)$ *and* $v^{(n-2)}(b, m)$ *are, respectively, strictly increasing and decreasing functions of* m *with ranges all of* $\mathbb{R}$.

We now prove our existence result.

Theorem 5.13. *Assume the hypotheses of Theorem* 5.11 *hold. Then,* (5.4), (5.5) *has a unique solution.*

Proof. The existence is immediate from either Theorem 5.11 or Theorem 5.12. Making use of Theorem 5.12, there exists a unique $m_0 \in \mathbb{R}$ such that $u^{(n-2)}(b, m) = v^{(n-2)}(b, m)$. Then

$$y(x) = \begin{cases} u(x, m), & a \le x \le b, \\ v(x, m), & b \le x \le c, \end{cases}$$

is a solution of (5.4), (5.5), and by Theorem 5.10, $y(x)$ is the unique solution. □

Bibliography

Agarwal R. P. (1986). *Boundary Value Problems for Higher Order Differential Equations* (World Scientific Publishing Co., Inc., Teaneck, NJ).

Bailey P. B., Shampine L. F. and Waltman P. E. (1968). *Nonlinear Two Point Boundary Value Problems* (Academic Press, New York).

Barr D. and Miletta P. (1974). An existence and uniqueness criterion for solutions of boundary value problems, *J. Differential Equations* **16**, pp. 460–471.

Barr D. and Sherman T. L. (1973). Existence and uniqueness of solutions of three-point boundary value problems, *J. Differential Equations* **13**, pp. 197–212.

Das K. M. and Lalli B. S. (1981). Boundary value problems for $y''' = f(x, y, y', y'')$, *J. Math. Anal. Appl.* **81**, pp. 300–307.

Ehrke J., Henderson J. and Kunkel C. (2008). Five-point boundary value problems for n-th order differential equations by solution matching, *Involve* **1**, No. 1, pp. 1–7.

Henderson J. (1983). Three-point boundary value problems for ordinary differential equations by matching solutions, *Nonlinear Anal.* **7**, 411–417.

Henderson J. (2009). Boundary value problems for third order differential equations by solution matching, *Electron. J. Qual. Theory Differ. Equ.*, Special Edition I, No. 14, pp. 1–9.

Henderson J. and Taunton D. (1993). Solutions of boundary value problems by matching methods, *Appl. Anal.* **49**, pp. 235–246.

Henderson J. and Tisdell C. C. (2005). Five-point boundary value problems for third-order differential equations by solution matching, *Math. Comput. Model. Dyn. Syst.* **42**, pp. 133–137.

Lakshmikantham V. and Murty K. N. (1991) Theory of differential inequalities and three-point boundary value problems, *Panamer. Math. J.* **1**, pp. 1–9.

Liu X. (2010). Boundary value problems with gaps in boundary conditions for differential equations by matching solutions, *Comm. Appl. Nonlinear Anal.* **17**, No. 3, pp. 81–88.

Liu X. (2011) Nonlocal boundary-value problems for N-th order ordinary differential equations by matching solutions, *Electron. J. Differential Equations*, No. 17, pp. 1–9.

Moorti V. R. G. and Garner J. B. (1978) Existence-uniqueness theorems for three-point boundary value problems for nth-order nonlinear differential equations, *J. Differential Equations* **29**, pp. 205–213.

Murty K. N. (1977). Three point boundary value problems, existence and uniqueness, *J. Math. Phys. Sci.* **11**, pp. 265–272.

Murty M. S. N. and Kumar G. S. (2007). Extension of Liapunov theory to five-point boundary value problems for third order differential equations, *Novi Sad J. Math.* **37**, pp. 85–92.

Rao D. R. K. S., Murty K. N. and Murty M. S. N. (1982). Extension of Liapunov theory to three-point boundary value problems, *Indian J. Pure Appl. Math.* **13**, No. 4, pp. 421–425.

Rao D. R. K. S., Murty K. N. and Rao A. S. (1981). On three-point boundary value problems associated with third order differential equations, *Nonlinear Anal.* **5**, pp. 669–673.

Chapter 6

Comparison of Smallest Eigenvalues

The theory of u_0-positive operators with respect to a cone in a Banach space is applied to BVP's for second order linear ODE's. The existence of a smallest positive eigenvalue is established, and then a comparison theorem for smallest positive eigenvalues is proven. The results are first obtained for eigenvalue problems satisfying 2-point conjugate BC's, followed by results for problems satisfying 3-point nonlocal BC's. The final section studies the solutions for multi-point BVP's by using the smallest positive eigenvalues for associated linear problems. The techniques involve properties of Green's functions, along with cone theoretic applications of results due to Krasnosel'skii and to Krein and Rutman.

6.1 Comparison Theorems for 2-Point Conjugate Eigenvalue Problems

For this section, we will concern ourselves with comparison results for the eigenvalue problems,

$$y'' + \lambda p(x)y = 0, \tag{6.1}$$

and

$$y'' + \sigma q(x)y = 0, \tag{6.2}$$

subject to the boundary conditions

$$y(a) = y(b) = 0, \tag{6.3}$$

where $p(x)$ and $q(x)$ are continuous nonnegative functions on $[a, b]$, and both $p(x)$ and $q(x)$ do not vanish identically on any subinterval $[\alpha, \beta] \subseteq [a, b]$.

Definition 6.1. A scalar λ_0 is an *eigenvalue* of (6.1), (6.3), if there exists a nontrivial solution $y(x)$ of (6.1), (6.3) on $[a, b]$ corresponding to $\lambda = \lambda_0$. $y(x)$ is called an *eigenvector* or *eigenfunction*.

Remark 6.1. Since $y'' = 0$ is disconjugate on $[a, b]$, it follows that $\lambda = 0$, $(\sigma = 0)$, is not an eigenvalue of (6.1), (6.3), ((6.2), (6.3)).

Our particular goal here is to compare the smallest eigenvalues λ and σ of (6.1), (6.3) and (6.2), (6.3), respectively, if $p(x) \leq q(x)$ on $[a, b]$.

Our techniques make use of the theory of u_0-positive operators with respect to an abstract cone in a suitable ordered Banach space. We will draw heavily on results from cone theory due to M.A. Krasnosel'skii.

Definition 6.2. Let $\mathcal{B}$ be a real Banach space. A closed nonempty subset $\mathcal{P}$ of $\mathcal{B}$ is said to be a *cone* provided

(i) $\alpha u + \beta v \in \mathcal{P}$, for all $u, v \in \mathcal{P}$ and all $\alpha, \beta \geq 0$; and

(ii) $u \in \mathcal{P}$ and $-u \in \mathcal{P}$ imply $u = 0$.

A cone $\mathcal{P}$ is said to be *reproducing*, provided $\mathcal{B} = \mathcal{P} - \mathcal{P}$; i.e., given $w \in \mathcal{B}$, there exists $u, v \in \mathcal{P}$ such that $w = u - v$. A cone $\mathcal{P}$ is said to be *solid*, if $\mathcal{P}^\circ \neq \emptyset$.

Remark 6.2. It is straightforward that every solid cone is reproducing. The converse is true if $\mathcal{B} = \mathbb{R}^n$.

Now, every cone $\mathcal{P} \subseteq \mathcal{B}$ generates a partial ordering, $\preceq$, on $\mathcal{B}$ and also on the space of bounded linear operators mapping $\mathcal{B}$ into itself.

Definition 6.3. Let $\mathcal{P}$ be a cone in a real Banach space $\mathcal{B}$. If $u, v \in \mathcal{B}$, we say $u \preceq v$ with respect to $\mathcal{P}$, if $v - u \in \mathcal{P}$. If both $M, N : \mathcal{B} \to \mathcal{B}$ are bounded linear operators, we say that $M \preceq N$ with respect to $\mathcal{P}$, if $Mu \preceq Nu$, for all $u \in \mathcal{P}$.

Finally, we say that a bounded linear operator $M : \mathcal{B} \to \mathcal{B}$ is u_0*-positive* with respect to $\mathcal{P}$, if $0 \neq u_0 \in \mathcal{P}$ and for each $0 \neq u \in \mathcal{P}$, there exist positive numbers $k_1(u)$ and $k_2(u)$ such that $k_1 u_0 \preceq Mu \preceq k_2 u_0$ with respect to $\mathcal{P}$.

Definition 6.4. Let X and Y be Banach spaces. A bounded linear operator $L : X \to Y$ is said to be *compact*, (or *completely continuous*), if for every bounded sequence $\{x_n\} \subset X$, $\{Lx_n\}$ has a convergent subsequence in Y.

Fundamental to our arguments of the section is the following theorem due to Krasnosel'skii.

Theorem 6.1. *Let $\mathcal{B}$ be a real Banach space and let $\mathcal{P} \subseteq \mathcal{B}$ be a reproducing cone. Let $L : \mathcal{B} \to \mathcal{B}$ be a compact u_0-positive linear operator. Then L has an essentially unique, (i.e., unique up to constant multiples), eigenvector* *in $\mathcal{P}$, and the corresponding eigenvalue is simple, positive and larger than the absolute value of any other eigenvalue.*

For our eventual comparison of eigenvalues λ and σ of (6.1), (6.3) and (6.2), (6.3), we will apply the next theorem due to Keener and Travis.

Theorem 6.2. *Let $\mathcal{B}$ be a real Banach space and let $\mathcal{P} \subseteq \mathcal{B}$ be a cone. Let both $M, N : \mathcal{B} \to \mathcal{B}$ be bonded linear operators and assume that at least one of the operators is u_0-positive. If $M \preceq N$,*

$$\lambda_1 u_1 \preceq M u_1, \quad \text{for some } u_1 \in \mathcal{P} \text{ and some } \lambda_1 > 0,$$

and

$$N u_2 \preceq \lambda_2 u_2, \quad \text{for some } u_2 \in \mathcal{P} \text{ and some } \lambda_2 > 0,$$

then $\lambda_1 \leq \lambda_2$. Further $\lambda_1 = \lambda_2$ implies u_1 is a scalar multiple of u_2.

In applying the two previous theorems, we will make use of the Green's function, $G(x, s)$, for

$$-y'' = 0, \tag{6.4}$$

satisfying (6.3).

Modifying the construction in Section 2.3, due to the minus sign,

$$G(x,s) = \begin{cases} \frac{(b-s)(x-a)}{b-a}, & a \leq x \leq s \leq b, \\ \frac{(b-x)(s-a)}{b-a}, & a \leq s \leq x \leq b, \end{cases}$$

and we know $G(x, s) > 0$ on $(a, b) \times (a, b)$ and satisfies properties such as, $G(x, s)$ satisfies the conditions (6.3) at $x = a$ and $x = b$, for each fixed s, etc. Moreover, from inspection,

$$\frac{\partial}{\partial x} G(a, s) > 0, \ a < s < b,$$

and

$$\frac{\partial}{\partial x} G(b, s) < 0, \ a < s < b.$$

Now, define the Banach space

$$\mathcal{B} = \{u \in C^{(1)}[a, b] \mid u \text{ satisfies } (6.3)\},$$

with norm

$$\|u\| = \max\left\{ \sup_{a \leq x \leq b} |u(x)|, \sup_{a \leq x \leq b} |u'(x)| \right\}.$$

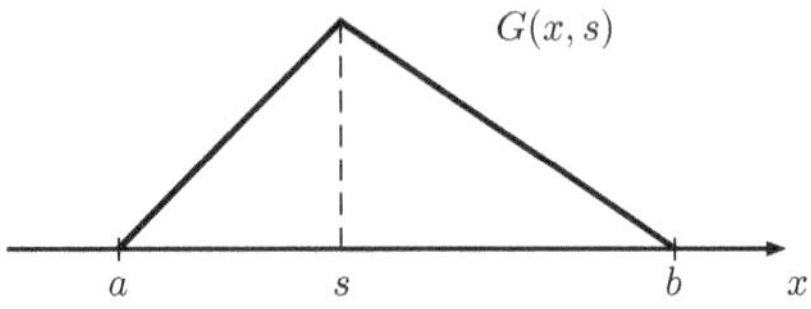

Fig. 6.1 The graph of $G(x, s)$.

Rather than study the differential equations (6.1) and (6.2) subject to (6.3), we seek eigenvalues of the linear operators $M, N : \mathcal{B} \to \mathcal{B}$ defined by

$$(Mu)(x) = \int_a^b G(x,s)p(s)u(s)\,ds, \ a \le x \le b, \tag{6.5}$$

$$(Nu)(x) = \int_a^b G(x,s)q(s)u(s)\,ds, \ a \le x \le b. \tag{6.6}$$

By mimicking part of the proof of Theorem 3.5, a routine application of the Arzelà-Ascoli Theorem shows that both M and N are compact operators.

Remark 6.3. Note, for example, if $0 \neq \Lambda$ is an eigenvalue of the above operator M with eigenvector u, (i.e., $Mu = \Lambda u$), then u is an eigenvector of (6.1), (6.3) with corresponding eigenvalue $\lambda = \frac{1}{\Lambda}$; in particular,

$$\begin{aligned} Mu = \Lambda u \ \text{ iff } u(x) &= \frac{1}{\Lambda}\int_a^b G(x,s)p(s)u(s)\,ds, \ a \le x \le b, \\ \text{iff } &\begin{cases} -u''(x) = \frac{1}{\Lambda}p(x)u(x), \ a \le x \le b, \\ u(a) = u(b) = 0. \end{cases} \end{aligned}$$

We now proceed to define a cone $\mathcal{P} \subseteq \mathcal{B}$ which is reproducing and then show M and N are both u_0-positive with respect to $\mathcal{P}$. Define

$$\mathcal{P} = \{u \in \mathcal{B} |\, u(x) \ge 0, \ a \le x \le b\}.$$

Exercise 36. Show $\mathcal{P}$ is a cone in $\mathcal{B}$.

In order to show that $\mathcal{P}$ is reproducing, it suffices to show $\mathcal{P}^\circ \neq \emptyset$. For that purpose, define the set,

$$\Omega = \{u \in \mathcal{B} |\, u(x) > 0, \ a < x < b, \ u'(a) > 0, \ \text{and } u'(b) < 0\}.$$

Theorem 6.3. *The cone $\mathcal{P} \subseteq \mathcal{B}$ is reproducing.*

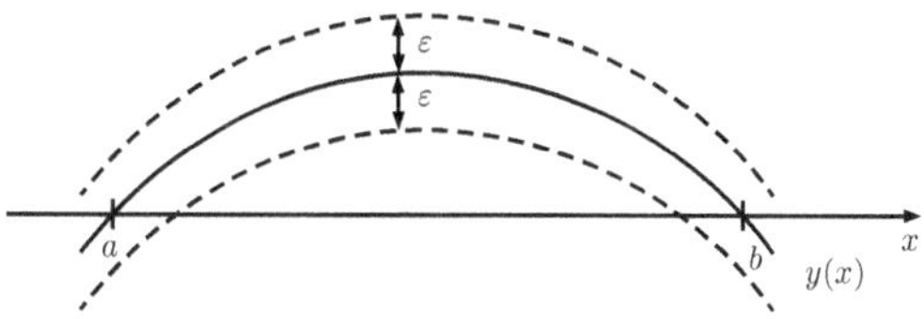

Fig. 6.2 The graph of $y(x)$.

Proof. In arguing $\mathcal{P}^\circ \neq \emptyset$, we show $\Omega \subseteq \mathcal{P}^0$. First, it is clear that $\Omega \subseteq \mathcal{P}$.

Now, let $y \in \Omega$. For $\varepsilon > 0$ sufficiently small, if $v \in S_\varepsilon(y) = \{u \in \mathcal{B} |\, \|u - y\| = \max\{\sup_{[a,b]} |u(x) - y(x)|, \sup_{[a,b]} |u'(x) - y'(x)|\} < \varepsilon\}$, it follows that $v(x) > 0$ on (a, b) and $v(a) = v(b) = 0$. Hence, $v \in \mathcal{P}$. Therefore, $S_\varepsilon(y) \subseteq \mathcal{P}$ so that $y \in \mathcal{P}^\circ$. But, $y \in \Omega$ was arbitrary and so $\Omega \subseteq \mathcal{P}^\circ$. As a consequence $\mathcal{P}$ is solid and, in turn by Remark 6.2, reproducing. □

Recall at this point that $p(x)$ and $q(x)$ are continuous nonnegative functions on $[a, b]$, each of which does not vanish identically on any subinterval $[\alpha, \beta] \subseteq [a, b]$.

Theorem 6.4. *The operators M and N are u_0-positive with respect to $\mathcal{P}$.*

Proof. We will make the argument for the operator M. For this, we will show $M : \mathcal{P}\backslash\{0\} \to \Omega$.

First, we exhibit that $M : \mathcal{P} \to \mathcal{P}$. Thus, let $u \in \mathcal{P}$. Since $G(x, s) > 0$ on $(a, b) \times (a, b)$ and $p(x) \geq 0$ on $[a, b]$, it follows that

$$(Mu)(x) = \int_a^b G(x, s)p(s)u(s)\, ds \geq 0, \;\; a \leq x \leq b.$$

Therefore, $Mu \in \mathcal{P}$ and so $M(\mathcal{P}) \subseteq \mathcal{P}$.

Next, let us choose $u \in \mathcal{P}\backslash\{0\}$. Then, there exists $[\alpha, \beta] \subseteq [a, b]$ such that $u(x) > 0$ and $p(x) > 0$, for all $\alpha \leq x \leq \beta$.

Then, for $a < x < b$,

$$\begin{aligned}(Mu)(x) &= \int_a^b G(x, s)p(s)u(s)\, ds \\ &\geq \int_\alpha^\beta G(x, s)p(s)u(s)\, ds \\ &> 0.\end{aligned}$$

Also, since $\frac{\partial}{\partial x}G(a,s) > 0$, $a < s < b$, we have

$$\begin{aligned}(Mu)'(a) &= \int_a^b \frac{\partial}{\partial x}G(a,s)p(s)u(s)\,ds \\ &\geq \int_\alpha^\beta \frac{\partial}{\partial x}G(a,s)p(s)u(s)\,ds \\ &> 0.\end{aligned}$$

Similarly, $(Mu)'(b) < 0$. As a consequence, $Mu \in \Omega$ and so, $M : \mathcal{P}\backslash\{0\} \to \Omega$.

To finish the proof, recall $\Omega \subseteq \mathcal{P}^\circ$ from Theorem 6.3. Now, choose any $u_0 \in \mathcal{P}^\circ$, (note that $u_0 \neq 0$), and let $u \in \mathcal{P}\backslash\{0\}$. Then $Mu \in \mathcal{P}^\circ$, from above. Choosing $k_1 > 0$ sufficiently small, $Mu - k_1u_0 \in \mathcal{P}^\circ$. Similarly, for $k_2 > 0$ sufficiently large, $u_0 - \frac{1}{k_2}Mu \in \mathcal{P}^\circ$. Thus $k_1u_0 \preceq Mu \preceq k_2u_0$ with respect to $\mathcal{P}$.

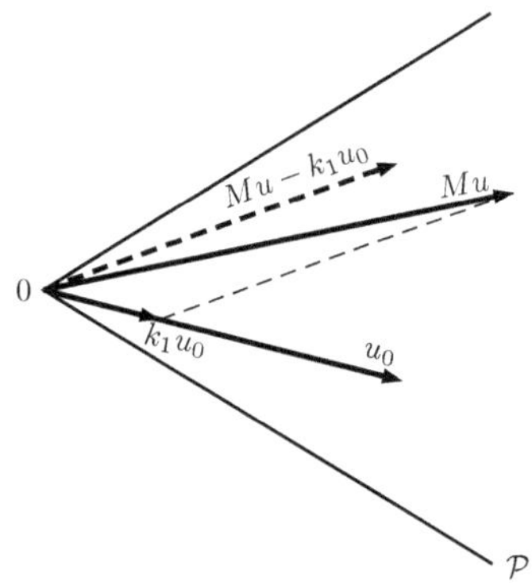

Fig. 6.3 $k_1u_0 \preceq Mu \preceq k_2u_0$ with respect to $\mathcal{P}$.

Therefore, M is u_0-positive with respect to $\mathcal{P}$. □

Corollary 6.1. *Let $\mathcal{B}$, $\mathcal{P}$, M (and N) be as above. Then M (and N) has an essentially unique eigenvector u in $\mathcal{P}^\circ$, and the corresponding eigenvalue is positive, simple and larger than the absolute value of any other eigenvalue.*

Proof. Theorems 6.3 and 6.4 satisfy the hypotheses of Theorem 6.1 which gives such an eigenvector u in $\mathcal{P}$ whose eigenvalue possesses the stated properties. The proof of Theorem 6.4 showed $Mu \in \Omega \subseteq \mathcal{P}^\circ$. But $Mu = \Lambda u$, where Λ is the appropriate positive eigenvalue. Hence $\Lambda u \in \mathcal{P}^\circ$. As a consequence, $u \in \mathcal{P}^\circ$, (in particular, $M\left(\frac{1}{\Lambda}u\right) = u \in \mathcal{P}^\circ$). □

Theorem 6.5. *Let $\mathcal{B}$, $\mathcal{P}$, M and N be as in Corollary* 6.1. *Assume moreover that $p(x) \leq q(x)$ on $[a, b]$. Let Λ and Σ be the eigenvalues given by Corollary* 6.1 *with respective essentially unique eigenvectors u_1 and u_2 corresponding to M and N, respectively. Then $\Lambda \leq \Sigma$. Moreover, $\Lambda = \Sigma$ if and only if $p(x) = q(x)$, for all $a \leq x \leq b$.*

Proof. $p(x) \leq q(x)$ on $[a, b]$ implies $M \preceq N$ with respect to $\mathcal{P}$. That $\Lambda \leq \Sigma$ then follows from Theorem 6.2 by taking $\lambda_1 = \Lambda$ and $\lambda_2 = \Sigma$.

For the last statement of the theorem, of course, if $p(x) \equiv q(x)$ on $[a, b]$, then $\Lambda = \Sigma$. On the other hand if $p(x) < q(x)$ on some subinterval $[\alpha, \beta] \subseteq [a, b]$, then one can argue as in Theorem 6.4 that $(N - M)u_1 \in \mathcal{P}^\circ$. Since $u_1 \in \mathcal{P}^\circ$ by Corollary 6.1, it can also be argued as in Theorem 6.4 that there exists $\varepsilon > 0$ such that

$$\varepsilon u_1 \preceq (N - M)u_1.$$

Then,

$$(\Lambda + \varepsilon)u_1 = Mu_1 + \varepsilon u_1 \preceq Nu_1.$$

But, we also have $Nu_2 = \Sigma u_2$, and so applying Theorem 6.2, (with N here playing the role of both M and N of Theorem 6.2), it follows that

$$\Lambda + \varepsilon \leq \Sigma.$$

In particular, $\Lambda < \Sigma$. As a consequence, if $\Lambda = \Sigma$, we have $p(x) = q(x)$, $a \leq x \leq b$. □

From Remark 6.3, we now state our comparison theorem for λ and σ as related to (6.1), (6.3) and (6.2), (6.3) .

Theorem 6.6. *Assume $p(x) \leq q(x)$ on $[a, b]$. Then, there exist smallest positive eigenvalues λ_0 and σ_0 of* (6.1), (6.3) *and* (6.2), (6.3), *respectively, each of which is simple and less than the absolute value of any other eigenvalue of the corresponding problem. Also, the eigenvectors corresponding to λ_0 and σ_0 may be chosen to belong to $\mathcal{P}^\circ$. Finally,*

$$\sigma_0 \leq \lambda_0,$$

and

$$\sigma_0 = \lambda_0, \text{ if and only if } p(x) = q(x), \ a \leq x \leq b.$$

6.2 Eigenvalue Comparisons for Nonlocal BVP's

We now consider the comparison of eigenvalues for the eigenvalue problems,

$$u'' + \lambda_1 p(x)u = 0, \tag{6.7}$$

$$u'' + \lambda_2 q(x)u = 0, \tag{6.8}$$

satisfying the nonlocal boundary conditions, for fixed $0 < \eta < 1$,

$$u(0) = 0, \quad u(\eta) - u(1) = 0, \tag{6.9}$$

where $p(x)$ and $q(x)$ are continuous, nonnegative functions on $[0,1]$, and both $p(x)$ and $q(x)$ do not vanish identically on any compact subinterval $[\alpha, \beta]$ of $[0,1]$. This comparison will be done when certain inequality comparisons are satisfied by $p(x)$ and $q(x)$.

Our techniques involve sign properties of a Green's function, followed by applications of Theorems 6.1 and 6.2 from the theory of u_0-positive operators with respect to a cone in a Banach space.

In applying these theorems, we will deal with integral operators whose kernel is the Green's function of $-u'' = 0$ satisfying (6.9). The Green's function is given by

$$G(x,s) = \begin{cases} x, & x \le s, s \le \eta, \\ s, & s \le x, s \le \eta, \\ \frac{1-s}{1-\eta}x, & x \le s, s \ge \eta, \\ s + \frac{\eta - s}{1-\eta}, & s \le x, s \ge \eta. \end{cases}$$

We note that $G(x,s) > 0$ on $(0,1)\times(0,1)$, and $\frac{\partial}{\partial x}G(x,s)|_{x=0} > 0, 0 < s < 1$.

To apply Theorems 6.1 and 6.2, we need to define a suitable Banach space $\mathcal{B}$ and a suitable cone $\mathcal{P}$ in $\mathcal{B}$. We define the Banach space $\mathcal{B}$ by

$$\mathcal{B} = \{u \in C^{(1)}[0,1] \mid u \text{ satisfies } (6.9)\}.$$

with the norm, $\|u\| = \sup_{0 \le x \le 1} |u'(x)|$, and the cone $\mathcal{P}$ is defined as

$$\mathcal{P} = \{u \in \mathcal{B} \mid u(x) \ge 0 \text{ on } [0,1]\}.$$

Note: For $u \in \mathcal{P}$ and for $0 \le x \le 1$,

$$|u(x)| = |u(x) - u(0)| = \left| \int_0^x u'(s)ds \right| \le \|u\|x \le \|u\|.$$

In particular, $\sup_{0 \le x \le 1} |u(x)| \le \|u\|$.

We now establish some preliminary lemmas which will be used to prove the main comparison theorems.

Lemma 6.1. *The cone $\mathcal{P}$ is a solid cone in $\mathcal{B}$, and hence, $\mathcal{P}$ is reproducing.*

Proof. It suffices to show that $\mathcal{P}^\circ \neq \emptyset$. Define $\Omega \subseteq \mathcal{P}^\circ$ by

$$\Omega = \{u \in \mathcal{B} \,|\, u(x) > 0 \text{ on } (0,1], u'(0) > 0\}.$$

We claim that $\Omega \subseteq \mathcal{P}^\circ$. To prove the claim, it is clear that $\Omega \subseteq \mathcal{P}$. Next, choose $u \in \Omega$. Then $u(x) > 0$ on $(0,1]$, and $u'(0) > 0$. Let

$$B_\varepsilon(u) = \{v \in \mathcal{B} \,|\, \|u - v\| < \varepsilon\}, \text{ where } \varepsilon > 0.$$

We wish to show that, for $\varepsilon > 0$ sufficiently small, $B_\varepsilon(u) \subseteq \mathcal{P}$. So, let $\varepsilon_0 > 0$ be such that $u'(0) - \varepsilon > 0$. Choose $v \in B_{\varepsilon_0}(u)$. Then $\sup_{0 \le x \le 1} |v'(x) - u'(x)| < \varepsilon_0$. So, $v'(0) > u'(0) - \varepsilon_0 > 0$. In addition, recalling $|v(x) - u(x)| \le \|v - u\| < \varepsilon_0$, we have $v(x) > 0$ on $(0,1]$. Thus, $v \in \Omega$ and hence $B_{\varepsilon_0}(u) \subseteq \Omega \subset \mathcal{P}$.

Therefore, $\Omega \subseteq \mathcal{P}^\circ$ and $\mathcal{P}$ is solid in $\mathcal{B}$. □

Next, we define two linear integral operators $M, N : \mathcal{B} \to \mathcal{B}$ by,

$$(Mu)(x) = \int_0^1 G(x,s)p(s)u(s)\,ds, \quad 0 \le x \le 1,$$

$$(Nu)(x) = \int_0^1 G(x,s)q(s)u(s)\,ds, \quad 0 \le x \le 1.$$

Lemma 6.2. *The bounded linear operators M and N are u_0-positive with respect to $\mathcal{P}$.*

Proof. We verify the statement for M only.

First, we show that $M : \mathcal{P} \setminus \{0\} \to \Omega \subseteq \mathcal{P}^\circ$. Let $u \in \mathcal{P}$. Then, $u(x) \ge 0$ on $[0,1]$. Since $G(x,s) > 0$ on $(0,1) \times (0,1)$ and $p(x) \ge 0$ on $[0,1]$, we have

$$(Mu)(x) = \int_0^1 G(x,s)p(s)u(s)\,ds \ge 0, \quad 0 \le x \le 1,$$

So, $Mu \in \mathcal{P}$, and $M : \mathcal{P} \to \mathcal{P}$.

Next, let $u \in \mathcal{P}\setminus\{0\}$. Then there exists $[\alpha, \beta] \subseteq [0,1]$ such that $u(x) > 0$ and $p(x) > 0$, for all $x \in [\alpha, \beta]$. Then, for all $0 < x \le 1$.

$$\begin{aligned}(Mu)(x) &= \int_0^1 G(x,s)p(s)u(s)\,ds \\ &\ge \int_\alpha^\beta G(x,s)p(s)u(s)\,ds \\ &> 0.\end{aligned}$$

Recall, $\frac{\partial}{\partial x}G(x,s)|_{x=0} > 0$, $0 < s < 1$, and so

$$\begin{aligned}(Mu)'(0) &= \int_0^1 \frac{\partial}{\partial x}G(0,s)p(s)u(s)\,ds \\ &\geq \int_\alpha^\beta \frac{\partial}{\partial x}G(0,s)p(s)u(s)\,ds \\ &> 0.\end{aligned}$$

So, $Mu \in \Omega$; i.e., $Mu \in \mathcal{P}^\circ$, and in particular $M : \mathcal{P} \setminus \{0\} \to \mathcal{P}^\circ$.

To complete the proof, choose any $u_0 \in \mathcal{P}^\circ$ (so $u_0 \neq 0$), and let $u \in \mathcal{P} \setminus \{0\}$. Then, from above, we have $Mu \in \Omega \subset \mathcal{P}^\circ$. Now choose $k_1 > 0$ sufficiently small that $Mu - k_1u_0 \in \mathcal{P}^\circ$. Similarly, choose $k_2 > 0$ sufficiently large that $u_0 - \frac{1}{k_2}Mu \in \mathcal{P}^\circ$. Thus,

$$k_1 u \preceq Mu \text{ wrt } \mathcal{P}, \text{ and } Mu \preceq k_2 u_0 \text{ wrt } \mathcal{P}.$$

Therefore, M is u_0-positive with respect to the cone $\mathcal{P}$. □

Lemma 6.3. *The operators M and N are compact.*

Proof. Again, we establish the statement for M only.

Let $L = \max_{0 \leq x \leq 1} p(x)$ and $k = \max_{(x,s)\in(0,1)^2} \left|\frac{\partial}{\partial x}G(x,s)\right|$. Let $\varepsilon > 0$ be given and let $\delta = \frac{\varepsilon}{kL}$. Then for $u, v \in \mathcal{B}$ with $\|u - v\| < \delta$, we have for each $0 \leq x \leq 1$,

$$\begin{aligned}|(Mu)'(x) - (Mv)'(x)| &= \left|\int_0^1 \frac{\partial}{\partial x}G(x,s)p(s)[u(s) - v(s)]\,ds\right| \\ &\leq \int_0^1 \left|\frac{\partial}{\partial x}G(x,s)\right| p(s)|u(s) - v(s)|\,ds \\ &\leq Lk\delta \\ &= \varepsilon.\end{aligned}$$

So, $\|Mu - Mv\| \leq \varepsilon$, and M is continuous.

Next, we recall that $M : \mathcal{P} \to \mathcal{P}$. Let $\{u_n\}$ be a bounded sequence in $\mathcal{P}$; say, $\|u_n\| \leq k_0$, for all n. Then, for each n, and $0 \leq x \leq 1$

$$|(Mu_n)'(x)| \leq \int_0^1 \left|\frac{\partial}{\partial x}G(x,s)\right| p(s)|u_n(s)|\,ds \leq kk_0L,$$

or $\|(Mu_n)'\| \leq kk_0L$ for all n.

From the piecewise continuity of $\frac{\partial}{\partial x}G(x,s)$ and the absolute continuity of the integral, $\{Mu_n\}$ is equicontinuous.

Hence, an application of the Arzelà-Ascoli Theorem yields that M is compact. □

Remark 6.4. If $\Lambda \neq 0$ is an eigenvalue of the operator M with eigenvector u; i.e., $Mu = \Lambda u$, then

$$\Lambda u = Mu = \int_0^1 G(x,s)p(s)u(s)\,ds,$$

iff

$$u(x) = \frac{1}{\Lambda} = \int_0^1 G(x,s)p(s)u(s)\,ds,$$

iff

$$\begin{cases} -u''(x) = \frac{1}{\Lambda}p(x)u(x), & 0 \le x \le 1, \\ u(0) = 0 = u(\eta) - u(1). \end{cases}$$

So, eigenvalues of (6.7), (6.9) are reciprocals of eigenvalues of M, and conversely. Similarly, eigenvalues of (6.8), (6.9) are reciprocals of eigenvalues of N, and conversely.

We now derive our comparison results.

Theorem 6.7. *Let $\mathcal{B}$, $\mathcal{P}$, M, N be as defined above. Then M (and N) has an essentially unique eigenvector $u \in \mathcal{P}^\circ$, and the corresponding eigenvalue is simple, positive and larger than the absolute value of any other eigenvalue.*

Proof. Since M is a compact linear operator which is u_0-positive wrt $\mathcal{P}$, M has an essentially unique eigenvector, say $u \in \mathcal{P}$, and eigenvalue Λ_1 with the above properties by Theorem 6.1. Since $u \neq 0$, $Mu \in \Omega \subseteq \mathcal{P}^\circ$. But

$$Mu = \Lambda_1 u, \text{ or } M\left(\frac{1}{\Lambda_1}u\right) = u.$$

Thus, $u = M\left(\frac{1}{\Lambda_1}u\right) \in \mathcal{P}^\circ$. □

Theorem 6.8. *Let $\mathcal{B}, \mathcal{P}, M, N$ be as defined above. Let $p(x) \le q(x)$ on $[0,1]$. Let Λ_1 and Λ_2 be the eigenvalues in Theorem 6.7 associated with M and N, respectively, and with respective essentially unique eigenvectors u_1 and u_2 in $\mathcal{P}^\circ$. Then $\Lambda_1 \le \Lambda_2$. Moreover, $\Lambda_1 = \Lambda_2$ iff $p(x) = q(x)$ on $[0,1]$.*

Proof. Let $p(x) \le q(x)$ on $[0,1]$ and recall that neither vanishes identically on any compact subinterval $[\alpha,\beta]$ of $[0,1]$. Then, for any $u \in \mathcal{P}$ and $x \in [0,1]$,

$$(Nu - Mu)(x) = \int_0^1 G(x,s)(q(s) - p(s))u(s)\,ds \ge 0.$$

In particular, $Nu - Mu \in \mathcal{P}$, for all $u \in \mathcal{P}$, or $Mu \preceq Nu$ wrt $\mathcal{P}$, for all $u \in \mathcal{P}$, or $M \preceq N$ wrt $\mathcal{P}$. By Theorem 6.2, we get $\Lambda_1 \leq \Lambda_2$.

Now, if $p(x) = q(x)$ on $[0,1]$, then naturally $\Lambda_1 = \Lambda_2$.

On the other hand, we want to show that $\Lambda_1 = \Lambda_2$ implies $p(x) = q(x)$ on $[0,1]$. We show here that, if $p(x) \neq q(x)$, then $\Lambda_1 \neq \Lambda_2$. Let $p(x) < q(x)$ on some subinterval $[\alpha, \beta] \subset [0,1]$.

Then, along the lines of previous arguments, it follows that $(N-M)u_1 \in \Omega \subseteq \mathcal{P}^\circ$. Since $u_1 \in \mathcal{P}^\circ$, we can find $\varepsilon > 0$ small so that, $(N-M)u_1 - \varepsilon u_1 \in \mathcal{P}$, i.e., $\Lambda_1 u_1 + \varepsilon u_1 = Mu_1 + \varepsilon u_1 \preceq Nu_1$, which implies $Nu_1 \succeq (\Omega_1 + \varepsilon)u_1$. Since $N \leq N$ and $Nu_2 = \Lambda_2 u_2$, we have by Theorem 6.2 that $\Lambda_1 + \varepsilon \leq \Lambda_2$, or $\Lambda_1 < \Lambda_2$. By contrapositive, if $\Lambda_1 = \Lambda_2$, then $p(x) = q(x)$ on $[0,1]$. □

Since the eigenvalues of M (and N) are reciprocals of eigenvalues of (6.7), (6.9) (and (6.8), (6.9), respectively), the following is immediate from Theorems 6.7 and 6.8.

Theorem 6.9. *Assume the hypotheses of Theorem* 6.8 *hold. Then there exist smallest positive eigenvalues* λ_1 *and* λ_2 *of* (6.7), (6.9) *and* (6.8), (6.9), *respectively, each of which is simple, positive and less than the absolute value of any other eigenvalue of the corresponding problems. Also, eigenfunctions corresponding to* λ_1 *and* λ_2 *may be chosen to belong to* $\mathcal{P}^\circ$. *Finally,* $\lambda_2 \leq \lambda_1$, *and* $\lambda_1 = \lambda_2$ *iff* $p(x) = q(x)$, $0 \leq x \leq 1$.

6.3 Solutions of Second Order Multi-Point BVP's

In this section, we are concerned with the existence of solutions of the BVP's consisting of one of the equations

$$u'' + f(t,u) = 0, \quad t \in (0,1), \tag{6.10}$$

and

$$u'' + g(t)f(t,u) = 0, \quad t \in (0,1), \tag{6.11}$$

and the multi-point BC

$$u(0) = u(1) = \sum_{i=1}^{m} a_i u(t_i), \tag{6.12}$$

where $f \in C([0,1] \times \mathbb{R}, \mathbb{R})$, $g \in C((0,1), \mathbb{R}^+)$ with $\mathbb{R}^+ = [0,\infty)$, $m \geq 1$ is an integer, $t_i \in (0,1)$, and $a_i \in \mathbb{R}^+$ for $i = 1,\ldots,m$. If a function $u \in C([0,1],\mathbb{R}) \cap C^{(2)}((0,1),\mathbb{R})$ satisfies Eq. (6.10) and BC (6.12), then $u(t)$ is called a *solution* of BVP (6.10), (6.12). Moreover, if $u(t)$ is *symmetric* on $[0,1]$, i.e., $u(t) = u(1-t)$ for $t \in [0,1]$, then it is said to be a *symmetric solution*. Finally, if a (symmetric) solution $u(t)$ of BVP (6.10), (6.12) satisfies that $u(t) > 0$ on $(0,1)$, then $u(t)$ is called a (symmetric) *positive solution*. Similar definitions also apply for BVP (6.11), (6.12) as well as for negative solutions of these problems. We observe that BC (6.12) includes the Dirichlet BC, i.e,

$$u(0) = u(1) = 0.$$

In this section, we need the following assumptions:

(A1) $\sum_{i=1}^{m} a_i < 1$.

(A2) There exist $\Lambda \geq 0$ and $\rho \geq 0$ such that

$$f(t,x) + \rho^2 x + \Lambda \geq 0 \quad \text{for } (t,x) \in [0,1] \times \mathbb{R}. \tag{6.13}$$

(A3) There exists $\rho \geq 0$ such that

$$x\big(f(t,x) + \rho^2 x\big) \geq 0 \quad \text{for } (t,x) \in [0,1] \times \mathbb{R}. \tag{6.14}$$

(A4) g satisfies

$$\int_0^1 K(s,s)g(s)ds < \infty, \tag{6.15}$$

and there exist c and d with $0 < c < d < 1$ such that

$$\int_c^d K(s,s)g(s)ds > 0, \tag{6.16}$$

where

$$K(t,s) = G(t,s) + \frac{1}{1 - \sum_{i=1}^{m} a_i} \sum_{i=1}^{m} a_i G(t_i, s) \tag{6.17}$$

with

$$G(t,s) = \begin{cases} t(1-s), & 0 \leq t \leq s \leq 1, \\ s(1-t), & 0 \leq s \leq t \leq 1; \end{cases} \tag{6.18}$$

6.3.1 *Preliminary lemmas*

For ρ given in (A2) or (A3), let ϕ_ρ and ψ_ρ be the unique solutions of the initial value problems

$$-u'' + \rho^2 u = 0, \quad u(0) = 0,\ u'(0) = 1,$$

and

$$-u'' + \rho^2 u = 0, \quad u(1) = 0,\ u'(1) = -1,$$

respectively. Then, for $\rho > 0$

$$\phi_\rho(t) = \frac{1}{\rho}\sinh(\rho t) \tag{6.19}$$

and

$$\psi_\rho(t) = \frac{1}{\rho}\sinh(\rho(1-t)), \tag{6.20}$$

and if $\rho = 0$, then

$$\phi_0(t) = t \quad \text{and} \quad \psi_0(t) = 1 - t.$$

Clearly,

$\phi_\rho(t)$ is increasing on $[0,1]$ and $\phi_\rho(t) > 0$ on $(0,1]$, and

$\psi_\rho(t)$ is decreasing on $[0,1]$ and $\psi_\rho(t) > 0$ on $[0,1)$.

Define

$$c_\rho = \phi_\rho'(0)\psi_\rho(0) - \phi_\rho(0)\psi_\rho'(0) \tag{6.21}$$

and

$$\Delta_\rho = c_\rho\left(c_\rho - \sum_{i=1}^m a_i(\phi_\rho(t_i) + \psi_\rho(t_i))\right). \tag{6.22}$$

When $\rho = 0$, it is easy to see that

$$\Delta_0 = 1 - \sum_{i=1}^m a_i > 0 \quad \text{and} \quad c_0 = 1.$$

Lemma 6.4. *Assume* (A1) *holds. Then* $\Delta_\rho > 0$.

Proof. In the following, we assume that $\rho > 0$. From (6.19), (6.20), and (6.21), we have $c_\rho = (\sinh\rho)/\rho$. Then, in view of (6.22),

$$\Delta_\rho = \frac{1}{\rho}c_\rho\left(\sinh\rho - \sum_{i=1}^m a_i\big(\sinh(\rho t_i) + \sinh(\rho(1-t_i))\big)\right).$$

Let

$$\kappa(x) = \sinh x - \sum_{i=1}^m a_i\big(\sinh(xt_i) + \sinh(x(1-t_i))\big) \quad \text{for } x \in \mathbb{R}.$$

Then

$$\kappa(0) = 0 \quad \text{and} \quad \Delta_\rho = \frac{1}{\rho}c_\rho\kappa(\rho).$$

Note that

$$\begin{aligned}\kappa'(x) &= \cosh x - \sum_{i=1}^{m} a_i\big(t_i \cosh(xt_i) + (1-t_i)\cosh(x(1-t_i))\big)\\ &\geq \cosh x - \sum_{i=1}^{m} a_i\big(t_i \cosh x + (1-t_i)\cosh x\big)\\ &= \left(1 - \sum_{i=1}^{m} a_i\right)\cosh x > 0.\end{aligned}$$

Then $\kappa(x)$ is increasing, and so $\kappa(\rho) > 0$. Consequently, $\Delta_\rho > 0$. This completes the proof of the lemma. □

By Lemma 6.19, (A1) implies that $\Delta_\rho > 0$. The following lemma is a special case of [Ma (2003), Lemma 2.2] and it can also be easily verified.

Lemma 6.5. *Assume* (A1) *holds. Then for any* $h \in L(0,1)$*, the BVP consisting of the equation*

$$u'' - \rho^2 u + h(t) = 0, \quad t \in (0,1),$$

and BC (6.12) *has a solution* $u(t)$ *if and only if*

$$u(t) = \int_0^1 J(t,s;\rho)h(s)ds,$$

where

$$J(t,s;\rho) = H(t,s;\rho) + \frac{c_\rho}{\Delta_\rho}\left(\phi_\rho(t) + \psi_\rho(t)\right)\sum_{i=0}^{m} a_i H(t_i,s;\rho) \tag{6.23}$$

with

$$H(t,s;\rho) = \frac{1}{c_\rho}\begin{cases}\phi_\rho(t)\psi_\rho(s), & 0 \leq t \leq s \leq 1,\\ \phi_\rho(s)\psi_\rho(t), & 0 \leq s \leq t \leq 1.\end{cases} \tag{6.24}$$

Remark 6.5. It is easy to see that if $\rho = 0$, then $J(t,s;\rho)$ reduces to $K(t,s)$ defined by (6.17) and $H(t,s;\rho)$ reduces to $G(t,s)$ defined by (6.18).

From here on, let

$$w_\rho(t) = \min\left\{\frac{\phi_\rho(t)}{\phi_\rho(1)}, \frac{\psi_\rho(t)}{\psi_\rho(0)}\right\} \quad \text{for } t \in [0,1], \tag{6.25}$$

and let a and b be any fixed numbers satisfying $0 < a < 1/2 < b < 1$. Define

$$d_\rho = e_\rho^{-1} \min_{t\in[a,b]} w_\rho(t), \tag{6.26}$$

where

$$e_\rho = 1 + \frac{c_\rho \sinh \rho}{\rho \Delta_\rho} \sum_{i=1}^{m} a_i. \tag{6.27}$$

Clearly, $0 < d_\rho < 1$ and

$$e_0 = \frac{1}{1 - \sum_{i=1}^{m} a_i}. \tag{6.28}$$

In the sequel, let X be the Banach space $C[0,1]$ equipped with the norm

$$||u|| = \max_{t \in [0,1]} |u(t)|.$$

Define a cone in X by

$$P = \{u \in X \ : \ u(t) \geq 0 \quad \text{for } t \in [0,1]\}. \tag{6.29}$$

We also define a smaller cone P_ρ by

$$P_\rho = \left\{ u \in P \ : \ \min_{t \in [a,b]} u(t) \geq d_\rho ||u|| \right\}. \tag{6.30}$$

Lemma 6.6. *We have the following:*

(i) $0 \leq J(t,s;\rho) \leq e_\rho H(s,s;\rho)$ *for* $(t,s) \in [0,1] \times [0,1]$,
(ii) $J(t,s;\rho) \geq w_\rho(t) H(s,s;\rho)$ *for* $(t,s) \in [0,1] \times [0,1]$, *and*
(iii) $J(t,s;\rho) \geq \min_{t \in [a,b]} w_\rho(t) H(s,s;\rho)$ *for* $(t,s) \in [a,b] \times [0,1]$,

where e_ρ *and* $w_\rho(t)$ *are defined by* (6.27) *and* (6.25), *respectively.*

Proof. We first prove (i). In view of the monotonicity of ϕ_ρ and ψ_ρ and from (6.24), $H(t,s;\rho) \leq H(s,s;\rho)$ for $(t,s) \in [0,1] \times [0,1]$. Note that

$$\max_{t \in [0,1]} (\phi_\rho(t) + \psi_\rho(t)) = \phi_\rho(0) + \psi_\rho(0) = \phi_\rho(1) + \psi_\rho(1) = \frac{\sinh \rho}{\rho}.$$

Then, from (6.23), it is easy to see that (i) holds.

We now prove (ii). From (6.24), it follows that

$$\frac{H(t,s;\rho)}{H(s,s;\rho)} = \begin{cases} \dfrac{\phi_\rho(t)}{\phi_\rho(s)}, & 0 \leq t \leq s \leq 1, \\ \dfrac{\psi_\rho(t)}{\psi_\rho(s)}, & 0 \leq s \leq t \leq 1, \end{cases}$$

$$\geq \begin{cases} \dfrac{\phi_\rho(t)}{\phi_\rho(1)}, & 0 \leq t \leq s \leq 1, \\ \dfrac{\psi_\rho(t)}{\psi_\rho(0)}, & 0 \leq s \leq t \leq 1. \end{cases}$$

Thus, $H(t,s;\rho) \geq w_\rho(t) H(s,s;\rho)$ for $(t,s) \in [0,1] \times [0,1]$, where $w_\rho(t)$ is defined by (6.25). Hence, from (6.23), (ii) holds.

Finally, (iii) readily follows from (ii), and this completes the proof of the lemma. □

In view of Remark 6.5, the following corollary to Lemma 6.6 is immediate.

Corollary 6.2. *We have the following:*

(i) $0 \leq K(t,s) \leq e_0\ G(s,s)$ *for* $(t,s) \in [0,1] \times [0,1]$,
(ii) $K(t,s) \geq w_0(t)G(s,s)$ *for* $(t,s) \in [0,1] \times [0,1]$, *and*
(iii) $K(t,s) \geq \min_{t\in[a,b]} w_0(t)G(s,s)$ *for* $(t,s) \in [a,b] \times [0,1]$,

where e_0 *and* $w_0(t)$ *are defined by* (6.28) *and* (6.25) *with* $\rho = 0$, *respectively.*

Define a linear operator $L_\rho : X \to X$ by

$$(L_\rho u)(t) = \int_0^1 J(t,s;\rho)u(s)ds. \tag{6.31}$$

When $\rho = 0$, we denote L_ρ by L. Then,

$$(Lu)(t) = \int_0^1 K(t,s)u(s)ds. \tag{6.32}$$

We also define a linear operator $M : X \to X$ by

$$(Mu)(t) = \int_0^1 K(t,s)g(s)u(s)ds. \tag{6.33}$$

We recall that λ is an *eigenvalue* of L with corresponding *eigenfunction* φ if φ is nontrivial and $L\varphi = \lambda\varphi$. The reciprocals of eigenvalues are called the *characteristic values* of L. The *radius of the spectrum* of L, denoted by $r(L)$, is given by the well known spectral radius formula $r(L) = \lim_{n\to\infty} ||L^n||^{1/n}$. Similar definitions and formulas hold for other linear operators such as L_ρ and M.

Recall also that in a Banach space X, a cone P is called a *total cone* if $P = \overline{X - X}$. The following well known Krein-Rutman theorem can be found in [Deimling (1985), Theorem 19.2] and [Zeidler (1986), Proposition 7.26].

Lemma 6.7. *Assume that P is a total cone in a real Banach space X. Let $L : X \to X$ be a compact linear operator with $L(P) \subseteq P$ and $r(L) > 0$. The $r(L)$ is an eigenvalue of L with a positive eigenfunction.*

The next two lemmas provide additional information about the operators L_ρ, L, and M.

Lemma 6.8. *Assume (A1) holds and let P and P_ρ be defined by (6.29) and (6.30), respectively. Then L_ρ maps P into P_ρ and is compact. In particular, L maps P into P_0 and is compact. In addition, if (A4) holds, then M maps P into P_0 and is compact.*

Proof. We first show that $L_\rho(P) \subseteq P_\rho$. For $u \in P$ and $t \in [0,1]$, Lemma 6.6(i) implies

$$(L_\rho u)(t) \le e_\rho \int_0^1 H(s,s;\rho)u(s)ds,$$

so

$$||L_\rho u|| \le e_\rho \int_0^1 H(s,s;\rho)u(s)ds.$$

On the other hand, by Lemma 6.6(iii),

$$(L_\rho u)(t) \ge \min_{t\in[a,b]} w_\rho(t) \int_0^1 H(s,s;\rho)u(s)ds \quad \text{for } t \in [a,b].$$

Thus, from (6.26), we see that $\min_{t\in[a,b]} L_\rho(t) \ge d_\rho ||L_\rho u||$, and so $L_\rho : P \to P_\rho$. A standard argument shows that L_ρ is compact and we omit the details here. In view of (6.15) in (A4), a similar argument shows that M maps P into P_0 and is compact. This completes the proof of the lemma. □

Lemma 6.9. *Assume* (A1) *holds. The spectral radius, $r(L)$, of L satisfies $r(L) > 0$. In addition, if* (A4) *holds, then the spectral radius, $r(M)$, of M satisfies $r(M) > 0$.*

Proof. To prove that $r(L) > 0$, let $u \in P_0$ and $t \in [a,b]$. By Corollary 6.2(iii), we have

$$(Lu)(t) \ge \min_{t\in[a,b]} w_0(t) \int_a^b G(s,s)u(s)ds \ge d_0||u|| \min_{t\in[a,b]} w_0(t) \int_a^b G(s,s)ds$$

and

$$\begin{aligned}(L^2u)(t) &= (L(Lu))(t) \\ &\ge \min_{t\in[a,b]} w_0(t) \int_a^b G(s,s)\left(d_0||u|| \min_{t\in[a,b]} w_0(t) \int_a^b G(s,s)ds\right) ds \\ &= d_0||u|| \left(\min_{t\in[a,b]} w_0(t)\right)^2 \left(\int_a^b G(s,s)ds\right)^2.\end{aligned}$$

By induction, we obtain that

$$(L^n u)(t) \ge d_0||u|| \left(\min_{t\in[a,b]} w_0(t)\right)^n \left(\int_a^b G(s,s)ds\right)^n.$$

Then,

$$||L^n||\,||u|| \ge ||L^n u|| \ge (L^n u)(t) \ge d_0||u|| \left(\min_{t\in[a,b]} w_0(t)\right)^n \left(\int_a^b G(s,s)ds\right)^n,$$

and so

$$||L^n|| \geq d_0 \left(\min_{t\in[a,b]} w_0(t) \right)^n \left(\int_a^b G(s,s)ds \right)^n .$$

Hence,

$$r(L) = \lim_{n\to\infty} ||L^n||^{1/n} \geq \min_{t\in[a,b]} w_0(t) \int_a^b G(s,s) > 0.$$

In view of condition (6.16) in (A4), we can prove that $r(M) > 0$ in a similar way. This completes the proof of the lemma. □

Lemma 6.10. *Assume (A1) holds. Then $r(L)$ is an eigenvalue of L with a positive eigenfunction φ_L. In addition, if (A4) holds, then $r(M)$ is an eigenvalue of M with a positive eigenfunction φ_M.*

Proof. Noting that P defined by (6.29) is a total cone, the conclusions readily follow from Lemmas 6.7, 6.8, and 6.9. □

In what follows, we let

$$\mu_L = \frac{1}{r(L)} \tag{6.34}$$

and

$$\mu_M = \frac{1}{r(M)}. \tag{6.35}$$

Clearly, μ_L and μ_M are the smallest positive characteristic values of L and M, respectively, satisfying $\varphi_L = \mu_L L \varphi_L$ and $\varphi_M = \mu_M M \varphi_M$.

Lemma 6.11. *Assume (A1) holds. Then*

$$\nu_\rho = \mu_L + \rho^2 \tag{6.36}$$

is the smallest positive characteristic value of L_ρ with a positive eigenfunction φ_L, where φ_L is given as in Lemma 6.10, i.e, $\varphi_L = \nu_\rho L_\rho \varphi_L$.

Proof. From $\varphi_L = \mu_L L\varphi$, it follows that

$$\varphi'' + \mu_L \varphi_L = 0 \quad \text{on } (0,1).$$

Thus,

$$\varphi_L'' - \rho^2 \varphi_L + (\mu_L + \rho^2)\varphi_L = 0 \quad \text{on } (0,1).$$

As a result,

$$\varphi_L = (\mu_L + \rho^2) \int_0^1 J(t,s;\rho)\varphi_L(s)ds = \nu_\rho L_\rho \varphi_L,$$

i.e., ν_ρ is a positive characteristic value of L_ρ with eigenfunction φ_L. Using the fact that μ_L is the smallest positive characteristic values of L, we can see that ν_ρ is the smallest positive characteristic value of L_ρ. This completes the proof of the lemma. □

We also need the following well known lemmas. We refer the reader to Theorem A.3.3(ix) and Lemma 2.5.1 in [Guo and Lakshmikantham (1988)], respectively, for their proofs. In the following, the bold $\mathbf{0}$ stands for the zero element in a Banach space X.

Lemma 6.12. *Let Ω be a bounded open set in a real Banach space X and $T : \overline{\Omega} \to X$ be compact. If there exists $u_0 \in X$, $u_0 \neq \mathbf{0}$, such that*

$$u - Tu \neq \tau u_0 \quad \textit{for all } u \in \partial\Omega \textit{ and } \tau \geq 0,$$

then the Leray-Schauder degree

$$\deg(I - T, \Omega, \mathbf{0}) = 0.$$

Lemma 6.13. *Let Ω be a bounded open set in a real Banach space X with $\mathbf{0} \in \Omega$ and $T : \overline{\Omega} \to X$ be compact. If*

$$Tu \neq \tau u \quad \textit{for all } u \in \partial\Omega \textit{ and } \tau \geq 1,$$

then the Leray-Schauder degree

$$\deg(I - T, \Omega, \mathbf{0}) = 1.$$

6.3.2 *Existence results*

In what follows, we let μ_L and μ_M be defined by (6.34) and (6.35), respectively. We first state the results for BVP (6.10), (6.12).

Theorem 6.10. *Assume* (A1) *and* (A2) *hold. If*

$$\liminf_{x \to 0} \min_{t \in [0,1]} \frac{f(t,x)}{|x|} > \mu_L + 2\rho^2 \tag{6.37}$$

and

$$\limsup_{x \to \infty} \max_{t \in [0,1]} \frac{f(t,x)}{x} < \mu_L, \tag{6.38}$$

then BVP (6.10), (6.12) *has at least one nontrivial solution. In addition, if $\Lambda = 0$ in* (A2), *then BVP* (6.10), (6.12) *has at least one positive solution.*

Moreover, if $f(t,x)$ is symmetric in t for $(t,x) \in [0,1] \times \mathbb{R}$, then all the solutions of BVP (6.10), (6.12) *are symmetric.*

Proof. From (6.37), there exists $R_1 > 0$ such that

$$f(t,x) \geq (\mu_L + 2\rho^2)|x| \quad \text{for } (t,x) \in [0,1] \times [-R_1, R_1].$$

Let

$$F(t,x) = f(t,x) + \rho^2 x \quad \text{for } (t,x) \in [0,1] \times \mathbb{R}. \tag{6.39}$$

Then,

$$F(t,x) \geq (\mu_L + 2\rho^2)|x| + \rho^2 x \geq \nu_\rho |x| \quad \text{for } (t,x) \in [0,1] \times [-R_1, R_1], \tag{6.40}$$

where ν_ρ is defined by (6.36). Define a compact operator $T : X \to X$ by

$$(Tu)(t) = \int_0^1 J(t,s;\rho) F(s,u(s)) ds, \tag{6.41}$$

where $J(t,s;\rho)$ is given by (6.23). By Lemma 6.5, a solution of BVP (6.10), (6.12) is equivalent to a fixed point of the operator T. Let

$$\Omega_1 = \{u \in X \ : \ ||u|| < R_1\}.$$

Then, for $u \in \overline{\Omega}_1$ and $t \in [0,1]$, from Lemma 6.6(i) and (6.40),

$$(Tu)(t) \geq \nu_\rho \int_0^1 J(t,s;\rho)|u(s)| ds \geq 0.$$

Thus,

$$T(\overline{\Omega}_1) \subseteq P, \tag{6.42}$$

where P is defined by (6.29). From (6.31) and (6.40), we see that

$$(Tu)(t) \geq \nu_\rho \int_0^1 J(t,s;\rho) u(s) ds = \nu_\rho L_\rho u(s) \tag{6.43}$$

for $u \in \overline{\Omega}_1$ and $t \in [0,1]$.

We now suppose that T has no fixed point on $\partial\Omega_1$. Otherwise, BVP (6.10), (6.12) has a nontrivial solution. Let φ_L be given as in Lemma 6.10. Then, by Lemma 6.11, $\varphi_L = \nu_\rho L_\rho \varphi_L$. In the following, we show that

$$u - Tu \neq \tau\varphi_L \quad \text{for all } u \in \partial\Omega_1 \text{ and } \tau \geq 0. \tag{6.44}$$

If this is not the case, then there exist $u^* \in \partial\Omega_1$ and $\tau^* \geq 0$ such that $u^* - Tu^* = \tau^*\varphi_L$ Thus, $\tau^* > 0$, and in view of (6.42),

$$u^* = Tu^* + \tau^*\varphi_L \geq \tau^*\varphi_L.$$

Define

$$\hat{\tau} = \sup\{\tau \ : \ u^* \geq \tau\varphi_L\}.$$

Then, $\hat{\tau} \geq \tau^* > 0$ and $u^* \geq \hat{\tau}\varphi_L$. As a result,

$$\nu_\rho L_\rho u^* \geq \hat{\tau}\nu_\rho L_\rho \varphi_L = \hat{\tau}\varphi_L.$$

Thus, from (6.43),

$$u^* = Tu^* + \tau^*\varphi_L \geq \nu_\rho L_\rho u^* + \tau^*\varphi_L \geq (\hat{\tau} + \tau^*)\varphi_L,$$

which contradicts the definition of $\hat{\tau}$. Therefore, (6.44) holds. By Lemma 6.12, it follows that

$$\deg(I - T, \Omega_1, \mathbf{0}) = 0. \tag{6.45}$$

Let $\tilde{\varphi} = \Lambda \int_0^1 J(t, s; \rho) ds$, where Λ is given in (mH2). Define $A : X \to X$ by

$$Au = T(u - \tilde{\varphi}) + \tilde{\varphi}.$$

Then, (A2) implies that

$$A(X) \subseteq P. \tag{6.46}$$

From (6.38) and (6.39), it is clear that

$$\limsup_{x\to\infty} \max_{t\in[0,1]} \frac{F(t,x)}{x} < \nu_\rho.$$

Hence, there exist $0 < \sigma < 1$ and $R_2 > R_1 + ||\tilde{\varphi}||$ such that

$$F(t, x) \le \sigma\nu_\rho x \quad \text{for } (t, x) \in [0, 1] \times [R_2, \infty). \tag{6.47}$$

Let $\tilde{L}u = \sigma\nu_\rho L_\rho u$ for $u \in X$. Then $\tilde{L} : X \to X$ is a bounded linear operator and $\tilde{L}(P) \subseteq P$. Let

$$C = ||\tilde{\varphi}|| + e_\rho \sup_{\substack{u\in X\\ ||u||\le R_2}} \int_0^1 H(s, s; \rho)|F(s, u(s))|ds. \tag{6.48}$$

Then, $0 < C < \infty$. Define

$$W = \{u \in P \ : \ u = sAu, \ 0 \le s \le 1\}.$$

We now show that W is bounded, i.e.,

$$\sup W < \infty. \tag{6.49}$$

For any $u \in W$, let

$$I_1^u = \{t \in [0, 1] \ : \ u(t) - \tilde{\varphi}(t) > R_2\},$$

$$I_2^u = [0, 1] \setminus I_1^u,$$

and

$$\tilde{u}(t) = \min\{u(t) - \tilde{\varphi}(t), R_2\}.$$

Since $u(t) - \tilde{\varphi}(t) \ge -\tilde{\varphi}(t) > -R_2$, we have that $||\tilde{u}|| \le R_2$. Moreover, from (6.47),

$$F(t, u(t) - \tilde{\varphi}(t)) \le \sigma\nu_\rho(u(t) - \tilde{\varphi}(t)) \le \sigma\nu_\rho u(t) \quad \text{for } t \in I_1^u.$$

Thus, from (6.41), Lemma 6.6(i), (6.48), and (6.31), it follows that

$$\begin{aligned}u(t) &= s(Au)(t) \le (T(u(t) - \tilde{\varphi}))(t) + \tilde{\varphi}(t) \\ &= \int_{I_1^u} J(t,s;\rho)F\big(s, u(s) - \tilde{\varphi}(s)\big)ds \\ &\quad + \int_{I_2} J(t,s;\rho)F\big(s, u(s) - \tilde{\varphi}(s)\big)ds + \tilde{\varphi}(t) \\ &\le \sigma\nu_\rho \int_0^1 J(t,s;\rho)u(s)ds + e_\rho \int_0^1 H(s,s;\rho)F(s,\tilde{u}(s))ds + \tilde{\varphi}(t) \\ &\le \sigma\nu_\rho (L_\rho u)(t) + C = \tilde{L}u(t) + C. \end{aligned} \tag{6.50}$$

By Lemma 6.26, ν_ρ is the smallest characteristic value of L_ρ. In view of the fact that $0 < \sigma < 1$, we see that the spectral radius, $r(\tilde{L})$, of $\tilde{L}$ satisfies $r(\tilde{L}) < 1$. Hence, the inverse operator $(I - \tilde{L})^{-1}$ exists and is given by (see, for example, [Swartz (1992), Chapter 23, Theorems 8 and 11])

$$(I - \tilde{L})^{-1} = \sum_{i=0}^{\infty} \tilde{L}^i.$$

Then, from $\tilde{L}(P) \subseteq P$, we obtain that $(I - \tilde{L})^{-1}(P) \subseteq P$. Now, from (6.50), we have $u(t) \le (I - \tilde{L})^{-1}(C)$ for $t \in [0,1]$. Hence, W is bounded, i.e., (6.49) holds.

Let $R_3 > \max\{R_2, \sup W + ||\tilde{\varphi}||\}$ be fixed. Define

$$\Omega_2 = \{u \in X \ : \ ||u|| < R_3\}.$$

Then,

$$Au \ne \tau u \quad \text{for all } u \in \partial\Omega_2 \text{ and } \tau \ge 1. \tag{6.51}$$

If this is not so, then there exists $u^* \in \partial\Omega_2$ and $\tau^* \ge 1$ such that $Au^* = \tau^* u^*$. Hence, $u^* = s^* A u^*$, where $s^* = 1/\tau^* \in [0,1]$. From (6.46), we see that $Au^* \in P$, and so $u^* \in P$, which implies $u^* \in W$. But this contradicts the assumption that $||u^*|| = R_3 > \sup W$. Thus, (6.51) holds.

By Lemma 6.13, it follows that

$$\deg(I - A, \Omega_2, \mathbf{0}) = 1. \tag{6.52}$$

Define a compact homotopy $\mathcal{H} : [0,1] \times X \to R$ by

$$\mathcal{H}(s,u) = T(u - s\tilde{\varphi}) + s\tilde{\varphi}.$$

We claim that

$$\mathcal{H}(s,u) \ne u \quad \text{for all } (s,u) \in [0,1] \times \partial\Omega_2. \tag{6.53}$$

In fact, if there exists $(s_0, u_0) \in [0,1] \times \partial\Omega_2$ such that $\mathcal{H}(s_0, u_0) = u_0$, then

$$T(u_0 - s_0\tilde{\varphi}) = u_0 - s_0\tilde{\varphi},$$

and so from the definition of A,

$$A(u_0 - s_0\tilde{\varphi} + \tilde{\varphi}) = T(u_0 - s_0\tilde{\varphi}) + \tilde{\varphi} = u_0 - s_0\tilde{\varphi} + \tilde{\varphi}.$$

Thus, $u_0 - s_0\tilde{\varphi} + \tilde{\varphi} \in W$. But this contradicts the fact that

$$||u_0 - s_0\tilde{\varphi} + \tilde{\varphi}|| \geq ||u_0|| - (1 - s_0)||\tilde{\varphi}|| \geq ||R_3|| - ||\tilde{\varphi}|| > \sup W.$$

Hence, (6.53) holds. From the homotopy invariance of the Leray-Schauder degree and (6.52), we have

$$\begin{aligned}\deg(I - T, \Omega_2, \mathbf{0}) &= \deg(I - \mathcal{H}(0, \cdot), \Omega_2, \mathbf{0}) \\ &= \deg(I - \mathcal{H}(1, \cdot), \Omega_2, \mathbf{0}) \\ &= \deg(I - A, \Omega_2, \mathbf{0}) = 1. \end{aligned} \tag{6.54}$$

By the additivity property of the Leray-Schauder degree, (6.45), and (6.54), we obtain

$$\deg(I - T, \Omega_2 \setminus \overline{\Omega}_1, \mathbf{0}) = 1.$$

Thus, from the solution property of the Leray-Schauder degree, T has at least one fixed point u in $\Omega_2 \setminus \overline{\Omega}_1$. Clearly, $u(t)$ is a nontrivial solution of BVP (6.10), (6.12).

We next suppose that $\Lambda = 0$ in (A2). Then $F(t, x)$ defined by (6.39) is nonnegative on $[0, 1] \times \mathbb{R}$. By Lemma 6.6(i),

$$0 \leq u(t) = \int_0^1 J(t, s; \rho)F(s, u(s))ds \leq e_\rho \int_0^1 H(s, s; \rho)F(s, u(s))ds$$

for $t \in [0, 1]$, so

$$||u|| \leq e_\rho \int_0^1 H(s, s; \rho)F(s, u(s))ds.$$

On the other hand, by Lemma 6.6(ii),

$$u(t) = \int_0^1 J(t, s; \rho)F(s, u(s))ds \geq w_\rho(t) \int_0^1 H(s, s; \rho)F(s, u(s))ds$$

for $t \in [0, 1]$, where $w_\rho(t)$ is defined by (6.25). Thus,

$$u(t) \geq e_\rho^{-1} w_\rho(t)||u|| \quad \text{for } t \in [0, 1]. \tag{6.55}$$

Since $R_1 < ||u|| < R_3$ and $w_\rho(t) > 0$ for $t \in (0, 1)$, $u(t)$ is a positive solution of BVP (6.10), (6.12).

Finally, we suppose that $f(t,x)$ is symmetric in t for $(t,x) \in [0,1] \times \mathbb{R}$. Then so is $F(t,x)$. From (6.23), we see that any solution $u(t)$ of BVP (6.10), (6.12) satisfies

$$\begin{aligned} u(t) &= \int_0^1 J(t,s;\rho)F(s,u(s))ds \\ &= \int_0^1 H(t,s;\rho)F(s,u(s))ds \\ &\quad + \frac{c_\rho}{\Delta_\rho}(\phi_\rho(t)+\psi_\rho(t))\sum_{i=0}^m a_i \int_0^1 H(t_i,s;\rho)ds \end{aligned}$$

for $t \in [0,1]$. From (6.19), (6.20), and (6.24), we have

$$\phi_\rho(1-t) = \psi_\rho(t), \ \psi_\rho(1-t) = \phi_\rho(t)$$

and

$$H(1-t,1-s;\rho) = H(t,s;\rho)$$

for $t, s \in [0,1]$. A change of variables then yields

$$\begin{aligned} u(1-t) &= \int_0^1 H(1-t,s;\rho)F(s,u(s))ds \\ &\quad + \frac{c_\rho}{\Delta_\rho}(\phi_\rho(1-t)+\psi_\rho(1-t))\sum_{i=0}^m a_i \int_0^1 H(t_i,s;\rho)ds \\ &= \int_0^1 H(1-t,1-s;\rho)F(s,u(s))ds \\ &\quad + \frac{c_\rho}{\Delta_\rho}(\phi_\rho(t)+\psi_\rho(t))\sum_{i=0}^m a_i \int_0^1 H(t_i,s;\rho)ds \\ &= \int_0^1 H(t,s;\rho)F(s,u(s))ds \\ &\quad + \frac{c_\rho}{\Delta_\rho}(\phi_\rho(t)+\psi_\rho(t))\sum_{i=0}^m a_i \int_0^1 H(t_i,s;\rho)ds \\ &= u(t) \end{aligned} \tag{6.56}$$

for $t \in [0,1]$. Hence, all the solutions of BVP (6.10), (6.12) are symmetric. This completes the proof of the theorem. □

For convenience, we introduce the following notations.

$$f_0 = \liminf_{x\to 0} \min_{t\in[0,1]} \frac{f(t,x)}{x}, \quad f^\infty = \limsup_{|x|\to\infty} \max_{t\in[0,1]} \frac{f(t,x)}{x},$$

$$\alpha = \frac{1}{e_0 \int_0^1 G(s,s)ds}, \quad \beta = \frac{1}{d_0 \min_{t\in[a,b]} w_0(t) \int_a^b G(s,s)ds}, \tag{6.57}$$

$$\xi = \frac{1}{e_\rho \int_0^1 H(s,s;\rho)ds}, \quad \eta = \frac{1}{\int_a^b J(1/2,s;\rho)ds}, \tag{6.58}$$

where $G(t,s)$, $J(t,s;\rho)$, $H(t,s;\rho)$, d_0, e_ρ, and e_0, and are defined by (6.18), (6.23), (6.24), (6.26) with $d_0 = 0$, (6.27), and (6.28), respectively.

As an application of Theorem 6.10, we have the following corollary.

Corollary 6.3. *Assume* (A1) *and* (A2) *hold. If*

$$\liminf_{x\to 0} \min_{t\in[0,1]} \frac{f(t,x)}{|x|} > \beta + 2\rho^2$$

and

$$\limsup_{x\to\infty} \max_{t\in[0,1]} \frac{f(t,x)}{x} < \alpha,$$

then the conclusions of Theorem 6.10 *hold.*

Lemma 6.14. *Let μ_L be defined by* (6.34)*. Then μ_L satisfies $\alpha \le \mu_L \le \beta$, where α and β are defined by* (6.57)*.*

Proof. Let φ_L be given as in Lemma 6.10; then,

$$\varphi_L(t) = \mu_L \int_0^1 K(t,s)\varphi_L(s)ds$$

for $t \in [0,1]$. By Corollary 6.2(i), we have

$$\varphi_L(t) \le \mu_L e_0 ||\varphi_L|| \int_0^1 G(s,s)ds \quad \text{on } [0,1].$$

Hence,

$$\mu_L \ge \frac{1}{e_0 \int_0^1 G(s,s)ds} = \alpha.$$

By Lemma 6.8, it is easy to see that $\min_{t\in[a,b]} \varphi_L(t) \ge d_0||\varphi_L||$. Corollary 6.2(iii) then implies

$$\begin{aligned}\varphi_L(t) &\ge \mu_L \min_{t\in[a,b]} w_0(t) \int_a^b G(s,s)\varphi_L(s)ds \\ &\ge \mu_L d_0 ||\varphi_L|| \min_{t\in[a,b]} w_0(t) \int_a^b G(s,s)ds\end{aligned}$$

for $t \in [a,b]$. Thus,

$$\mu_L \le \frac{1}{d_0 \min_{t\in[a,b]} w_0(t) \int_a^b G(s,s)ds} = \beta.$$

This completes the proof of the lemma. □

Proof of Corollary 6.3. In view of Lemma 6.14, Corollary 6.3 readily follows from Theorem 6.10. □

Theorem 6.11. *Assume* (A1) *and* (A3) *hold. If*

$$f^{\infty} < \mu_L < f_0,$$

then BVP (6.10), (6.12) *has at least one positive solution and one negative solution.*

Moreover, if $f(t,x)$ *is symmetric in* t *for* $(t,x) \in [0,1] \times \mathbb{R}$, *then all the solutions of* BVP (6.10), (6.12) *are symmetric.*

Proof. Let

$$F_1(t,x) = \begin{cases} f(t,x) + \rho^2 x, & x \geq 0, \\ -(f(t,x) + \rho^2 x), & x < 0. \end{cases} \tag{6.59}$$

From (6.14), it is easy to see that $F_1 : [0,1] \times \mathbb{R} \to \mathbb{R}$ is continuous and nonnegative. Since $f_0 > \mu_L$, there exists $R_1 > 0$ such that

$$f(t,x) \geq \mu_L x \quad \text{for } (t,x) \in [0,1] \times [0, R_1]$$

and

$$f(t,x) \leq \mu_L x \quad \text{for } (t,x) \in [0,1] \times [-R_1, 0].$$

Then, (6.59) implies that

$$F_1(t,x) \geq \nu_\rho |x| \quad \text{for } (t,x) \in [0,1] \times [-R_1, R_1],$$

where ν_ρ is defined by (6.36). Define a compact operator $T_1 : X \to X$ by

$$(T_1 u)(t) = \int_0^1 J(t,s;\rho) F_1(s, u(s)) \tag{6.60}$$

and let

$$\Omega_1 = \{u \in X \; : \; ||u|| < R_1\}.$$

An argument similar to the one used in deriving (6.45) in the proof of Theorem 6.10 yields

$$\deg(I - T_1, \Omega_1, \mathbf{0}) = 0. \tag{6.61}$$

It is clear that

$$T_1(X) \subseteq P, \tag{6.62}$$

where P is defined by (6.29). From the fact that $f^{\infty} < \mu_L$ and (6.59), it follows that

$$\limsup_{x \to \infty} \max_{t \in [0,1]} \frac{F_1(t,x)}{x} < \nu_\rho.$$

Thus, there exists $0 < \sigma < 1$ and $R_2 > R_1$ such that

$$F_1(t,x) \le \sigma\nu_\rho x \quad \text{for } (t,x) \in [0,1] \times [R_2, \infty).$$

Define

$$W = \{t \in [0,1] \ : \ u = sT_1u, \ 0 \le s \le 1\}.$$

An argument similar to the one used to obtain (6.49) in the proof of Theorem 6.10 (in which $\tilde{\varphi} = 0$) shows that W is bounded.

Let $R_3 > \max\{R_2, \sup W\}$ be fixed. Define

$$\Omega_2 = \{u \in X \ : \ ||u|| < R_3\}.$$

To show

$$T_1u \ne \tau u \quad \text{for all } u \in \partial\Omega_2 \text{ and } \tau \ge 1, \tag{6.63}$$

suppose there exists $u^* \in \partial\Omega_2$ and $\tau^* \ge 1$ such that $T_1u^* = \tau^*u^*$. Then, $u^* = s^*T_1u^*$, where $s^* = 1/\tau^* \in [0,1]$. From (6.62), we see that $T_1u^* \in P$, and so $u^* \in P$, which implies $u^* \in W$. But this contradicts the assumption that $||u^*|| = R_3 > \sup W$. Thus, (6.63) holds. By Lemma 6.13, it follows that

$$\deg(I - T_1, \Omega_2, \mathbf{0}) = 1. \tag{6.64}$$

By the additivity property of the Leray-Schauder degree, (6.61), and (6.64), we obtain

$$\deg(I - T_1, \Omega_2 \setminus \overline{\Omega}_1, \mathbf{0}) = 1.$$

Thus, from the solution property of the Leray-Schauder degree, T_1 has at least one fixed point u_1 in $\Omega_2 \setminus \overline{\Omega}_1$. Then

$$u_1(t) = \int_0^1 J(t,s;\rho)F_1(s,u_1(s)) \quad \text{for } t \in [0,1].$$

As in the proof of Theorem 6.10, (6.55) with $u(t)$ replaced by $u_1(t)$ still holds. In view of the fact that $R_1 < ||u_1|| < R_3$ and $w_\rho(t) > 0$ for $t \in (0,1)$, we have $u_1(t) > 0$ on $(0,1)$. Therefore, from (6.59), $F_1(t,u_1(t)) = f(t,u_1(t)) + \rho^2 u_1(t)$, and so $u_1(t)$ is a positive solution of BVP (6.10), (6.12).

Let

$$F_2(t,x) = \begin{cases} -(f(t,-x) + \rho^2(-x)), & x \ge 0, \\ f(t,-x) + \rho^2(-x), & x < 0. \end{cases} \tag{6.65}$$

Then, from (6.14), $F_2 : [0,1] \times \mathbb{R} \to \mathbb{R}$ is continuous and nonnegative. Define a compact operator $T_2 : X \to X$ by

$$(T_2u)(t) = \int_0^1 J(t,s;\rho)F_2(s,u(s)). \tag{6.66}$$

By a similar argument to the above, we see that T_2 has a fixed point $v(t)$ satisfying $v(t) > 0$ on $(0, 1)$. From (6.65) and

$$v(t) = \int_0^1 J(t, s; \rho) F_2(s, v(s)) ds,$$

we obtain

$$-v(t) = \int_0^1 J(t, s; \rho)\big(f(s, -v(s)) + \rho^2(-v(s))\big) ds.$$

Therefore, $u_2(t) := -v(t)$ is a negative solution of BVP (6.10), (6.12).

Finally, if $f(t, x)$ is symmetric in t for $(t, x) \in [0, 1] \times \mathbb{R}$, then as in deriving (6.56) in the proof of Theorem 6.10, we see that all solutions of BVP (6.10), (6.12) are symmetric. This completes the proof of Theorem 6.11. □

The following corollary is an immediate consequence of Theorem 6.11.

Corollary 6.4. *Assume* (*A*1) *and* (*A*3) *hold. If*

$$\frac{f^\infty}{\alpha} < 1 < \frac{f_0}{\beta}, \tag{6.67}$$

then the conclusions of Theorem 6.11 *hold.*

Proof. In view of Lemma 6.14 and (6.67), we have $f^\infty < \mu_L < f_0$. Corollary 6.4 readily follows from Theorem 6.11. □

We now introduce the following conditions that describe the "smallness" and "largeness" of $f(t, x) + \rho^2 x$ on fixed "slabs" of $[0, 1] \times \mathbb{R}$.

(B1) There exists $p_1 > 0$ such that

$$|f(t, x) + \rho^2 x| < p_1 \xi \quad \text{for } (t, x) \in [0, 1] \times [-p_1, p_1].$$

(B2) There exists $p_2 > 0$ such that

$$|f(t, x) + \rho^2 x| > p_2 \eta \quad \text{for } (t, |x|) \in [a, b] \times [p_2 d_\rho, p_2],$$

where d_ρ is defined by (6.26).

The following results provide sufficient conditions for the existence of multiple solutions of BVP (6.10), (6.12).

Theorem 6.12. *Assume* (A1) *and* (A3) *hold together with one of the following conditions:*

(C1) $f_0 > \mu_L$ *and* (B1) *holds;*
(C2) $f^\infty < \mu_L$ *and* (B2) *holds.*

Then BVP (6.10), (6.12) *has at least one positive solution and one negative solution.*

Moreover, if $f(t,x)$ is symmetric in t for $(t,x) \in [0,1] \times \mathbb{R}$, then all the solutions of BVP (6.10), (6.12) *are symmetric.*

Corollary 6.5. *Assume* (A1) *and* (A3) *hold together with one of the following conditions:*

(C1)* $f_0 > \beta$ *and* (B1) *holds;*
(C2)* $f^\infty < \alpha$ *and* (B2) *holds.*

Then the conclusions of Theorem 6.12 hold.

Theorem 6.13. *Assume* (A1) *and* (A3) *hold together with one of the following conditions:*

(D1) $f_0 > \mu_L$, (B1) *and* (B2) *hold with* $p_1 < p_2$*;*
(D2) $f^\infty < \mu_L$, (B1) *and* (B2) *hold with* $p_1 < p_2$.

Then BVP (6.10), (6.12) *has at least two positive solutions and two negative solutions.*

Moreover, if $f(t,x)$ is symmetric in t for $(t,x) \in [0,1] \times \mathbb{R}$, then all the solutions of BVP (6.10), (6.12) *are symmetric.*

Corollary 6.6. *Assume* (A1) *and* (A3) *hold together with one of the following conditions:*

(D1)* $f_0 > \beta$, (B1) *and* (B2) *hold with* $p_1 < p_2$*;*
(D2)* $f^\infty < \alpha$, (B1) *and* (B2) *hold with* $p_1 < p_2$.

Then the conclusions of Theorem 6.13 hold.

The following two lemmas are taken from [Guo and Lakshmikantham (1988), Corollary 2.3.1. and Lemma 2.3.1], respectively.

Lemma 6.15. *Let X be a Banach space and $P \subseteq X$ be a cone in E. Assume that Ω is a bounded open subset of X and that $T : P \cap \overline{\Omega} \to P$ is compact. If there exists $u_0 \in P \setminus \{\mathbf{0}\}$ such that*

$$u - Tu \neq \tau u_0 \quad \text{for all } u \in P \cap \partial\Omega \text{ and } \tau \geq 0,$$

then the fixed point index

$$i(T, P \cap \Omega, P) = 0.$$

Lemma 6.16. *Let X be a Banach space and $P \subseteq X$ be a cone in E. Assume that Ω is a bounded open subset of X with $\mathbf{0} \in \Omega$ and that $T : P \cap \overline{\Omega} \to P$ is compact. If*

$$Tu \neq \tau u \quad \text{for all } u \in P \cap \partial\Omega \text{ and } \tau \geq 1,$$

then the fixed point index

$$i(T, P \cap \Omega, P) = 1.$$

Remark 6.6. The following observations are made about the proofs of the results in this subsection.

(a) Theorem 6.10 was proved by a topological degree argument instead of a fixed point index argument because the operator T defined by (6.41) is not cone-preserving in general.
(b) Although Theorem 6.11 was proved using topological degree theory, it could also be proved using the fixed point index and Lemmas 6.15 and 6.16.
(c) In the following, Theorems 6.12 and 6.13 will be proved using the fixed point index and Lemmas 6.15 and 6.16.

Proof of Theorem 6.12. Let P_ρ and T be defined by (6.30) and (6.41), respectively. In view of (A3), as in the proof of Lemma 6.8, it is easy to show that $T(P_\rho) \subseteq P_\rho$.

We first assume (C1) holds. Since $f_0 > \mu_L$, there exists $0 < R_1 < p_1$ such that

$$F(t,x) \geq \nu_\rho x \quad \text{for } (t,x) \in [0,1] \times [0, R_1].$$

Let

$$\Omega_1 = \{u \in X \ : \ ||u|| < R_1\}.$$

For $u \in P_\rho \cap \partial\Omega_1$, we have

$$(Tu)(t) \geq \nu_\rho \int_0^1 J(t,s;\rho)u(s)ds = \nu_\rho L_\rho u(s)ds$$

for $t \in [0,1]$. Using a similar argument as in deriving (6.44), we have that

$$u - Tu \neq \tau\varphi_L \quad \text{for all } u \in P_\rho \cap \partial\Omega_1 \text{ and } \tau \geq 0.$$

By Lemma 6.15,

$$i(T, P_\rho \cap \Omega_1, P_\rho) = 0. \tag{6.68}$$

Let

$$\Omega_2 = \{u \in X \ : \ ||u|| < p_1\}.$$

From (A3), Lemma 6.6(i), (B1), and (6.58), it follows that for any $u \in P_\rho \cap \partial\Omega_2$, we have

$$\begin{aligned} 0 \le Tu(t) &\le e_\rho \int_0^1 H(s,s;\rho)F(s,u(s))ds \\ &< p_1 \xi e_\rho \int_0^1 H(s,s;\rho)ds = p_1 \end{aligned}$$

for $t \in [0,1]$. Hence, $||Tu|| < ||u||$. By Lemma 6.16, we have that

$$i(T, P_\rho \cap \Omega_2, P_\rho) = 1. \tag{6.69}$$

By the additivity property of the fixed point index, (6.68), and (6.69), we can conclude that

$$i(T, P_\rho \cap (\Omega_2 \setminus \overline{\Omega}_1), P_\rho) = 1. \tag{6.70}$$

Thus, from the solution property of the fixed point index, T has at least one fixed point u_1 in $P_\rho \cap (\Omega_2 \setminus \overline{\Omega}_1)$. Clearly, $u_1(t)$ is a positive solution of BVP (6.10), (6.12).

Now define a compact operator $T_3 : X \to X$ by

$$(T_3 u)(t) = \int_0^1 J(t,s;\rho)F_3(s,u(s))ds, \tag{6.71}$$

where

$$F_3(t,x) = -\big(f(t,-x) + \rho^2(-x)\big), \quad x \in \mathbb{R}. \tag{6.72}$$

By an argument similar to the one above, we see that T_3 has a fixed point v satisfying $v(t) > 0$ on $(0,1)$. Hence,

$$-v(t) = \int_0^1 J(t,s;\rho)\big(f(s,-v(s)) + \rho^2(-v(s))\big)ds.$$

Therefore, $u_2(t) := -v(t)$ is a negative solution of BVP (6.10), (6.12).

Next, we assume that (C2) holds. Since $f^\infty < \mu_L$, there exists $R_2 > p_2$ such that (see (6.47))

$$F(t,s) \le \sigma \nu_\rho x \quad \text{for } (t,x) \in [0,1] \times [R_2, \infty).$$

Define

$$W = \{u \in P \ : \ u = sTu, \ 0 \le s \le 1\}.$$

An argument similar to the one used to obtain (6.49) (in which $\tilde{\varphi} = 0$), we see that W is bounded.

Let $R_3 > \max\{R_2, \sup W\}$ be fixed. Define

$$\Omega_3 = \{u \in X \ : \ ||u|| < R_3\}.$$

As in deriving (6.63), we have that

$$Tu \neq \tau\varphi_L \quad \text{for all } u \in P_\rho \cap \partial\Omega_3 \text{ and } \tau \geq 1.$$

By Lemma 6.16,

$$i(T, P_\rho \cap \Omega_3, P_\rho) = 1. \tag{6.73}$$

Let

$$\Omega_4 = \{u \in X \ : \ ||u|| < p_2\}.$$

Then, for $u \in P_\rho \cap \partial\Omega_4$, $\min_{t\in[a,b]} u(t) \geq d_\rho ||u|| = d_\rho p_2$, and so from (B2) and (6.58), it follows that

$$\begin{aligned}(Tu)(1/2) &\geq \int_a^b J(1/2, s; \rho) F(s, u(s)) ds \\ &> p_2 \eta \int_0^1 J(1/2, s; \rho) ds = p_2.\end{aligned}$$

Hence, $||Tu|| > ||u||$. By Lemma 6.15, we have that

$$i(T, P_\rho \cap \Omega_4, P_\rho) = 0. \tag{6.74}$$

By (6.69) and (6.70), we obtain that

$$i(T, P_\rho \cap (\Omega_3 \setminus \overline{\Omega}_4), P_\rho) = 1.$$

Thus, from the solution property of the fixed point index, T has at least one fixed point u_1 in $P_\rho \cap (\Omega_3 \setminus \overline{\Omega}_4)$. Clearly, $u_1(t)$ is a positive solution of BVP (6.10), (6.12).

The proof for the existence of a negative solution is similar to the previous case and hence is omitted.

Finally, if $f(t, x)$ is symmetric in t for $(t, x) \in [0, 1] \times \mathbb{R}$, then as deriving (6.56), we can conclude that all the solutions of BVP (6.10), (6.12) are symmetric. This completes the proof of the theorem. □

Proof of Corollary 6.5. In view of Lemma 6.14, Corollary 6.5 readily follows from Theorem 6.12. □

Proof of Theorem 6.13. We only prove the case when (D1) holds. Let P_ρ and T be defined by (6.30) and (6.41), respectively. Then, $T(P_\rho) \subseteq P_\rho$. Since $f_0 > \mu_L$, (B1), and (B2) hold, so as in the proof of Theorem 6.12, it follows that (6.68), (6.69), and (6.74) hold, where Ω_1, Ω_2, and Ω_4 are defined as before. Thus, we have that (6.70) holds and

$$i(T, P_\rho \cap (\Omega_4 \setminus \overline{\Omega}_2), P_\rho) = -1.$$

Therefore, T has at least two fixed points, u_1 in $P_\rho \cap (\Omega_2 \setminus \overline{\Omega}_2)$ and v_1 in $P_\rho \cap (\Omega_4 \setminus \overline{\Omega}_2)$. Clearly, $u_1(t)$ and $v_1(t)$ are positive solutions of BVP (6.10), (6.12).

Let T_3 and F_3 be defined by (6.71) and (6.72), respectively. Using a similar argument to the one used in Theorem 6.12 to show the existence of negative solutions, we can show that BVP (6.10), (6.12) has at least two negative solutions.

Finally, the symmetry of all solutions of BVP (6.10), (6.11) can be proved as before and hence is omitted. This completes the proof of the theorem. □

Proof of Corollary 6.6. In view of Lemma 6.14, Corollary 6.6 readily follows from Theorem 6.13. □

Now, we state some parallel results for BVP (6.11), (6.12).

Theorem 6.14. *Assume* (A1), (A2) *with* $\rho = 0$, *and* (A4) *hold. If*

$$\liminf_{x \to 0} \min_{t \in [0,1]} \frac{f(t,x)}{|x|} > \mu_M$$

and

$$\limsup_{x \to \infty} \max_{t \in [0,1]} \frac{f(t,x)}{x} < \mu_M,$$

then BVP (6.11), (6.12) *has at least one nontrivial solution. In addition, if* $\Lambda = 0$ *in* (A2), *then BVP* (6.11), (6.12) *has at least one positive solution.*

Moreover, if $g(t)$ *and* $f(t,x)$ *are symmetric in* t *for* $(t,x) \in [0,1] \times \mathbb{R}$, *then all the solutions of BVP* (6.11), (6.12) *are symmetric.*

Theorem 6.15. *Assume* (A1), (A3) *with* $\rho = 0$, *and* (A4) *hold. If*

$$f^\infty < \mu_M < f_0,$$

then BVP (6.11), (6.12) *has at least one positive solution and one negative solution.*

Moreover, if $g(t)$ *and* $f(t,x)$ *are symmetric in* t *for* $(t,x) \in [0,1] \times \mathbb{R}$, *then all the solutions of BVP* (6.11), (6.12) *are symmetric.*

The proofs of Theorems 6.14 and 6.15 are similar to those of Theorems 6.10 and 6.11, and we omit the details here.

Results similar to Theorems 6.12, 6.13, and all the corollaries can be easily formulated for BVP (6.11), (6.12), and we leave this to the interested reader. (Under the approach present in this section, we need $\rho = 0$ in the corresponding results for BVP (6.11), (6.12).)

Finally, the following example illustrate one of the results in this section.

Example 6.1. Consider the BVP consisting of the equation

$$u'' + (g_1(t) - 4)\, u + g_2(t) u^{1/3} = 0, \quad t \in (0,1), \tag{6.75}$$

and the 3-point BC

$$u(0) = u(1) = \frac{1}{2} u\left(\frac{1}{2}\right), \tag{6.76}$$

where $g_1, g_2 \in C([0,1], \mathbb{R}^+)$ and $g_2(t) > 0$ on $[0,1]$.

We claim that if

$$||g_1|| + ||g_2|| < \frac{16(\sinh 2)(\sinh 2 - \sinh 1)}{(3\sinh 2 - 2\sinh 1)(2\cosh 2 - \sinh 2)}, \tag{6.77}$$

then BVP (6.75), (6.76) has at least one positive solution and one negative solution. In addition, if $g_1(t)$ and $g_2(t)$ are symmetric in t for $t \in [0,1]$, then all the solutions of BVP (6.75), (6.76) are symmetric.

Here, $m = 1$, $a_1 = t_1 = 1/2$, $\rho = 2$, and $f(t,x) = (g_1(t) - 4)x + g_2(t)x^{1/3}$, then it is easy to see that (A1) and (A3) hold. Moreover, for ξ defined in (6.58), by a simple calculation, we have

$$\xi = \frac{16(\sinh 2)(\sinh 2 - \sinh 1)}{(3\sinh 2 - 2\sinh 1)(2\cosh 2 - \sinh 2)}.$$

With this ξ, $\rho = 2$ and $p_1 = 1$, it is clear that if (3.33) holds, then

$$\begin{aligned} |f(t,x) + \rho^2 x| &= |g_1(t)x + g_2(t)x^{1/3}| \\ &\le ||g_1|| + ||g_2|| < \xi \quad \text{for } (t,x) \in [0,1] \times [-p_1, p_1], \end{aligned}$$

i.e., (B1) holds. Since $f_0 = \infty$, $f_0 > \beta$, where β is defined in (6.57) with $a = 1/4$ and $b = 3/4$. Thus (C1)* of Corollary 6.5 holds. The conclusion then follows from Corollary 6.5.

Bibliography

Chyan C. J., Davis J. M., Henderson J. and Yin W. K. C. (1998). Eigenvalue comparisons for differential equations on a measure chain, *Electron. J. Differential Equations* No. 35, pp. 1–7.

Davis J. M., Eloe P. W. and Henderson J. (1999). Comparison of eigenvalues for discrete Lidstone boundary value problems, *Dynam. Systems Appl.* **8**, No. 3–4, pp. 381–388.

Deimling K. (1985). *Nonlinear Functional Analysis* (Spring-Verlag, New York).

Eloe P. W. and Henderson J. (1989). Comparison of eigenvalues for a class of two-point boundary value problems, *Appl. Anal.* **34**, pp. 25–34.

Eloe P. W. and Henderson J. (1992). Comparison of eigenvalues for a class of multipoint boundary value problem, *Recent Trends in Differential Equations*, pp. 179–188, World Sci. Ser. Appl. Anal., Vol. 1 (World Scientific Publishing, River Edge, NJ).

Eloe P. W. and Henderson J. (1994). Comparison of eigenvalues for a system of two-point boundary value problems, *Inequalities and Applications*, pp. 187–196, World Sci. Ser. Appl. Anal., Vol. 1 (World Scientific Publishing, River Edge, NJ).

Eloe P. W. and Neugebauer J. T. (2014). Existence and comparison of smallest eigenvalues for a fractional boundary-value problem, *Electron. J. Differential Equations*, No. 43, pp. 1–10.

Eloe P. W. and Neugebauer J. T. (2016). Smallest eigenvalues for a right focal boundary value problem, *Fract. Cal. Appl. Anal.* **19**, pp. 11–18.

Gentry R. D. and Travis C. C. (1976). Existence and comparison of eigenvalues of nth order linear differential equations, *Bull. Amer. Math. Soc.* **82**, No. 2, pp. 350–352.

Gentry R. D. and Travis C. C. (1976). Comparison of eigenvalues associated with linear differential equations of arbitrary order, *Trans. Amer. Math. Soc.* **223**, pp. 167–179.

Gentry R. D. and Travis C. C. (1980). The existence and extremal characterization of eigenvalues for an nth order multiple point boundary value problem, *Ann. Mat. Pura Appl.* **126**, No. 4, pp. 223–232.

Graef J. R. and Kong L (2008). Solutions of second order multi-point boundary value problems, *Math. Proc. Camb. Phil. Soc.* **145**, pp. 489–510.

Guo D. and Lakshmikantham V. (1988). *Nonlinear Problems in Abstract Cones*, (Academic Press, Orlando).

Hankerson D. and Henderson J. (1989). Comparison of eigenvalues for n-point boundary value problems for difference equations, *Differential Equations* (Colorado Springs, CO), pp. 203–208, Lecture Notes in Pure and Appl. Math., Vol. 127 (Dekker, New York).

Hankerson D. and Peterson A. C. (1988). Comparison theorems for eigenvalue problems for nth order differential equations, *Proc. Amer. Math. Soc.* **104**, No. 4, pp. 1204–1211.

Hankerson D. and Peterson A. C. (1990). Comparison of eigenvalues for focal point problems for nth order difference equations, *Differential Integral Equations* **3**, No. 2, pp. 363–380.

Henderson J. and Kosmatov N. (2014) Eigenvalue comparison for fractional boundary value problems with Caputo derivative, *Fract. Calc. Appl. Anal.* **17**, No. 3, pp. 872–880.

Henderson J. and Prasad K. R. (1999). Comparison of eigenvalues for Lidstone boundary value problems on a measure chain *Comput. Math. Appl.* **38**, No. 11–12, pp. 55–62.

Karna B. (2004). Eigenvalue comparisons for m-point boundary value problems, *Comm. Appl. Nonlinear Anal.* **11**, No. 1, pp. 73–83.

Karna B. (2005). Eigenvalue comparisons for three-point boundary value problems, *Comm. Appl. Nonlinear Anal.* **12**, No. 3, pp. 83–91.

Kaufmann E. R. (1994). Comparison of eigenvalues for eigenvalue problems for a right disfocal operator, *Panamer. Math. J.* **4**, pp. 103–124.

Keener M. S. and Travis C. C. (1978). Positive cones and focal points for a class of nth-order differential equations, *Trans. Amer. Math. Soc.* **237**, pp. 331–351.

Krasnosel'skii M. A. (1964). *Positve Solutions of Operator Equations* (Fizmatgiz, Moscow, 1962); English transl. (Noordhoff, Groningen).

Krein M. G. and Rutman M. A. (1962). *Linear Operators Leaving a Cone Invariant in a Banach Space*, American Mathematical Society Translations, Series, Vol. 1 (American Mathematical Society, Providence).

Ma R. (2003). Existence of positive solutions for superlinear semipositone m-point boundary value problems, *Proc. Edinburgh Math. Soc.* **46** pp. 279–292.

Nelms C. (2015). Comparison of smallest eigenvalue for certain fifth order boundary value problems, *Adv. Dyn. Syst. Appl.* **10**, No. 1, pp. 77–84.

Neugebauer J. T. (2011). Methods of extending lower order problems to higher order problems in the context of smallest eigenvalue comparisons, *Electron. J. Qual. Theory Differ. Equ.* **99**, pp. 1–16.

Swartz C. (1992). *An Introduction to Functional Analysis* (Marcel Dekker, New York).

Travis C. C. (1973). Comparison of eigenvalues for linear differential equations order $2n$, *Trans. Amer. Math. Soc.* **177**, pp. 363–374.

Zeidler E. (1986). *Nonlinear Functional Analysis and its Applications I: Fixed-Point Theorems* (Springer-Verlag, New York).

Chapter 7

BVP's for Functional Differential Equations

In this chapter, we first obtain the existence of solutions for a second order functional differential equation satisfying an initial function value and final boundary value via an application of the Leray-Schauder Nonlinear Alternative, then we study periodic solutions for several first order functional differential equations.

7.1 Existence of Solutions for BVP's for Second Order Functional Differential Equations

In many applications, one assumes the system under consideration is governed by a principle of causality; that is, the future state of the system is independent of the past states and is determined solely by the present. If it is also assumed that the system is governed by an equation involving the state and the rate of change of the state, then, generally, one is considering either ordinary or partial differential equations. However, under closer scrutiny, it becomes apparent that the principle of causality is often only a first approximation to the true situation and that a more realistic model would include some of the past states of the system. Also, in some problems, it is meaningless not to have dependence on the past.

The simplest type of past dependence in a differential equation is that in which the past dependence is through the state variable and not the derivative of the state variable, the so-called retarded functional differential equations or retarded differential-difference equations such as

$$x'(t) = f(t, x(t), x(t-r)).$$

For example, the differential-difference equation,

$$x'(t) = -\alpha x(t-1)[1 + x(t)]$$

has been used in the study of the distribution of primes, with variants of this equation used as models in the theory of growth of a single species.

In this section, we will apply the Topological Transversality Method of Granas in the form of the Leray-Schauder Nonlinear Alternative to obtain solutions of

$$\begin{cases} x''(t) = f(t, x(t+\theta), x'(t)), & 0 \le t \le T, \\ x(s) = \varphi(s), & -r \le s \le 0, \\ x(T) = A, \end{cases}$$

where $-r \le \theta \le 0$ is fixed.

For $r > 0$, let $C[-r, 0]$ be the continuous real-valued functions such that, for all $\varphi \in C[-r, 0]$, we define the norm

$$\|\varphi\|_{[-r,0]} = \sup_{-r \le \theta \le 0} |\varphi(\theta)|,$$

which gives the usual Banach space.

Let $T > 0$ be fixed. Then, for each continuous $x : [-r, T] \to \mathbb{R}$ and each $t \in [0, T]$, we denote by x_t the element of $C[-r, 0]$ (i.e., $x_t : [-r, 0] \to \mathbb{R}$ is continuous), defined by

$$x_t(\theta) = x(t+\theta), \quad -r \le \theta \le 0.$$

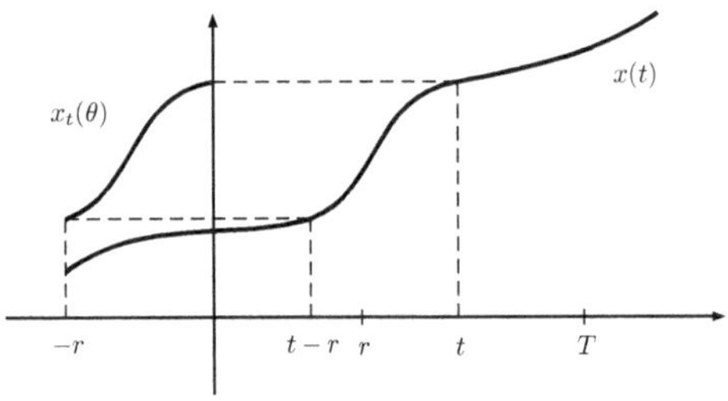

Fig. 7.1 The graphs of $x(t)$ and $x_t(\theta)$.

If I is a compact interval of $\mathbb{R}$, we will let $C(I, \mathbb{R})$ and $C^{(1)}(I, \mathbb{R})$ denote the usual spaces with respective norms,

$$\|x\|_I = \sup_{t \in I} |x(t)|, \quad \|x\|_1 = \max\{\|x\|_I, \|x'\|_I\}.$$

We consider solutions of BVP's for

$$\begin{cases} x''(t) = f(t, x_t, x'(t)), \quad 0 \le t \le T, & (7.1) \\ x_0 = \varphi, \quad \text{(i.e., } x(0+s) = \varphi(s), \ -r \le s \le 0), & \\ x(T) = A, & (7.2) \end{cases}$$

where $f : [0, T] \times C[-r, 0] \times \mathbb{R} \to \mathbb{R}$ is continuous, $\varphi \in C[-r, 0]$, and $A \in \mathbb{R}$.

By a *solution* x of (7.1), (7.2), we mean $x \in C([-r,T],\mathbb{R}) \cap C^{(2)}([0,T],\mathbb{R})$, and x satisfies (7.1) and (7.2).

If $G(t,s)$ is the Green's function for

$$x'' = 0, \quad x(0) = x(t) = 0,$$

that is, (from Section 2.3 of Chapter 2),

$$G(t,s) = \frac{1}{T}\begin{cases} (t-T)s, & 0 \le s \le t \le T, \\ t(s-T), & 0 \le t \le s \le T, \end{cases}$$

and then by Theorem 3.1, (7.1), (7.2) is equivalent to

$$x(t) = \begin{cases} \int_0^T G(t,s) f(s, x_s, x'(s))\, ds + \frac{A-\varphi(0)}{T} t + \varphi(0), & 0 \le t \le T, \\ \varphi(t), & -r \le t \le 0. \end{cases} \tag{7.3}$$

As a consequence, to find a solution of (7.1), (7.2) (equivalently (7.3)), it suffices to exhibit a fixed point of the mapping S defined by

$$(Sl)(t) = \begin{cases} \varphi(t), & -r \le t \le 0, \\ \int_0^T G(t,s) f(s, l_s, l'(s))\, ds + \frac{A-\varphi(0)}{T} t + \varphi(0), & 0 \le t \le T, \end{cases}$$

which we shall do by using a type of homotopy or degree method.

Theorem 7.1 (Leray-Schauder Nonlinear Alternative). *Let C be a convex subset of a normed linear space E and assume $0 \in C$. Let $F : C \to C$ be a completely continuous operator and let*

$$\mathcal{E}(F) = \{x \in C \mid x = \lambda F x, \text{ for some } 0 < \lambda < 1\}.$$

Then, either $\mathcal{E}(F)$ is unbounded, or F has a fixed point.

We will apply the above Alternative Theorem in obtaining solutions of (7.1), (7.2).

Theorem 7.2. *Let $f : [0,T] \times C[-r,0] \times \mathbb{R} \to \mathbb{R}$ be continuous. Suppose, there exists $M > 0$ such that $\|x\|_1 \le M$, for all solutions x of*

$$\begin{cases} x''(t) = \lambda f(t, x_t, x'(t)), \quad 0 \le t \le T, \quad 0 \le \lambda \le 1, & (7.1_\lambda) \\ x_0 = \varphi, & \\ x(T) = \lambda A, & (7.2_\lambda) \end{cases}$$

for every $0 \le \lambda \le 1$. Then, the BVP (7.1_1), (7.2_1) (i.e., (7.1), (7.2)), has at least one solution.

Proof. We will first establish the result for the case $\varphi(0) = 0$.

Let $E = C^{(1)}([0,T],\mathbb{R})$ be equipped with the $\|\cdot\|_1$ norm, and let the convex subset C be defined by

$$C = \{x \in E \,|\, x(0) = 0\}.$$

Of course, $0 \in C$. Next, define an operator $F : C \to E$ by

$$(Fx)(t) = \int_0^T G(t,s) f(s, x_s, x'(s))\, ds + \frac{A}{T}t, \quad 0 \le t \le T,$$

where

$$x_s(\theta) = \begin{cases} x(s+\theta), & s+\theta \ge 0, \\ \varphi(s+\theta), & s+\theta < 0. \end{cases}$$

Note that $(Fx)(0) = 0$, for all $x \in C$, so that $Fx \in C$. Hence, $F : C \to C$.

We now proceed to show that F is completely continuous; in particular, that F is continuous and maps bounded subsets of C into precompact sets.

The continuity of f along with the integral yields the continuity of F.

Next, choose a bounded $B \subseteq C$. Then, there exists $b \ge 0$ such that $\|x\|_1 \le b$, for $x \in B$; i.e., $\|x\|_{[0,T]} \le b$ and $\|x'\|_{[0,T]} \le b$, for all $x \in B$. If follows that for each $x \in B$ and $t_1, t_2 \in [0,T]$,

$$|x(t_1) - x(t_2)| \le b|t_1 - t_2|,$$

so that B is a uniformly equicontinuous family in the "function value" sense.

Now, let $\widehat{B} \subset C[-r,0]$ be defined by $\widehat{B} = \{x_t | x \in B\}$. We wish to show that $\widehat{B}$ itself is precompact by showing there exists a compact $D \subset C[-r,0]$ such that $\widehat{B} \subseteq D \subset C[-r,0]$.

To carry this out, it suffices to show that $\widehat{B}$ is uniformly bounded and equicontinuous.

So, let $x \in B$ and $t \in [0,T]$. Then,

$$\sup_{-r\le\theta\le0} |x_t(\theta)| = \sup_{-r\le\theta\le0} |x(t+\theta)| \le \max\{b, \|\varphi\|_{[-r,0]}\},$$

and hence $\widehat{B}$ is uniformly bounded.

For the equicontinuity of $\widehat{B}$, let $\theta_1, \theta_2 \in [-r,0]$ and consider for any $x_t \in \widehat{B}$,

$$|x_t(\theta_1) - x_t(\theta_2)| = \begin{cases} |x(t+\theta_1) - x(t+\theta_2)|, & t+\theta_1 \ge 0,\ t+\theta_2 \ge 0, \ \text{(a)} \\ |\varphi(t+\theta_1) - x(t+\theta_2)|, & t+\theta_1 < 0,\ t+\theta_2 \ge 0, \ \text{(b)} \\ |\varphi(t+\theta_1) - \varphi(t+\theta_2)|, & t+\theta_1 < 0,\ t+\theta_2 < 0, \ \text{(c)} \\ |x(t+\theta_1) - \varphi(t+\theta_2)|, & t+\theta_1 \ge 0,\ t+\theta_2 < 0. \ \text{(d)} \end{cases}$$

Let $\varepsilon > 0$ be given.

For Case (a), $|x(t+\theta_1) - x(t+\theta_2)| \leq b|\theta_1 - \theta_2| < \varepsilon$, for all $|\theta_1 - \theta_2| < \delta \equiv \frac{\varepsilon}{b}$.

For Case (b), $|\varphi(t+\theta_1) - x(t+\theta_2)| \leq |\varphi(t+\theta_1) - \varphi(0)| + |x(0) - x(t+\theta_2)|$. Since φ is uniformly continuous on $[-r, 0]$ and B is uniformly equicontinuous in the "function value" sense, there exist $\delta_1(\varepsilon)$ and $\delta_2(\varepsilon)$ such that $|\varphi(s_1) - \varphi(s_2)| < \frac{\varepsilon}{2}$ when $|s_1 - s_2| < \delta_1$, and $|x(r_1) - x(r_2)| < \frac{\varepsilon}{2}$ when $|r_1 - r_2| < \delta_2$. So choose $\delta(\varepsilon) = \min\{\delta_1, \delta_2\}$. Then, for $|\theta_1 - \theta_2| < \delta$, we have $|t + \theta_1 - 0| < \delta_1$ and $|t + \theta_2 - 0| < \delta_2$ in this case, and so

$$|\varphi(t+\theta_1) - \varphi(0)| + |x(0) - x(t+\theta_2)| < \frac{\varepsilon}{2} + \frac{\varepsilon}{2} = \varepsilon.$$

Case (d) is analogous to (b).

For Case (c), the situation is independent of x, and so for $|\theta_1 - \theta_2|$ small, by the uniform continuity of φ, $|x_t(\theta_1) - x_t(\theta_2)| < \varepsilon$.

It follows from the four cases that, for each $\varepsilon > 0$, there exists $\delta(\varepsilon) > 0$ such that $|\theta_1 - \theta_2| < \delta$ implies $|x_t(\theta_1) - x_t(\theta_2)| < \varepsilon$, for all $x_t \in \widehat{B}$. In particular, $\widehat{B}$ is uniformly equicontinuous.

We can then apply the Arzela-Ascoli Theorem to conclude that $\overline{\widehat{B}}$ is compact. That is, there exists a compact $D \subset C[-r, 0]$ such that $\widehat{B} \subseteq D \subset C[-r, 0]$.

Now we choose a bounded sequence $\{x_\nu\}$ in C. (This sequence is like the bounded set B above.) Say, $\|x_\nu\|_1 \leq b$, for all $\nu \in \mathbb{N}$. Also, there exists a compact $D_1 \subset C[-r, 0]$ such that $\{(x_\nu)_t\} \subseteq D_1 \subset C[-r, 0]$ for all $t \in [0, T]$.

Now by Tychonoff's Theorem, the space $X = [0, T] \times D_1 \times [-b_1, b_1]$ is compact in $[0, T] \times C[-r, 0] \times \mathbb{R}$. Recall from Exercise 26 of Chapter 2 that

$$\int_0^T |G(t,s)| ds \leq \frac{T^2}{8} \quad \text{and} \quad \int_0^T \left| \frac{\partial}{\partial t} G(t,s) \right| ds \leq \frac{T}{2}.$$

Set $\theta = \max_{(t,u,v) \in X} |f(t, u, v)|$ and $K = \max\{\frac{\theta T^2}{8}, \frac{\theta T}{2}\}$. Then, for each $\nu \in \mathbb{N}$ and $t \in [0, T]$,

$$\begin{aligned} |(Fx_\nu)(t)| &\leq \int_0^T |G(t,s)||f(s, (x_\nu)_s, x_\nu'(s))| \, ds + |A| \\ &\leq \frac{\theta T^2}{8} + |A| \leq K + |A|, \end{aligned}$$

so that $\|Fx_\nu\|_{[0,T]} \leq K + |A|$, and also

$$\begin{aligned}|(Fx_\nu)'(t)| &\leq \int_0^T \left|\frac{\partial}{\partial t}G(t,s)\right| |f(s,(x_\nu)_s, x_\nu'(s))|\,ds + \left|\frac{A}{T}\right| \\ &\leq \frac{\theta T}{2} + \left|\frac{A}{T}\right| \leq K + \left|\frac{A}{T}\right|,\end{aligned}$$

so that $\|(Fx_\nu)'\|_{[0,T]} \leq K + |\frac{A}{T}|$. Therefore, $\{Fx_\nu\}$ is uniformly bounded in $\|\cdot\|_1$-norm. The sequence $\{Fx_\nu\|$ is also equicontinuous with respect to the $\|\cdot\|_1$-norm, because for each $t_1, t_2 \in [0,T]$ and for each ν,

$$|(Fx_\nu)(t_1) - (Fx_\nu)(t_2)| = \left|\int_{t_1}^{t_2} (Fx_\nu)'(s)\,ds\right| \leq \overline{K}|t_1 - t_2|,$$

where $\overline{K} = K + |\frac{A}{T}|$, and

$$\begin{aligned}|(Fx_\nu)'(t_1) - (Fx_\nu)'(t_2)| &= \left|\int_{t_1}^{t_2} (Fx_\nu)''(s)\,ds\right| \\ &= \left|\int_{t_1}^{t_2} f(s,(x_\nu)_s, x_\nu'(s))\,ds\right| \\ &\leq \theta|t_1 - t_2|.\end{aligned}$$

By the Arzela-Ascoli Theorem, $\{Fx_\nu\}$ has a subsequence that converges in $\|\cdot\|_1$-norm, and we conclude that F is completely continuous.

Finally by the assumptions on (7.1_λ), (7.2_λ), the set $\mathcal{E}(F) = \{x \in C \mid x = \lambda Fx,\ 0 < \lambda < 1\}$ is bounded. By Theorem 7.1, F has a fixed point $x \in C$. It follows that the function

$$z(t) = \begin{cases} x(t), & t \in [0,T], \\ \varphi(t), & t \in [-r,0], \end{cases}$$

is a solution of (7.1), (7.2) for this case of $\varphi(0) = 0$.

For the case $\varphi(0) \neq 0$, we make the transformation $y = x - \varphi(0)$. Then the BVP (7.1), (7.2) is transformed to

$$\begin{cases} y''(t) = x''(t) = f(t, x_t, x'(t)) = f(t, y_t + \varphi(0), y'(t)) \\ \qquad := \tilde{f}(t, y_t, y'(t)), \quad 0 \leq t \leq T, & (\widetilde{7.1}) \\ y_0 = x_0 - \varphi(0) = \varphi - \varphi(0) := \widetilde{\varphi}, \\ y(T) = A - \varphi(0) := \widetilde{A}, & (\widetilde{7.2}) \end{cases}$$

and we note that $\widetilde{\varphi}(0) = 0$.

By the first case of the proof, $(\widetilde{7.1})$, $(\widetilde{7.2})$ has a solution $y(t)$. Then $x(t) := y(t) + \varphi(0))$ is a solution of (7.1), (7.2), because

$$\begin{cases} x''(t) = y''(t) = \widetilde{f}(t, y_t, y'(t)) = f(t, y_t + \varphi(0), y'(t)) \\ \qquad = f(t, x_t, x'(t)), \quad 0 \leq t \leq T, \\ x_0 = y_0 + \varphi(0) = \varphi - \varphi(0) + \varphi(0) = \varphi, \\ x(T) = y(T) + \varphi(0) = A - \varphi(0) + \varphi(0) = A. \end{cases}$$

The proof is complete. □

7.2 Periodic Solutions for First Order Functional Differential Equations

Let $T > 0$ be fixed. In this section, we are concerned with the existence of T-periodic solutions of the first order functional differential equations

$$u'(t) = a(t)g(u(t))u(t) - b(t)f(u(t - \tau(t))) \tag{7.4}$$

and

$$u'(t) = a(t)g(u(t))u(t) - \lambda b(t)f(u(t - \tau(t))), \tag{7.5}$$

where $a, b, \tau \in C(\mathbb{R}, \mathbb{R})$ are T-periodic functions, $f, g \in C(\mathbb{R}, \mathbb{R})$ and λ is a positive parameter. These equations, or their variations, appear in a number of applications, such as in the model of blood cell productions in an animal [Glass and Mackey (1988); Wazewska-Czyzewska and Lasota (1988)], the control of testosterone levels in the blood stream [Murray (1989)], and so on. When $g(x)$ is constant or bounded, the existence of T-periodic solutions of (7.4) and (7.5) has been extensively investigated in the literature. When $g(x)$ is not necessarily bounded, papers [Jin and Wang (2010); Wang (2004)] studied the existence of one and two positive T-periodic solutions of (3.11) by using the fixed point index theory. No result is obtained in [Jin and Wang (2010); Wang (2004)] for the existence of three or more T-periodic solutions.

In this section, we first establish the existence of at least one T-periodic solution of (7.4) by assuming the existence of a pair of lower and upper solutions. Then, applying our result, we derive several explicit conditions for the existence of T-periodic solutions of (7.4) and (7.5) (without assuming the existence of lower and upper solutions). In particular, our Corollary 7.2 provides sufficient conditions for the existence of multiple (even infinitely many) T-periodic solutions of (7.4).

Let the Banach space

$$X := \{u(t) \in C(\mathbb{R}, \mathbb{R}) \ : \ u(t) = u(t+T) \quad \text{for } t \in \mathbb{R}\}$$

be equipped with the norm $||u|| = \sup_{t\in[0,T]} |u(t)|$.

Definition 7.1. A function $\alpha \in X \cap C^{(1)}(\mathbb{R}, \mathbb{R})$ is said to be a *lower solution* of (7.4) if

$$\alpha'(t) \geq a(t)g(\alpha(t))\alpha(t) - b(t)f(\alpha(t - \tau(t))).$$

Similarly, a function $\beta \in X \cap C^1(\mathbb{R}, \mathbb{R})$ is said to be an *upper solution* of (7.4) if

$$\beta'(t) \leq a(t)g(\beta(t))\beta(t) - b(t)f(\beta(t - \tau(t))). \tag{7.6}$$

Now, we state our first existence result.

Theorem 7.3. *Assume that*

(E1) *$f(x)$ is nondecreasing in x;*
(E2) *(7.4) has a lower solution $\alpha(t)$ and an upper solution $\beta(t)$ satisfying*

$$\alpha(t) \leq \beta(t) \quad \text{for } t \in \mathbb{R}. \tag{7.7}$$

Moreover, $a(t)g(u(t)) \geq 0$ and $a(t)g(u(t)) \not\equiv 0$ on $\mathbb{R}$ for $u \in X$ with $\alpha(t) \leq u(t) \leq \beta(t)$.

Then, (7.4) *has at least one T-periodic solution $u(t)$ satisfying*

$$\alpha(t) \leq u(t) \leq \beta(t) \quad \text{for } t \in \mathbb{R}. \tag{7.8}$$

The following corollaries are consequences of Theorem 7.3 and provide explicit conditions for the existence of T-periodic solutions of (7.4) and (7.5).

Corollary 7.1. *Assume that* (E1) *holds and there exist $-\infty < \delta_1 < \delta_2 < \infty$ such that*

$$a(t)g(\delta_1)\delta_1 \leq b(t)f(\delta_1) \quad \text{and} \quad a(t)g(\delta_2)\delta_2 \geq b(t)f(\delta_2) \quad \text{on } \mathbb{R}. \tag{7.9}$$

Moreover, $a(t)g(u(t)) \geq 0$ and $a(t)g(u(t)) \not\equiv 0$ on $\mathbb{R}$ for $u \in X$ with $\delta_1 \leq u(t) \leq \delta_2$. Then, (7.4) *has at least one T-periodic solution $u(t)$ satisfying $\delta_1 \leq u(t) \leq \delta_2$ on $\mathbb{R}$.*

Corollary 7.2. *Let $k \in \{1, 2, \ldots\}$. Assume that* (E1) *holds and there exist $\mu_i, \eta_i \in \mathbb{R}$, $i = 1, \ldots, k$, such that*

$$\mu_1 < \eta_1 < \mu_2 < \eta_2 < \ldots < \mu_k < \eta_k,$$

and

$$a(t)g(\mu_i)\mu_i \leq b(t)f(\mu_i) \quad \text{and} \quad a(t)g(\eta_i)\eta_i \geq b(t)f(\eta_i) \quad \text{on } \mathbb{R}.$$

Moreover, $a(t)g(u(t)) \geq 0$ and $a(t)g(u(t)) \not\equiv 0$ on $\mathbb{R}$ for $u \in X$ with $\mu_i \leq u(t) \leq \eta_i$, $i = 1, \ldots, k$. Then, (7.4) *has at least k T-periodic solutions $u_i(t)$ satisfying $\mu_i \leq u_i(t) \leq \eta_i$ on $\mathbb{R}$ for $i = 1, \ldots, k$.*

Corollary 7.3. *Assume that* (E1) *holds and there exist $-\infty < \delta_1 < \delta_2 < \infty$ such that $b(t)f(\delta_1) \geq 0$, $b(t)f(\delta_2) \geq 0$, and $\underline{\lambda} < 1 < \overline{\lambda}$, where*

$$\underline{\lambda} = \frac{\delta_1 g(\delta_1)}{f(\delta_1)} \max_{t \in [0,T]} \frac{a(t)}{b(t)} \quad \text{and} \quad \overline{\lambda} = \frac{\delta_2 g(\delta_2)}{f(\delta_2)} \min_{t \in [0,T]} \frac{a(t)}{b(t)}. \tag{7.10}$$

Moreover, $a(t)g(u(t)) \geq 0$ and $a(t)g(u(t)) \not\equiv 0$ on $\mathbb{R}$ for $u \in X$ with $\delta_1 \leq u(t) \leq \delta_2$. Then, for each $\underline{\lambda} \leq \lambda \leq \overline{\lambda}$, (7.5) *has at least one T-periodic solution $u(t)$ satisfying $\delta_1 \leq u(t) \leq \delta_2$ on $\mathbb{R}$.*

Corollary 7.4. *Assume that* (E1) *holds, $f(0) > 0$, and $g(x) = e^x$. Then, for each $L > 0$, there exists $\lambda^* = \frac{L(1-\sigma)}{f(L)\int_0^T b(s)ds} > 0$ such that* (7.5) *has at least one positive T-periodic solution $u(t)$ satisfying $0 < ||u|| \leq L$ for each $0 < \lambda \leq \lambda^*$, where*

$$\sigma = e^{-\int_0^T a(s)ds} < 1. \tag{7.11}$$

Remark 7.1. In Corollaries 7.1–7.3, if $\delta_1 > 0$ and $\lambda_1 > 0$, the corresponding solutions are obviously positive.

Proof of Theorem 7.3. Let α and β be given as in (E2). For any $u \in X$, define $p_{\alpha,\beta}$, $g_{\alpha,\beta}$, and $b(t)f_{\alpha,\beta}$ by

$$p_{\alpha,\beta}(u(t)) = \max\{\alpha(t), \min\{u(t), \beta(t)\}\}, \tag{7.12}$$

$$g_{\alpha,\beta} = g(p_{\alpha,\beta}(u(t))), \tag{7.13}$$

and

$$b(t)f_{\alpha,\beta}(u(t)) = b(t)f(p_{\alpha,\beta}(u(t))) - \frac{u(t+\tau(t)) - p_{\alpha,\beta}(u(t+\tau(t)))}{1 + u^2(t+\tau(t))}. \tag{7.14}$$

Then,

$$\begin{cases} p_{\alpha,\beta}(u(t)) = \alpha(t) & \text{if } u(t) \leq \alpha(t), \\ p_{\alpha,\beta}(u(t)) = \beta(t) & \text{if } u(t) \geq \beta(t), \\ \alpha(t) \leq p_{\alpha,\beta}(u(t)) \leq \beta(t) & \text{for all } u \in X, \end{cases} \tag{7.15}$$

and $g_{\alpha,\beta}, f_{\alpha,\beta} : X \to X$ are continuous, and $g_{\alpha,\beta}(X)$ and $f_{\alpha,\beta}(X)$ are bounded. Consider the modified equation

$$u'(t) = a(t)g_{\alpha,\beta}(u(t))u(t) - b(t)f_{\alpha,\beta}(u(t-\tau(t))). \tag{7.16}$$

Define an operator $T_{\alpha,\beta} : X \to X$ by

$$T_{\alpha,\beta}u(t) = \int_t^{t+T} G_{\alpha,\beta}(t,s;u)b(s)f_{\alpha,\beta}(u(s-\tau(s)))ds, \tag{7.17}$$

where

$$G_{\alpha,\beta}(t,s;u) = \frac{\exp(-\int_t^s a(\tau)g_{\alpha,\beta}(u(\tau))d\tau)}{1-\exp(-\int_0^T a(\tau)g_{\alpha,\beta}(u(\tau))d\tau)}. \tag{7.18}$$

Then, it is easy to see that $T : X \to X$ is continuous and $T(X)$ is relatively compact. Moreover, as in the proof of [Jin and Wang (2010), Lemma 2.3], we can show that a T-periodic solution of (7.16) is equivalent to a fixed point of T in X. Now, applying the Schauder fixed point theorem, we conclude that (7.16) has a T-periodic solution $u \in X$.

Now, to finish the proof, it suffices to show that $u(t)$ satisfies (7.8). Assume, to the contrary, that $u(t^*) > \beta(t^*)$ for some $t^* \in \mathbb{R}$. Without loss of generality, we may assume that $w(t) := u(t) - \beta(t)$ is maximized at t^*. Then, $u'(t^*) = \beta'(t^*)$. On the other hand, from (7.6), (7.13)–(7.16), (E1), and (E2), we have

$$\begin{aligned}
u'(t^*) &= a(t^*)g_{\alpha,\beta}(u(t^*))u(t^*) - b(t^*)f_{\alpha,\beta}(u(t^*-\tau(t^*))) \\
&= a(t^*)g(p_{\alpha,\beta}(u(t^*)))u(t^*) - b(t^*)f(p_{\alpha,\beta}(u(t^*-\tau(t^*)))) \\
&\quad + \frac{u(t^*) - p_{\alpha,\beta}(u(t^*))}{1+u^2(t^*)} \\
&\geq a(t^*)g(\beta(t^*))u(t^*) - b(t^*)f(\beta(t^*-\tau(t^*))) + \frac{u(t^*)-\beta(t^*)}{1+u^2(t^*)} \\
&\geq a(t^*)g(\beta(t^*))\beta(t^*) - b(t^*)f(\beta(t^*-\tau(t^*))) + \frac{u(t^*)-\beta(t^*)}{1+u^2(t^*)} \\
&\geq \beta'(t^*) + \frac{u(t^*)-\beta(t^*)}{1+u^2(t^*)} > \beta'(t^*).
\end{aligned}$$

This is a contradiction. Thus, $u(t) \leq \beta(t)$ for $t \in \mathbb{R}$. Similarly, we can show that $u(t) \geq \alpha(t)$ on $\mathbb{R}$. Hence, (7.8) holds. This completes the proof of the theorem. □

Proof of Corollary 7.1. Let $\alpha(t) = \delta_1$ and $\beta(t) = \delta_2$ for $t \in \mathbb{R}$. Then, in view of (7.9), it is easy to see that $\alpha(t)$ and $\beta(t)$ are lower and upper solutions of (7.4), respectively, such that (E2) holds. The conclusion now follows directly from Theorem 7.3. □

Proof of Corollary 7.2. Corollary 7.2 follows readily from Corollary 7.1. □

Proof of Corollary 7.3. Let $\alpha(t) = \delta_1$ and $\beta(t) = \delta_2$ for $t \in \mathbb{R}$. Then, in view of (7.10), it is easy to see that, for $\underline{\lambda} \leq \lambda \leq \overline{\lambda}$, $\alpha(t)$ and $\beta(t)$ are lower and upper solutions of (7.5), respectively, and $a(t)g(u(t)) \geq 0$ and $a(t)g(u(t)) \not\equiv 0$ on $\mathbb{R}$ for $u \in X$ with $\alpha(t) \leq u(t) \leq \beta(t)$. Now applying Theorem 7.3 to Eq. (7.5) obtains the desired conclusion. □

Proof of Corollary 7.4. Let $g(x) = e^x$. For any $L > 0$, let $p_{0,L}$, $g_{0,L}$, and $G_{0,L}$ be defined by (7.12), (7.13), and (7.18) with $\alpha = 0$ and $\beta = L$, respectively. Clearly,

$$1 \leq g_{0,L}(u) \leq e^L \quad \text{for } u \in X,$$

and

$$\frac{\sigma^{e^L}}{1 - \sigma^{e^L}} \leq G_{0,L}(t, s; u) \leq \frac{1}{1 - \sigma} \quad \text{for } t \leq s \leq t + T \text{ and } u \in X, \tag{7.19}$$

where σ is defined in (7.11). For any $\lambda > 0$, define a compact operator $S_\lambda u(t) : X \to X$ by

$$(S_\lambda u)(t) = \lambda f(L) \int_t^{t+T} G_{0,L}(t, s; u) b(s) ds.$$

An application of the Schauder fixed point theorem shows that S_λ has a fixed point $\beta_\lambda \in X$. Then,

$$\beta_\lambda(t) = \lambda f(L) \int_t^{t+T} G_{0,L}(t, s; u) b(s) ds.$$

From (7.19), we have

$$\beta_\lambda(t) \geq \frac{\lambda f(L) \sigma^{e^L}}{1 - \sigma^{e^L}} \int_0^T b(s) ds > 0 \quad \text{if } \lambda > 0,$$

and

$$\beta_\lambda(t) \leq \frac{\lambda f(L)}{1 - \sigma} \int_0^T b(s) ds \leq L \quad \text{if } \lambda \leq \lambda^* := \frac{L(1 - \sigma)}{f(L) \int_0^T b(s) ds}. \tag{7.20}$$

Then, for $0 < \lambda \leq \lambda^*$, $f(\beta_\lambda(t - \tau(t))) \leq f(L)$ by (E1), and thus

$$\begin{aligned}
\beta_\lambda'(t) &= a(t) g_{0,L}(\beta_\lambda(t)) \beta_\lambda(t) - \lambda f(L) b(t) \\
&= a(t) e^{\beta_\lambda(t)} \beta_\lambda(t) - \lambda f(L) b(t) \\
&\leq a(t) e^{\beta_\lambda(t)} \beta_\lambda(t) - \lambda f(\beta_\lambda(t - \tau(t))) b(t).
\end{aligned}$$

This implies that $\beta_\lambda(t)$ is an upper solution of (7.5). On the other hand, it is clear that $\alpha(t) \equiv 0$ is a lower solution of (7.5) satisfying $\alpha(t) \le \beta_\lambda(t)$ on $\mathbb{R}$. Then by Theorem 7.3, (7.5) has a solution $u(t)$ satisfying $0 \le u(t) \le \beta_\lambda(t)$ on $\mathbb{R}$ for $0 < \lambda \le \lambda^*$. Finally, from the assumption that $f(0) > 0$ and (7.20), it is easy to see that $u(t)$ is positive and $0 < ||u|| \le L$. This completes the proof of the corollary. □

We conclude this section with the following three examples.

Example 7.1. Let $f, g \in C(\mathbb{R}, \mathbb{R})$ satisfy that $f(x)$ is nondecreasing in x,

$$f(x) = \begin{cases} 2 & \text{for } x \in (-\infty, 1), \\ 2x - n + 1 & \text{for } x \in [2n-1, 2n], \ n = 1, 2, \ldots, \end{cases} \tag{7.21}$$

and

$$g(x) = \begin{cases} 1 & \text{for } x \in (-\infty, 1), \\ 2x - 4n + 3 & \text{for } x \in [2n-1, 2n], \ n = 1, 2, \ldots. \end{cases} \tag{7.22}$$

We claim that the functional differential equation

$$u'(t) = g(u(t))u(t) - f(u(t - \sin t)) \tag{7.23}$$

has an infinite number positive 2π-periodic solutions $u_i(t)$ satisfying $2i-1 \le u_i(t) \le 2i$ on $\mathbb{R}$ for $i = 1, 2, \ldots$.

In fact, with $a(t) = b(t) = 1$, $T = 2\pi$, and $\tau(t) = \sin t$, it is clear that (7.23) is of the form (7.4). By the assumption, (E1) obviously holds. Let $\mu_i = 2i - 1$ and $\eta_i = 2i$ for $i = 1, 2, \ldots$. Then,

$$\mu_1 < \eta_1 < \mu_2 < \eta_2 < \cdots < \mu_i < \eta_i < \cdots.$$

Moreover, from (7.21) and (7.22), we see that

$$g(\mu_i)\mu_i \le f(\mu_i) \quad \text{and} \quad g(\eta_i)\eta_i \ge f(\eta_i) \quad \text{on } \mathbb{R}, \ i = 1, 2, \ldots,$$

and $g(u(t)) \ge 0$ and $g(u(t)) \not\equiv 0$ on $\mathbb{R}$ for $u \in X$ with $\mu_i \le u(t) \le \eta_i$, $i = 1, 2 \ldots$. Thus, all the assumptions of Corollary 7.2 are satisfied with any $k \in \{1, 2, \ldots\}$. The conclusion then follows from Corollary 7.2.

Example 7.2. We claim that for each $1/2 < \lambda < e/2$, the functional differential equation

$$u'(t) = e^{|\cos(u(t))|}u(t) - 2\lambda u(t - \sin t) \tag{7.24}$$

has an infinitely many 2π-periodic solutions $u_k(t)$, $k = 1, 2, \ldots$, satisfying

$$k\pi + \frac{\pi}{2} \le u_k(t) \le k\pi + \pi \quad \text{for } t \in \mathbb{R}. \tag{7.25}$$

In fact, with $a(t) = b(t) = 1$, $T = 2\pi$, $\tau(t) = \sin t$, $f(x) = 2x$, and $g(x) = e^{|\cos x|}$, we see that (7.25) is of the form (7.5) and (E1) holds.

For any $k \in \mathbb{N}$, let $\delta_1 = k\pi + \pi/2$ and $\delta_2 = k\pi + \pi$. Then, $\delta_1 < \delta_2$, $f(\delta_1) > 0$, and $f(\delta_2) > 0$. For $\underline{\lambda}$ and $\overline{\lambda}$ defined in (7.10), by a simple calculation, we have $\underline{\lambda} = 1/2$ and $\overline{\lambda} = e/2$. Clearly, $g(u(t)) > 0$ on $\mathbb{R}$ for $u \in X$ with $\delta_1 \leq u(t) \leq \delta_2$. Thus, all the assumptions of Corollary 7.3 are satisfied. By Corollary 7.3, we see that for each $1/2 < \lambda < e/2$, (7.24) has a 2π-periodic solution $u_k(t)$ satisfying (7.25) for any $k \in \mathbb{N}$.

Example 7.3. We claim that, for any $L > 0$, then exists $\lambda^* = \frac{L(1-e^{-2\pi})}{2\pi(L^3+1)} > 0$ such that the functional differential equation

$$u'(t) = e^{u(t)}u(t) - \lambda(u^3(t - \cos(t)) + 1) \tag{7.26}$$

has at least one positive 2π-periodic solution $u(t)$ satisfying $0 < ||u|| \leq L$ for each $0 < \lambda \leq \lambda^*$.

In fact, with $a(t) = b(t) = 1$, $T = 2\pi$, $\tau(t) = \cos t$, $f(x) = x^3 + 1$, and $g(x) = e^x$, it is clear that (7.26) is of the form (7.5), (E1) holds, and $f(0) > 0$. Moreover, from (7.11), we have $\sigma = e^{-2\pi}$. The conclusion now follows from Corollary 7.4.

7.3 Periodic Solutions for Functional Differential Equations with Sign-Changing Nonlinearities

Let $T > 0$ be fixed. In this section, we study the existence of nontrivial T-periodic solutions of the first order functional differential equation

$$u'(t) = -a(t)u(t) + b(t)f(u(t - \tau(t))), \tag{7.27}$$

where $\tau \in C(\mathbb{R}, \mathbb{R})$ and $a, b \in C(\mathbb{R}, \mathbb{R}^+)$ with $\mathbb{R}^+ = [0, \infty)$ are T periodic functions, and $f \in C(\mathbb{R}, \mathbb{R})$. As by-products of our results, we also derive conditions for the existence of nontrivial T-periodic solutions of the eigenvalue problem

$$u'(t) = -a(t)u(t) + \lambda b(t)f(u(t - \tau(t))), \tag{7.28}$$

where λ is a positive parameter. Here, by a nontrivial *T-periodic solution* of Eq. (7.27), we mean a nontrivial function $u \in C^1(\mathbb{R}, \mathbb{R})$ such that $u(t+T) = u(t)$ for $t \in \mathbb{R}$ and $u(t)$ satisfies (7.27) on $\mathbb{R}$. A similar definition also applies to Eq. (7.28). We assume throughout, and without further mention, that the following assumption holds.

(H) The function $g(t) := t - \tau(t)$ is strictly increasing on $\mathbb{R}$,

$$\int_0^T a(v)dv > 0, \quad \text{and} \quad \int_0^T b(v)dv > 0.$$

In this section, by means of topological degree theory, we derive new criteria for the existence of nontrivial T-periodic solutions of Eqs. (7.27) and (7.28) when f is a sign-changing function and not necessarily bounded from below. Our existence conditions are determined by the relationship between the behavior of the quotient $f(x)/x$ for x near 0 and $\pm\infty$ and the smallest positive characteristic value (given by (7.54) below) of a related linear operator $\mathcal{M}$ defined by (7.46) in Subsection 7.3.1. Here, we comment that the linear operator $\mathcal{M}$ plays a very important role in the proofs of our results and its construction is nontrivial.

7.3.1 *Preliminary results*

Let $(X, ||\cdot||)$ be a real Banach space and $\mathcal{L} : X \to X$ be a linear operator. We recall that λ is an *eigenvalue* of $\mathcal{L}$ with a corresponding *eigenvector* ϕ if ϕ is nontrivial and $\mathcal{L}\phi = \lambda\phi$. The reciprocals of eigenvalues are called the *characteristic values* of $\mathcal{L}$. The *spectral radius* of $\mathcal{L}$, denoted by $r_{\mathcal{L}}$, is given by the well known spectral radius formula $r_{\mathcal{L}} = \lim_{n\to\infty} ||\mathcal{L}^n||^{1/n}$. Recall also that a cone P in X is called a *total cone* if $X = \overline{P - P}$.

Let X^* be the dual space of X, P be a total cone in X, and P^* be the dual cone of P, i.e.,

$$P^* = \{l \in X^* \ : \ l(u) \geq 0 \text{ for all } u \in P\}.$$

Let $\mathcal{L}, \mathcal{M} : X \to X$ be two linear compact operators such that $\mathcal{L}(P) \subseteq P$ and $\mathcal{M}(P) \subseteq P$. If their spectral radii $r_{\mathcal{L}}$ and $r_{\mathcal{M}}$ are positive, then by Lemma 6.7, there exist $\phi_{\mathcal{L}}, \phi_{\mathcal{M}} \in P \setminus \{\mathbf{0}\}$ such that

$$\mathcal{L}\phi_{\mathcal{L}} = r_{\mathcal{L}}\phi_{\mathcal{L}} \quad \text{and} \quad \mathcal{M}\phi_{\mathcal{M}} = r_{\mathcal{M}}\phi_{\mathcal{M}}. \tag{7.29}$$

Assume there exists $h \in P^* \setminus \{\mathbf{0}\}$ such that

$$\mathcal{L}^* h = r_{\mathcal{M}} h, \tag{7.30}$$

where $\mathcal{L}^*$ is the dual operator of $\mathcal{L}$. Choose $\delta > 0$ and define

$$P(h, \delta) = \{u \in P \ : \ h(u) \geq \delta||u||\}. \tag{7.31}$$

Then, $P(h, \delta)$ is a cone in X.

The following two lemmas are crucial in the proofs of our theorems. From here on, for any $R > 0$, let $B(\mathbf{0}, R) = \{u \in X \ : \ ||u|| < R\}$ be the open ball of X centered at $\mathbf{0}$ with radius R.

Lemma 7.1. *Assume that the following conditions hold:*

(F1) *there exist* $\phi_{\mathcal{L}}$, $\phi_{\mathcal{M}} \in P \setminus \{\mathbf{0}\}$ *and* $h \in P^* \setminus \{\mathbf{0}\}$ *such that* (7.29) *and* (7.30) *hold and* $\mathcal{L}(P) \subseteq P(h, \delta)$;

(F2) $\mathcal{H} : X \to P$ *is a continuous operator and satisfies*

$$\lim_{||u|| \to \infty} \frac{||\mathcal{H}u||}{||u||} = 0;$$

(F3) $\mathcal{F} : X \to X$ *is a bounded continuous operator and there exists* $u_0 \in X$ *such that* $\mathcal{F}u + \mathcal{H}u + u_0 \in P$ *for all* $u \in X$;

(F4) *there exist* $v_0 \in X$ *and* $\epsilon > 0$ *such that*

$$\mathcal{L}\mathcal{F}u \geq r_{\mathcal{M}}^{-1}(1+\epsilon)\mathcal{L}u - \mathcal{L}\mathcal{H}u - v_0 \quad \textit{for all } u \in X.$$

Let $\mathcal{T} = \mathcal{L}\mathcal{F}$. *Then there exists* $R > 0$ *such that the Leray-Schauder degree*

$$\deg(\mathcal{I} - \mathcal{T}, B(\mathbf{0}, R), \mathbf{0}) = 0.$$

Lemma 7.2. *Assume that* (F1) *and the following conditions hold:*

(F2)* $\mathcal{H} : X \to P$ *is a continuous operator and satisfies*

$$\lim_{||u|| \to 0} \frac{||\mathcal{H}u||}{||u||} = 0;$$

(F3)* $\mathcal{F} : X \to X$ *is a bounded continuous operator and there exists* $r_1 > 0$ *such that*

$$\mathcal{F}u + \mathcal{H}u \in P \quad \textit{for all } u \in X \textit{ with } ||u|| < r_1;$$

(F4)* *there exist* $\epsilon > 0$ *and* $r_2 > 0$ *such that*

$$\mathcal{L}\mathcal{F}u \geq r_{\mathcal{M}}^{-1}(1+\epsilon)\mathcal{L}u \quad \textit{for all } u \in X \textit{ with } ||u|| < r_2.$$

Let $\mathcal{T} = \mathcal{L}\mathcal{F}$. *Then there exists* $0 < R < \min\{r_1, r_2\}$ *such that the Leray-Schauder degree*

$$\deg(\mathcal{I} - \mathcal{T}, B(\mathbf{0}, R), \mathbf{0}) = 0.$$

Lemma 7.1 is a generalization of [Han and Wu (2007), Theorem 2.1] and it is proved in [Liu *et al.* (2009), Lemma 2.5] for the case when $\mathcal{L}$ and $\mathcal{M}$ are two specific linear operators, but the proof there also works for any general linear operators $\mathcal{L}$ and $\mathcal{M}$ satisfying (7.29) and (7.30). Lemma 7.2 generalizes [Graef and Kong (2010), Lemma 3.5]. In what follows, we only give the proof of Lemma 7.2.

Proof of Lemma 7.2. For any $\nu > 0$ satisfying

$$\nu(\delta^{-1} r_{\mathcal{M}} ||h|| + ||\mathcal{L}||) < 1, \tag{7.32}$$

from (F2)*, there exists $r_3 > 0$ such that

$$||\mathcal{H}u|| \leq \nu ||u|| \quad \text{for all } u \in X \text{ with } ||u|| < r_3. \tag{7.33}$$

We claim that there exists $0 < R < \min\{r_1, r_2, r_3\}$ such that

$$u - \mathcal{T}u \neq \tau \phi_{\mathcal{L}} \quad \text{for all } u \in \partial B(\mathbf{0}, R) \text{ and } \tau \geq 0. \tag{7.34}$$

If this is not the case, then, for all $0 < R < \min\{r_1, r_2, r_3\}$, there exist $u_1 \in \partial B(\mathbf{0}, R)$ and $\tau_1 \geq 0$ such that

$$u_1 - \mathcal{L}\mathcal{F}u_1 = \tau_1 \phi_{\mathcal{L}}. \tag{7.35}$$

Then, from (7.30) and (F4)*, we have

$$\begin{aligned} h(u_1) &= h(\mathcal{L}\mathcal{F}u_1) + \tau_1 h(\phi_{\mathcal{L}}) \\ &\geq h(\mathcal{L}\mathcal{F}u_1) \\ &\geq r_{\mathcal{M}}^{-1}(1+\epsilon) h(\mathcal{L}u_1) \\ &= r_{\mathcal{M}}^{-1}(1+\epsilon)(\mathcal{L}^* h)(u_1) \\ &= r_{\mathcal{M}}^{-1}(1+\epsilon) r_{\mathcal{M}} h(u_1) \\ &= (1+\epsilon) h(u_1). \end{aligned}$$

Hence, $h(u_1) \leq 0$. This, together with (7.30) and (7.33), implies that

$$\begin{aligned} h(u_1 + \mathcal{L}\mathcal{H}u_1) &= h(u_1) + h(\mathcal{L}\mathcal{H}u_1) \\ &= h(u_1) + (\mathcal{L}^* h)(\mathcal{H}u_1) \\ &\leq (\mathcal{L}^* h)(\mathcal{H}u_1) \\ &= r_{\mathcal{M}} h(\mathcal{H}u_1) \\ &\leq \nu r_{\mathcal{M}} ||h|| \, ||u_1||, \end{aligned} \tag{7.36}$$

From (7.29) and (7.35), we see that

$$\begin{aligned} u_1 + \mathcal{L}\mathcal{H}u_1 &= \mathcal{L}\mathcal{F}u_1 + \mathcal{L}\mathcal{H}u_1 + \tau_1 \phi_{\mathcal{L}} \\ &= \mathcal{L}(\mathcal{F}u_1 + \mathcal{H}u_1) + \tau_1 r_{\mathcal{L}}^{-1} \mathcal{L}\phi_{\mathcal{L}}. \end{aligned}$$

In view of (F1) and (F3)*, we see that $u_1 + \mathcal{L}\mathcal{H}u_1 \in P(h, \delta)$. Thus, by (7.31), we have

$$h(u_1 + \mathcal{L}\mathcal{H}u_1) \geq \delta ||u_1 + \mathcal{L}\mathcal{H}u_1|| \geq \delta ||u_1|| - \delta ||\mathcal{L}\mathcal{H}u_1||,$$

and so

$$||u_1|| \leq \delta^{-1} h(u_1 + \mathcal{L}\mathcal{H}u_1) + ||\mathcal{L}\mathcal{H}u_1||.$$

Hence, from (3.16) and (3.19),

$$\begin{aligned}R = ||u_1|| &\le \delta^{-1}\nu r_{\mathcal{M}}||h||\ ||u_1|| + \nu||\mathcal{L}||\ ||u_1|| \\ &= \nu(\delta^{-1}r_{\mathcal{M}}||h|| + ||\mathcal{L}||)R.\end{aligned}$$

Thus,

$$\nu(\delta^{-1}r_{\mathcal{M}}||h|| + ||\mathcal{L}||) \ge 1,$$

which contradicts (7.32). Therefore, there exists $0 < R < \min\{r_1, r_2, r_3\}$ such that (7.34) holds. Note that the operator $\mathcal{T}$ is compact. The conclusion now readily follows Lemma 6.12, and this completes the proof of the lemma. □

Now, we define

$$G(t,s) = \frac{\exp(\int_t^s a(v)dv)}{\exp(\int_0^T a(v)dv) - 1},$$

$$c = \frac{1}{\exp(\int_0^T a(v)dv) - 1} \quad \text{and} \quad d = \frac{\exp(2\int_0^T a(v)dv)}{\exp(\int_0^T a(v)dv) - 1}. \tag{7.37}$$

Then, it is easy to see that $G(t+T, s+T) = G(t,s)$, $d > c > 0$,

$$c \le G(t,s) \le d \quad \text{if } t \le s \le t+T, \tag{7.38}$$

$$c \le G(t,s) \le d \quad \text{if } -\tau(0) \le t \le s \le T - \tau(0), \tag{7.39}$$

and

$$c \le G(t,s) \le d \quad \text{if } -\tau(0) \le t \le T - \tau(0) \le s \le 2T - \tau(0). \tag{7.40}$$

The following lemma can be directly verified.

Lemma 7.3. *The function $u(t)$ is a T-periodic solution of the equation*

$$u' = -a(t)u + k(t), \tag{7.41}$$

if and only if

$$u(t) = \int_t^{t+T} G(t,s)k(s)ds,$$

where $k \in C(\mathbb{R}, \mathbb{R})$ is a T-periodic function.

In what follows we let the Banach space X be defined by

$$X = \{u \in C(\mathbb{R}, \mathbb{R}) \; : \; u(t+T) = u(t) \text{ for } t \in \mathbb{R}\} \tag{7.42}$$

equipped with the norm $||u|| = \sup_{t\in\mathbb{R}} |u(t)|$. Define a cone P in X by

$$P = \{u \in X \; : \; u(t) \geq 0 \text{ on } \mathbb{R}\}, \tag{7.43}$$

and a subcone K of P by

$$K = \{u \in P \; : \; u(t) \geq \sigma ||u|| \text{ on } \mathbb{R}\}, \tag{7.44}$$

where $\sigma = c/d$. Let the linear operators $\mathcal{L}, \mathcal{M} : X \to X$ be defined by

$$(\mathcal{L}u)(t) = \int_t^{t+T} G(t,s)b(s)u(g(s))ds \tag{7.45}$$

and

$$\begin{aligned}(\mathcal{M}u)(t) = & \int_0^{g^{-1}(t)} G(g(s),t)b(s)u(s)ds \\ & + \int_{g^{-1}(t)}^{T} G(g(s),t+T)b(s)u(s)ds,\end{aligned} \tag{7.46}$$

where $g^{-1}(t)$ is the inverse function of $g(t)$.

The next two lemmas provide some useful information about the operators $\mathcal{L}$ and $\mathcal{M}$.

Lemma 7.4. *The operators $\mathcal{L}$ and $\mathcal{M}$ map P into K and are compact.*

Proof. We first show that $\mathcal{L}(P) \subseteq K$. For $u \in P$ and $t \in \mathbb{R}$, from (7.38) and (7.45), we have and

$$c\int_0^T b(s)u(g(s))ds \leq \mathcal{L}u(t) \leq d\int_0^T b(s)u(g(s))ds.$$

As a result, $(\mathcal{L}u)(t) \geq (c/d)||\mathcal{L}u|| = \sigma||\mathcal{L}u||$. Thus, $\mathcal{L}(P) \subseteq K$.

Next, we show that $\mathcal{M}$ maps P into K. To this end, we first prove that

$$(\mathcal{M}u)(t+T) = (\mathcal{M}u)(t) \quad \text{for any } u \in X \text{ and } t \in \mathbb{R}. \tag{7.47}$$

In fact, for any $u \in X$ and $t \in \mathbb{R}$, from (7.46),

$$\begin{aligned}(\mathcal{M}u)(t+T) = & \int_0^{g^{-1}(t+T)} G(g(s),t+T)b(s)u(s)ds \\ & + \int_{g^{-1}(t+T)}^{T} G(g(s),t+2T)b(s)u(s)ds \\ = & (I_1u)(t) + (I_2u)(t),\end{aligned}$$

where

$$(I_1u)(t)) = \int_0^{g^{-1}(t+T)} G(g(s), t+T)b(s)u(s)ds$$

and

$$(I_2u)(t)) = \int_{g^{-1}(t+T)}^{T} G(g(s), t+2T)b(s)u(s)ds.$$

Since $g^{-1}(t+T) = g^{-1}(t) + T$, $g(s+T) = g(s) + T$, and $G(t+T, s+T) = G(t, s)$ for any $t, s \in \mathbb{R}$, we have

$$\begin{aligned}
&(I_1u)(t)) = \int_0^{g^{-1}(t)+T} G(g(s), t+T)b(s)u(s)ds \\
&= \int_0^{g^{-1}(t)} G(g(s), t+T)b(s)u(s)ds + \int_{g^{-1}(t)}^{T} G(g(s), t+T)b(s)u(s)ds \\
&\quad + \int_T^{g^{-1}(t)+T} G(g(s), t+T)b(s)u(s)ds \\
&= \int_0^{g^{-1}(t)} G(g(s), t+T)b(s)u(s)ds + \int_{g^{-1}(t)}^{T} G(g(s), t+T)b(s)u(s)ds \\
&\quad + \int_0^{g^{-1}(t)} G(g(s)+T, t+T)b(s+T)u(s+T)ds \\
&= \int_0^{g^{-1}(t)} G(g(s), t+T)b(s)u(s)ds + \int_{g^{-1}(t)}^{T} G(g(s), t+T)b(s)u(s)ds \\
&\quad + \int_0^{g^{-1}(t)} G(g(s), t)b(s)u(s)ds
\end{aligned}$$

and

$$\begin{aligned}
(I_2u)(t)) &= \int_{g^{-1}(t)+T}^{T} G(g(s), t+2T)b(s)u(s)ds \\
&= \int_{g^{-1}(t)}^{0} G(g(s)+T, t+2T)b(s+T)u(s+T)ds \\
&= \int_{g^{-1}(t)}^{0} G(g(s), t+T)b(s)u(s)ds.
\end{aligned}$$

Thus,

$$\begin{aligned}
&(\mathcal{M}u)(t+T) = (I_1u)(t)) + (I_2u)(t)) \\
&= \int_{g^{-1}(t)}^{T} G(g(s), t+T)b(s)u(s)ds + \int_0^{g^{-1}(t)} G(g(s), t)b(s)u(s)ds \\
&= (\mathcal{M}u)(t),
\end{aligned}$$

i.e., (7.47) holds. Hence, $\mathcal{M}(X) \subseteq X$. Consequently, for $u \in P$ and $t \in \mathbb{R}$, we have

$$\begin{aligned}(\mathcal{M}u)(t) &\geq \min_{v\in[-\tau(0),T-\tau(0)]} (\mathcal{M}u)(v) \\ &= \min_{v\in[-\tau(0),T-\tau(0)]} \left(\int_0^{g^{-1}(v)} G(g(s),v)b(s)u(s)ds \right. \\ &\quad \left. + \int_{g^{-1}(v)}^T G(g(s),v+T)b(s)u(s)ds \right) \end{aligned} \tag{7.48}$$

and

$$\begin{aligned}(\mathcal{M}u)(t) &\leq \max_{v\in[-\tau(0),T-\tau(0)]} \mathcal{M}u(v) \\ &= \max_{v\in[-\tau(0),T-\tau(0)]} \left(\int_0^{g^{-1}(v)} G(g(s),v)b(s)u(s)ds \right. \\ &\quad \left. + \int_{g^{-1}(v)}^T G(g(s),v+T)b(s)u(s)ds \right). \end{aligned} \tag{7.49}$$

Note that

$$0 \leq g^{-1}(v) \leq T \iff -\tau(0) = g(0) \leq v \leq g(T) = T - \tau(0). \tag{7.50}$$

$$0 \leq s \leq g^{-1}(v) \iff -\tau(0) = g(0) \leq g(s) \leq v, \tag{7.51}$$

$$g^{-1}(v) \leq s \leq T \iff v \leq g(s) \leq g(T) = T - \tau(0). \tag{7.52}$$

Then, for $u \in P$ and $t \in \mathbb{R}$, from (7.39), (7.40), (7.48), and (7.49), it follows that

$$\begin{aligned}(\mathcal{M}u)(t) &\geq \min_{v\in[-\tau(0),T-\tau(0)]} (\mathcal{M}u)(v) \\ &\geq c\int_0^{g^{-1}(v)} b(s)u(s)ds + c\int_{g^{-1}(v)}^T b(s)u(s)ds \\ &= c\int_0^T b(s)u(s)ds\end{aligned}$$

and

$$\begin{aligned}\mathcal{M}u(t) &\leq \max_{v\in[-\tau(0),T-\tau(0)]} \mathcal{M}u(v) \\ &\leq d\int_0^{g^{-1}(v)} b(s)u(s)ds + d\int_{g^{-1}(v)}^T b(s)u(s)ds \\ &= d\int_0^T b(s)u(s)ds,\end{aligned}$$

from which we have $(\mathcal{M}u)(t) \geq (c/d)||\mathcal{M}u|| = \sigma||\mathcal{M}u||$. Therefore, $\mathcal{M}(P) \subseteq K$.

Finally, standard arguments can be used to show that $\mathcal{L}$ and $\mathcal{M}$ are compact and we omit the details here. This completes the proof of the lemma. □

Lemma 7.5. *We have the following:*

(a) *The spectral radius, $r_{\mathcal{L}}$, of $\mathcal{L}$ satisfies $r_{\mathcal{L}} > 0$. Moreover, $r_{\mathcal{L}}$ is an eigenvalue of $\mathcal{L}$ with an eigenvector $\phi_{\mathcal{L}} \in P$.*
(b) *The spectral radius, $r_{\mathcal{M}}$, of $\mathcal{M}$ satisfies $r_{\mathcal{M}} > 0$. Moreover, $r_{\mathcal{M}}$ is an eigenvalue of $\mathcal{M}$ with an eigenvector $\phi_{\mathcal{M}} \in P$.*

Proof. The ideas of the proof for parts (a) and (b) are essentially the same. In the following, we only prove part (b). Let $u \in K$ and $t \in \mathbb{R}$. Noting (7.50)–(7.52), from (7.39), (7.40), and (7.48), we see that

$$\begin{aligned}(\mathcal{M}u)(t) &\geq \min_{v\in[-\tau(0),T-\tau(0)]}(Mu)(v)\\ &\geq c\int_0^{g^{-1}(v)} b(s)u(s)ds + c\int_{g^{-1}(v)}^{T} b(s)u(s)ds\\ &= c\int_0^T b(s)u(s)ds \geq \sigma||u||c\int_0^T b(s)ds\end{aligned}$$

and

$$\begin{aligned}(\mathcal{M}^2u)(t) &= (\mathcal{M}(\mathcal{M}u))(t) \geq \min_{v\in[-\tau(0),T-\tau(0)]}(\mathcal{M}(\mathcal{M}u))(v))\\ &\geq c\int_0^{g^{-1}(v)} b(s)\left(c\sigma||u||\int_0^T b(s)ds\right)ds\\ &\quad +c\int_{g^{-1}(v)}^{T} b(s)\left(c\sigma||u||\int_0^T b(s)ds\right)ds\\ &= \sigma||u||\left(c\int_0^T b(s)ds\right)^2.\end{aligned}$$

By induction, we obtain that

$$(\mathcal{M}^n u)(t) \geq \sigma||u||\left(c\int_0^T b(s)ds\right)^n.$$

Then,

$$||\mathcal{M}^n||\ ||u|| \geq ||\mathcal{M}^n u|| \geq (\mathcal{M}^n u)(t) \geq \sigma||u||\left(c\int_0^T b(s)ds\right)^n,$$

and so

$$||\mathcal{M}^n|| \geq \sigma \left(c \int_0^T b(s)ds \right)^n .$$

Hence,

$$r_{\mathcal{M}} = \lim_{n\to\infty} ||\mathcal{M}^n||^{1/n} \geq c \int_0^T b(s) > 0.$$

Now, in view of the fact that the cone P defined by (7.43) is a total cone and that $r_{\mathcal{M}} > 0$, the "moreover" part of part (b) readily follows from Lemmas 6.7 and 7.4. This completes the proof of the lemma. □

7.3.2 *Existence results*

For convenience, we introduce the following notations:

$$f_0 = \liminf_{x\to 0^+} \frac{f(x)}{x}, \quad f_\infty = \liminf_{x\to\infty} \frac{f(x)}{x},$$

$$F_0 = \limsup_{x\to 0} \left| \frac{f(x)}{x} \right|, \quad F_\infty = \limsup_{|x|\to\infty} \left| \frac{f(x)}{x} \right|,$$

$$\xi = \frac{1}{d\int_0^T b(s)ds}, \quad \eta = \frac{1}{c\sigma \int_0^T b(s)ds}. \tag{7.53}$$

In the rest of this subsection, we also let

$$\mu_{\mathcal{M}} = \frac{1}{r_{\mathcal{M}}}, \tag{7.54}$$

where $r_{\mathcal{M}}$ is given in Lemma 7.5 (b). Clearly, $\mu_{\mathcal{M}}$ is the smallest positive characteristic value of $\mathcal{M}$ satisfying $\phi_{\mathcal{M}} = \mu_{\mathcal{M}} M \phi_{\mathcal{M}}$, and as we will see by Lemma 7.6 in Subsection 7.3.3 below, $\xi \leq \mu_{\mathcal{M}} \leq \eta$.

We need the following assumptions.

(G1) There exist a constant $M \geq 0$ and a function $\alpha \in C(\mathbb{R}, \mathbb{R}^+)$ such that α is even and nondecreasing on $\mathbb{R}^+$,

$$f(x) \geq -M - \alpha(x) \quad \text{for all } x \in \mathbb{R}, \tag{7.55}$$

and

$$\lim_{x\to\infty} \frac{\alpha(x)}{x} = 0. \tag{7.56}$$

(G2) There exist a constant $0 < r < 1$ and a function $\beta \in C(\mathbb{R}, \mathbb{R}^+)$ such that β is even and nondecreasing on $\mathbb{R}^+$,

$$f(x) \geq -\beta(x) \quad \text{for all } x \in [-r, 0], \tag{7.57}$$

and

$$\lim_{x \to 0} \frac{\beta(x)}{x} = 0. \tag{7.58}$$

Remark 7.2. Here, we want to emphasize that, in (B1), we assume that $f(x)$ is bounded from below by $-M - \alpha(x)$ for all $x \in \mathbb{R}$; however in (G2), we only require that $f(x)$ is bounded from below by $-\beta(x)$ for x in a small left-neighborhood of 0.

We first state our existence results for Eq. (7.27).

Theorem 7.4. *Assume* (G1) *holds. If*

$$F_0 < \mu_{\mathcal{M}} < f_\infty,$$

then Eq. (7.27) *has at least one nontrivial T-periodic solution.*

Theorem 7.5. *Assume* (G2) *holds. If*

$$F_\infty < \mu_{\mathcal{M}} < f_0,$$

then Eq. (7.27) *has at least one nontrivial T-periodic solution.*

Corollary 7.5. *Assume* (G1) *holds. If*

$$\frac{F_0}{\xi} < 1 < \frac{f_\infty}{\eta},$$

then Eq. (7.27) *has at least one nontrivial T-periodic solution.*

Corollary 7.6. *Assume* (G2) *holds. If*

$$\frac{F_\infty}{\xi} < 1 < \frac{f_0}{\eta},$$

then Eq. (7.27) *has at least one nontrivial T-periodic solution.*

Next, we state our existence results for Eq. (7.28); they are immediate consequences of the above results.

Corollary 7.7. *Assume* (G1) *holds. If*

$$\frac{\mu_{\mathcal{M}}}{f_\infty} < \lambda < \frac{\mu_{\mathcal{M}}}{F_0},$$

then Eq. (7.28) *has at least one nontrivial T-periodic solution.*

Corollary 7.8. *Assume* (G2) *holds. If*

$$\frac{\mu_{\mathcal{M}}}{f_0} < \lambda < \frac{\mu_{\mathcal{M}}}{F_\infty},$$

then Eq. (7.28) *has at least one nontrivial T-periodic solution.*

Corollary 7.9. *Assume* (G1) *holds. If*

$$\frac{\eta}{f_\infty} < \lambda < \frac{\xi}{F_0},$$

then Eq. (7.28) *has at least one nontrivial T-periodic solution.*

Corollary 7.10. *Assume* (G2) *holds. If*

$$\frac{\eta}{f_0} < \lambda < \frac{\xi}{F_\infty},$$

then Eq. (7.28) *has at least one nontrivial T-periodic solution.*

We conclude this subsection with the following two simple examples.

Example 7.4. Let

$$f(x) = \begin{cases} \sum_{i=1}^{n} \gamma_i x^i, & x \in [-1, \infty), \\ \sum_{i=1}^{n} (-1)^i \gamma_i - |x|^\theta \ln(1+|x|) + \ln 2, & x \in (-\infty, -1], \end{cases} \tag{7.59}$$

where $n \geq 1$ is an integer, $\gamma_i \in \mathbb{R}$ with $0 \leq \gamma_1 < (e^{2\pi} - 1)/(2\pi e^{4\pi})$ and $\gamma_n > 0$, and $0 < \theta < 1$. Clearly, $f \in C(\mathbb{R}, \mathbb{R})$.

We claim that functional differential equation

$$u'(t) = -u(t) + f(u(t - \sin t)) \tag{7.60}$$

has a nontrivial 2π-periodic solution.

In fact, with $a(t) = b(t) = 1$, $T = 2\pi$, $\tau(t) = \sin t$, and $g(t) = t - \sin t$, it is easy to see that Eq. (7.60) is of the form of Eq. (7.27) and assumption (H) is satisfied. Moreover, from (7.53), we have

$$\xi = \frac{e^{2\pi} - 1}{2\pi e^{4\pi}} \quad \text{and} \quad \eta = \frac{e^{4\pi}(e^{2\pi} - 1)}{2\pi}. \tag{7.61}$$

Let

$$M = \sum_{i=1}^{n} |\gamma_i| + \ln 2 \quad \text{and} \quad \alpha(x) = |x|^\theta \ln(1 + |x|).$$

Then, in view of (7.59), we have

$$f(x) \geq -M - \alpha(x) \quad \text{for all } x \in \mathbb{R}$$

and

$$\lim_{x\to\infty}\frac{\alpha(x)}{x}=\lim_{x\to\infty}\frac{\ln(1+x)}{x^{1-\theta}}=0.$$

Thus, (G1) holds. From (7.59), we have

$$F_0=\limsup_{x\to 0}\left|\frac{f(x)}{x}\right|=\gamma_1<\xi\quad\text{and}\quad f_\infty=\liminf_{x\to\infty}\frac{f(x)}{x}=\infty.$$

Hence, $F_0/\xi<1<f_\infty/\eta$. The conclusion then follows from Corollary 7.5.

Example 7.5. Let f be defined as in Example 7.4. Then, we claim that for any λ satisfying $0<\lambda<(e^{2\pi}-1)/(2\pi\gamma_1 e^{4\pi})$, the eigenvalue problem

$$u'(t)=-u(t)+\lambda f(u(t-\sin t))$$

has a nontrivial 2π-periodic solution.

In fact, as in Example 7.4, (G1) holds. Moreover, ξ and η are given by (7.61), $F_0=\gamma_1$, and $f_\infty=\infty$. Note that

$$0<\lambda<\frac{e^{2\pi}-1}{2\pi\gamma_1 e^{4\pi}}\quad\Longleftrightarrow\quad\frac{\eta}{f_\infty}<\lambda<\frac{\xi}{F_0}.$$

The conclusion then readily follows from Corollary 7.9.

Notice that in these examples, for x negative, the function f is negative and unbounded from below.

7.3.3 *Proofs of the existence results*

Let X, P, K, $\mathcal{L}$, and $\mathcal{M}$ be defined by (7.42)–(7.46), respectively. By Lemma 7.4, $\mathcal{L}$ and $\mathcal{M}$ map P into K and are compact. Define operators $\mathcal{F},\mathcal{T}:X\to X$ by

$$(\mathcal{F}u)(t)=f(u(t))\tag{7.62}$$

and

$$(\mathcal{T}u)(t)=(\mathcal{L}\mathcal{F}u)(t)=\int_t^{t+T}G(t,s)b(s)f(u(g(s)))ds.\tag{7.63}$$

Then $\mathcal{F}:X\to X$ is bounded and $\mathcal{T}:X\to X$ is compact. Moreover, by Lemma 7.3, a T-periodic solution of Eq. (7.27) is equivalent to a fixed point of the operator $\mathcal{T}$ in X.

Proof of Theorem 7.4. We first verify that conditions (F1)–(F4) of Lemma 7.1 are satisfied.

By Lemma 7.5, there exist $\phi_{\mathcal{L}}, \phi_{\mathcal{M}} \in P \setminus \{\mathbf{0}\}$ such that (7.29) holds. To show that (7.30) holds, we let

$$h(u) = \int_0^T b(t)\phi_{\mathcal{M}}(t)u(g(t))dt, \; u \in X. \tag{7.64}$$

Then, $h \in P^* \setminus \{\mathbf{0}\}$ and

$$\begin{aligned}(\mathcal{L}^*h)(u) &= h(\mathcal{L}u)\\ &= \int_0^T b(t)\phi_{\mathcal{M}}(t)\mathcal{L}u(g(t))dt\\ &= \int_0^T b(t)\phi_{\mathcal{M}}(t)\left(\int_{g(t)}^{g(t)+T} G(g(t),s)b(s)u(g(s))ds\right)dt\\ &= \int_0^T \int_{g(t)}^{g(t)+T} G(g(t),s)b(s)u(g(s))b(t)\phi_{\mathcal{M}}(t)dsdt.\end{aligned}$$

Interchanging the order of integration and noting that $g(T) = g(0)+T$, we have

$$\begin{aligned}(\mathcal{L}^*h)(u) = &\int_{g(0)}^{g(T)} b(s)u(g(s))\left(\int_0^{g^{-1}(s)} G(g(t),s)b(t)\phi_{\mathcal{M}}(t)dt\right)ds\\ &+ \int_{g(0)+T}^{g(T)+T} b(s)u(g(s))\left(\int_{g^{-1}(s-T)}^T G(g(t),s)b(t)\phi_{\mathcal{M}}(t)dt\right)ds.\end{aligned}$$

Letting $s = v+T$ in the second term and noting that $g(v+T) = g(v)+T$, we obtain

$$\begin{aligned}(\mathcal{L}^*h)(u) = &\int_{g(0)}^{g(T)} b(s)u(g(s))\left(\int_0^{g^{-1}(s)} G(g(t),s)b(t)\phi_{\mathcal{M}}(t)dt\right)ds\\ &+ \int_{g(0)}^{g(T)} b(v+T)u(g(v)+T)\left(\int_{g^{-1}(v)}^T G(g(t),v+T)b(t)\phi_{\mathcal{M}}(t)dt\right)dv\\ = &\int_{g(0)}^{g(T)} b(s)u(g(s))\left(\int_0^{g^{-1}(s)} G(g(t),s)b(t)\phi_{\mathcal{M}}(t)dt\right)ds\\ &+ \int_{g(0)}^{g(T)} b(v)u(g(v))\left(\int_{g^{-1}(v)}^T G(g(t),v+T)b(t)\phi_{\mathcal{M}}(t)dt\right)dv\\ = &\int_{g(0)}^{g(T)} b(s)u(g(s))\left(\int_0^{g^{-1}(s)} G(g(t),s)b(t)\phi_{\mathcal{M}}(t)dt\right.\\ &\left.+ \int_{g^{-1}(s)}^T G(g(t),s+T)b(t)\phi_{\mathcal{M}}(t)dt\right)ds.\end{aligned}$$

Thus, in view of (7.29), (7.46), (7.64), and the fact that $g(T) = g(0) + T$, we have

$$\begin{aligned}(\mathcal{L}^*h)(u) &= \int_{g(0)}^{g(T)} b(s)u(g(s))\mathcal{M}\phi_{\mathcal{M}}(s)ds \\ &= r_{\mathcal{M}} \int_{g(0)}^{g(T)} b(s)\phi_{\mathcal{M}}(s)u(g(s))ds \\ &= r_{\mathcal{M}} \int_0^T b(s)\phi_{\mathcal{M}}(s)u(g(s))ds = r_{\mathcal{M}}h(u),\end{aligned}$$

i.e., h satisfies (7.30). Since $\phi_{\mathcal{M}}(s) \geq \sigma||\phi_{\mathcal{M}}|| > 0$ on $\mathbb{R}$, there exists $\delta_1 > 0$ such that

$$\phi_{\mathcal{M}}(s) \geq \delta_1 G(t,s) \quad \text{for } t, s \in \mathbb{R}. \tag{7.65}$$

Let $\delta = r_{\mathcal{M}}\delta_1$. For any $u \in P$ and $t \in \mathbb{R}$, from (7.45), (7.64), and (7.65), it follows that

$$\begin{aligned}h(\mathcal{L}u) = r_{\mathcal{M}}h(u) &= r_{\mathcal{M}} \int_0^T b(s)\phi_{\mathcal{M}}(s)u(g(s))ds \\ &= r_{\mathcal{M}} \int_t^{t+T} b(s)\phi_{\mathcal{M}}(s)u(g(s))ds \\ &\geq r_{\mathcal{M}}\delta_1 \int_t^{t+T} G(t,s)b(s)u(g(s))ds \\ &= \delta(\mathcal{L}u)(t).\end{aligned}$$

Hence, $h(\mathcal{L}u) \geq \delta||\mathcal{L}u||$, i.e., $\mathcal{L}(P) \subseteq P(h,\delta)$. Therefore, condition (F1) of Lemma 7.1 holds.

Since α is nondecreasing on $\mathbb{R}^+$, we have

$$\alpha(u) \leq \alpha(||u||) \quad \text{for all } u \in P.$$

Then, from the fact that α is even, it follows that

$$\alpha(u) \leq \alpha(||u||) \quad \text{for all } u \in X.$$

Thus,

$$||\alpha(u)|| \leq \alpha(||u||) \quad \text{for all } u \in X.$$

From (7.56), we see that

$$\lim_{||u||\to\infty} \frac{||\alpha(u)||}{||u||} = 0 \quad \text{for any} \quad u \in X.$$

Letting $\mathcal{H}u = \alpha(u)$ for $u \in X$, we have that condition (F2) in Lemma 7.1 holds.

With $\mathcal{F}$ defined by (7.62) and $u_0 = M$, from (7.55), we have $\mathcal{F}u + \mathcal{H}u + u_0 \in P$ for all $u \in X$. Hence, (F3) of Lemma 7.1 holds.

Since $f_\infty > \mu_\mathcal{M}$, there exist $\epsilon > 0$ and $N > 0$ such that

$$f(x) \geq \mu_\mathcal{M}(1+\epsilon)x \quad \text{for } x \geq N.$$

Then, in view of (7.55), there exists $\rho > 0$ such that

$$f(x) \geq \mu_\mathcal{M}(1+\epsilon)x - \alpha(x) - \rho \quad \text{for all } x \in \mathbb{R}.$$

From (7.54) and (7.62), we have

$$\mathcal{F}u \geq \mu_\mathcal{M}(1+\epsilon)u - \alpha(u) - \rho = r_\mathcal{M}^{-1}(1+\epsilon)u - \mathcal{H}u - \rho \quad \text{for all } u \in X.$$

Thus,

$$\mathcal{L}\mathcal{F}u \geq r_\mathcal{M}^{-1}(1+\epsilon)\mathcal{L}u - \mathcal{L}\mathcal{H}u - \mathcal{L}\rho \quad \text{for all } u \in X.$$

Then, (F4) of Lemma 7.1 holds with $v_0 = \mathcal{L}\rho$.

We have verified that all the conditions of Lemma 7.1 hold, so there exists $R_1 > 0$ such that

$$\deg(\mathcal{I} - \mathcal{T}, B(\mathbf{0}, R_1), \mathbf{0}) = 0. \tag{7.66}$$

Next, since $F_0 < \mu_\mathcal{M}$, there exist $0 < \nu < 1$ and $0 < R_2 < R_1$ such that

$$|f(x)| \leq \mu_\mathcal{M}(1-\nu)|x| \quad \text{for } |x| < R_2. \tag{7.67}$$

We claim that

$$\mathcal{T}u \neq \tau u \quad \text{for all } u \in \partial B(\mathbf{0}, R_2) \text{ and } \tau \geq 1. \tag{7.68}$$

If this is not the case, then there exist $\bar{u} \in \partial B(\mathbf{0}, R_2)$ and $\bar{\tau} \geq 1$ such that $\mathcal{T}\bar{u} = \bar{\tau}\bar{u}$. It follows that $\bar{u} = \bar{s}\mathcal{T}\bar{u}$, where $\bar{s} = 1/\bar{\tau}$. Clearly, $\bar{s} \in (0, 1]$. Then, from (7.45), (7.63), and (7.67), we have

$$\begin{aligned} |\bar{u}(t)| = \bar{s}|\mathcal{T}\bar{u}(t)| &\leq \int_t^{t+T} G(t,s)b(s)|f(\bar{u}(g(s)))|ds \\ &\leq \mu_\mathcal{M}(1-\nu)\int_t^{t+T} G(t,s)b(s)|\bar{u}(g(s))|ds \\ &= \mu_\mathcal{M}(1-\nu)\mathcal{L}|\bar{u}(t)|. \end{aligned}$$

Consequently,

$$\begin{aligned} h(|\bar{u}|) \leq \mu_\mathcal{M}(1-\nu)h(\mathcal{L}|\bar{u}|) &= \mu_\mathcal{M}(1-\nu)(\mathcal{L}^*h)(|\bar{u}|) \\ &= r_\mathcal{M}^{-1}(1-\nu)r_\mathcal{M}h(|\bar{u}|) \\ &= (1-\nu)h(|\bar{u}|). \end{aligned}$$

Thus, $h(|\bar{u}|) \le 0$. On the other hand, in view of the fact that $\phi_{\mathcal{M}}(t) > 0$ and $\|\bar{u}\| = R_2 > 0$, by (7.64), $h(|\bar{u}|) > 0$. This contradiction implies (7.68) holds. Now Lemma 6.13 implies

$$\deg(\mathcal{I} - \mathcal{T}, B(\mathbf{0}, R_2), \mathbf{0}) = 1, \tag{7.69}$$

so by the additivity property of the Leray-Schauder degree, (7.66), and (7.69), we have

$$\deg(\mathcal{I} - \mathcal{T}, B(\mathbf{0}, R_1) \setminus \overline{B(\mathbf{0}, R_2)}) = -1.$$

Then, from the solution property of the Leray-Schauder degree, $\mathcal{T}$ has at least one fixed point u in $B(\mathbf{0}, R_1) \setminus \overline{B(\mathbf{0}, R_2)}$. Clearly, $u(t)$ is a nontrivial solution of Eq. (7.27). This complete the proof of the theorem. □

Proof of Theorem 7.5. We first verify that conditions (F1) and (F2)*–(F4)* of Lemma 7.2 are satisfied.

As in the proof of Theorem 7.4, there exist $\phi_{\mathcal{L}}, \phi_{\mathcal{M}} \in P \setminus \{\mathbf{0}\}$ and $h \in P^* \setminus \{\mathbf{0}\}$ defined by (7.64) such that (F1) holds.

From the fact that β is even and nondecreasing on $\mathbb{R}^+$, it is easy to see that

$$\beta(u) \le \beta(||u||) \quad \text{for all } u \in X.$$

Thus,

$$||\beta(u)|| \le \beta(||u||) \quad \text{for all } u \in X.$$

This, together with (7.58), implies that

$$\lim_{||u|| \to 0} \frac{||\beta(u)||}{||u||} = 0 \quad \text{for any} \quad u \in X.$$

Let $\mathcal{H}u = \beta(u)$ for $u \in X$. Then, (F2)* of Lemma 7.2 holds.

Since $f_0 > \mu_{\mathcal{M}}$, there exist $\epsilon > 0$ and $0 < \zeta_1 < 1$ such that

$$f(x) \ge \mu_{\mathcal{M}}(1+\epsilon)x = r_{\mathcal{M}}^{-1}(1+\epsilon)x \ge 0 \quad \text{for } x \in [0, \zeta_1]. \tag{7.70}$$

Let r be given in (G2) and F be defined by (7.62). Now, in view of (7.57) and (7.70), we see that (F3)* of Lemma 7.2 holds with $r_1 = \min\{r, \zeta_1\}$.

From (7.58), there exists $0 < \zeta_2 < \min\{r, \zeta_1\}$ such that

$$-\beta(x) \ge r_{\mathcal{M}}^{-1}(1+\epsilon)x \quad \text{for } x \in [-\zeta_2, 0].$$

Then, from (7.57),

$$f(x) \ge r_{\mathcal{M}}^{-1}(1+\epsilon)x \quad \text{for } x \in [-\zeta_2, 0]. \tag{7.71}$$

From (7.70) and (7.71), we have

$$f(x) \geq r_{\mathcal{M}}^{-1}(1+\epsilon)x \quad \text{for } x \in [-\zeta_2, \zeta_2],$$

which clearly implies that

$$\mathcal{L}\mathcal{F}u \geq r_{\mathcal{M}}^{-1}(1+\epsilon)\mathcal{L}u \quad \text{for all } u \in X \text{ with } ||u|| < \zeta_2.$$

Hence, (F4)* of Lemma 7.2 holds with $r_2 = \zeta_2$.

We have verified that all the conditions of Lemma 7.2 hold, so there exists $R_3 > 0$ such that

$$\deg(\mathcal{I} - \mathcal{T}, B(\mathbf{0}, R_3), \mathbf{0}) = 0. \tag{7.72}$$

Next, since $F_\infty < \mu_{\mathcal{M}}$, there exist $0 < \tilde{\nu} < 1$ and $\bar{R} > R_3$ such that

$$|f(x)| \leq \mu_{\mathcal{M}}(1-\tilde{\nu})|x| = r_{\mathcal{M}}^{-1}(1-\tilde{\nu})|x| \quad \text{for } |x| \in (\bar{R}, \infty). \tag{7.73}$$

Let

$$\mathcal{C} = \max_{|x| \leq \bar{R}} |f(x)| \sup_{t \in \mathbb{R}} \int_t^{t+T} G(t,s)b(s)ds. \tag{7.74}$$

Then $0 < \mathcal{C} < \infty$. Choose R_4 large enough so that

$$R_4 > \max\{\bar{R},\ \tilde{\nu}^{-1}\mathcal{C}\}. \tag{7.75}$$

We claim that

$$\mathcal{T}u \neq \tau u \quad \text{for all } u \in \partial B(\mathbf{0}, R_4) \text{ and } \tau \geq 1. \tag{7.76}$$

If this is not the case, then there exist $\tilde{u} \in \partial B(\mathbf{0}, R_4)$ and $\tilde{\tau} \geq 1$ such that $\mathcal{T}\tilde{u} = \tilde{\tau}\tilde{u}$. It follows that $\tilde{u} = \tilde{s}\mathcal{T}\tilde{u}$, where $\tilde{s} = 1/\tilde{\tau}$. Clearly, $\tilde{s} \in (0,1]$. Since $\tilde{u}$ is T-periodic, we may assume that $R_4 = ||\tilde{u}|| = |\tilde{u}(\tilde{t})|$ for some $\tilde{t} \in [0, T]$. Let

$$J_1(\tilde{u}) = \{t \in [\tilde{t}, \tilde{t}+T] \ : \ |\tilde{u}(g(t))| > \bar{R}\},$$

$$J_2(\tilde{u}) = [\tilde{t}, \tilde{t}+T] \setminus J_1(\tilde{u}),$$

and

$$p(\tilde{u}(t)) = \min\{|\tilde{u}(g(t))|, \bar{R}\} \quad \text{for} \quad t \in [\tilde{t}, \tilde{t}+T].$$

Then, from (7.45), (7.63), (7.73), and (7.74), we have

$$
\begin{aligned}
R_4 &= |\tilde{u}(\tilde{t})| = \tilde{s}|(\mathcal{T}\tilde{u})(\tilde{t})| \\
&\leq \int_{\tilde{t}}^{\tilde{t}+T} G(\tilde{t},s)b(s)|f(\tilde{u}(g(s)))|ds \\
&= \int_{J_1(\tilde{u})} G(\tilde{t},s)b(s)|f(\tilde{u}(g(s)))|ds + \int_{J_2(\tilde{u})} G(\tilde{t},s)b(s)|f(\tilde{u}(g(s)))|ds \\
&\leq r_{\mathcal{M}}^{-1}(1-\tilde{\nu}) \int_{J_1(\tilde{u})} G(\tilde{t},s)b(s)|\tilde{u}(g(s))|ds \\
&\quad + \int_{J_2(\tilde{u})} G(\tilde{t},s)b(s)|f(p(\tilde{u}(s)))|ds \\
&\leq r_{\mathcal{M}}^{-1}(1-\tilde{\nu}) \int_{\tilde{t}}^{\tilde{t}+T} G(\tilde{t},s)b(s)|\tilde{u}(g(s))|ds \\
&\quad + \int_{\tilde{t}}^{\tilde{t}+T} G(\tilde{t},s)b(s)|f(p(\tilde{u}(s)))|ds \\
&\leq r_{\mathcal{M}}^{-1}(1-\tilde{\nu})(\mathcal{L}|u|)(\tilde{t}) + \mathcal{C} = r_{\mathcal{M}}^{-1}(1-\tilde{\nu})\mathcal{L}R_4 + \mathcal{C}.
\end{aligned}
$$

Hence, for h defined by (7.64),

$$
\begin{aligned}
h(R_4) &\leq r_{\mathcal{M}}^{-1}(1-\tilde{\nu})h(\mathcal{L}R_4) + h(\mathcal{C}) \\
&= r_{\mathcal{M}}^{-1}(1-\tilde{\nu})(\mathcal{L}^*h)(R_4) + h(\mathcal{C}) \\
&= r_{\mathcal{M}}^{-1}(1-\tilde{\nu})r_{\mathcal{M}}h(R_4) + h(\mathcal{C}) \\
&= (1-\tilde{\nu})h(R_4) + h(\mathcal{C}),
\end{aligned}
$$

which implies

$$(\tilde{\nu}R_4 - \mathcal{C})h(1) \leq 0.$$

In view of the fact that $h(1) > 0$, it follows that $R_4 \leq \tilde{\nu}^{-1}\mathcal{C}$. This contradicts (7.75) and so (7.76) holds. By Lemma 6.13, we have

$$\deg(\mathcal{I} - \mathcal{T}, B(\mathbf{0}, R_4), \mathbf{0}) = 1. \tag{7.77}$$

By the additivity property of the Leray-Schauder degree, (7.72), and (7.77), we obtain

$$\deg(\mathcal{I} - \mathcal{T}, B(\mathbf{0}, R_4) \setminus \overline{B(\mathbf{0}, R_3)}) = 1.$$

Thus, from the solution property of the Leray-Schauder degree, $\mathcal{T}$ has at least one fixed point u in $B(\mathbf{0}, R_4) \setminus \overline{B(\mathbf{0}, R_3)}$. Clearly, $u(t)$ is a nontrivial solution of Eq. (7.27), and this completes the proof of the theorem. □

Lemma 7.6. *Let $\mu_{\mathcal{M}}$ be defined by (7.54). Then $\xi \leq \mu_{\mathcal{M}} \leq \eta$, where ξ and η are given by (7.53).*

Proof. Let $\phi_{\mathcal{M}}$ be given as in Lemma 7.5 (b). Then $\phi_{\mathcal{M}}(t) = \mu_{\mathcal{M}}\mathcal{M}\phi_{\mathcal{M}}(t)$, i.e.,

$$\phi_{\mathcal{M}}(t) = \mu_{\mathcal{M}}\left(\int_0^{g^{-1}(t)} G(g(s),t)b(s)\phi_{\mathcal{M}}(s)ds \right.$$
$$\left. + \int_{g^{-1}(t)}^{T} G(g(s),t+T)b(s)\phi_{\mathcal{M}}(s)ds\right).$$

Thus, in view of (7.50)–(7.52) and from (7.39) and (7.40), it follows that

$$\begin{aligned}\phi_{\mathcal{M}}(t) &\le \mu_{\mathcal{M}} \max_{v\in[-\tau(0),T-\tau(0)]}(\mathcal{M}\phi_{\mathcal{M}})(v)\\ &\le \mu_{\mathcal{M}}\left(d\|\phi_{\mathcal{M}}\|\int_0^{g^{-1}(v)} b(s)ds + d\|\phi_{\mathcal{M}}\|\int_{g^{-1}(v)}^{T} b(s)ds\right)\\ &= \mu_{\mathcal{M}} d\|\phi_{\mathcal{M}}\|\int_0^T b(s)ds \quad \text{on } \mathbb{R}.\end{aligned}$$

Hence,

$$\mu_{\mathcal{M}} \ge \frac{1}{d\int_0^T b(s)ds} = \xi.$$

On the other hand, since $\phi_{\mathcal{M}} \in K$, we have $\phi_{\mathcal{M}}(t) \ge \sigma\|\phi_{\mathcal{M}}\|$ on $\mathbb{R}$, and then again from (7.39) and (7.40), we have

$$\begin{aligned}\phi_{\mathcal{M}}(t) &\ge \mu_{\mathcal{M}} \min_{v\in[-\tau(0),T-\tau(0)]}(\mathcal{M}\phi_{\mathcal{M}})(v)\\ &\ge \mu_{\mathcal{M}}\left(c\sigma\|\phi_{\mathcal{M}}\|\int_0^{g^{-1}(v)} b(s)ds + c\sigma\|\phi_{\mathcal{M}}\|\int_{g^{-1}(v)}^{T} b(s)ds\right)\\ &= \mu_{\mathcal{M}} c\sigma\|\phi_{\mathcal{M}}\|\int_0^T b(s)ds \quad \text{on } \mathbb{R}.\end{aligned}$$

Thus,

$$\mu_{\mathcal{M}} \le \frac{1}{c\sigma\int_0^T b(s)ds} = \eta.$$

This completes the proof of the lemma. □

Proof of Corollary 7.5. The conclusion follows from Theorem 7.4 and Lemma 7.6. □

Proof of Corollary 7.6. The conclusion follows from Theorem 7.5 and Lemma 7.6. □

Finally, by virtue of Lemma 7.6, Corollaries 7.7–7.10 are direct applications of Theorems 7.4–7.5 and Corollaries 7.5–7.6 with f in Eq. (7.27) replaced by λf. We omit the proofs here.

Bibliography

Boucherif A. and Al-Malki N. (2007). Nonlinear three-point third-order boundary value problems, *Appl. Math. Comput.* **190**, No. 2, pp. 1168–1177.

Glass L. and Mackey M. C. (1988). *From Clocks to Chaos*, Princeton University Press, Princeton.

Graef J. R. and Kong L. (2011). Existence of multiple periodic solutions of first order functional differential equations, *Math. Comput. Modelling* **54**, pp. 2962–2968,

Graef J. R. and Kong L. (2011). Periodic solutions of first order functional differential equations, *Appl. Math. Lett.* **24**, pp. 1981–1985.

Graef J. R. and Kong L. (2010). Periodic solutions for functional differential equations with sign-changing nonlinearities, *Proc. Royal Soc. Edinburgh Sect. A* **140**, pp. 597–616.

Granas A. and Dugundji J. (2003). *Fixed Point Theory* (Springer, New York).

Granas A., Guenther R. B. and Lee J. W. (1980). Applications of topological transversality to differential equations. I. Some nonlinear diffusion problems, *Pacific J. Math.* **89**, No. 1, pp. 53–67.

Granas A., Guenther R. B. and Lee J. W. (1983). Topological transversality. II. Applications to the Neumann problem for $y'' = f(x, y, y')$, *Pacific J. Math.* **104**, No. 1, pp. 95–109.

Granas A., Guenther R. B. and Lee J. W. (1985). Nonlinear boundary value problems for ordinary differential equations, *Dissertationes Math.* (*Rozprawy Mat.*) **244**, pp. 1–128.

Han G. and Wu Y. (2007). Nontrivial solutions of singular two-point boundary value problems with sign-changing nonlinear terms, *J. Math. Anal. Appl.* **325**, pp. 1327–1338.

Henderson J. (1995). *Boundary Value Problems for Functional Differential Equations* (World Scientific, River Edge, NJ).

Jin Z. and Wang H. (2010). A note on positive periodic solutions of delayed differential equations, *Appl. Math. Lett.* **23**, pp. 581–584.

Liu L., Liu B. and Wu Y. (2009). Nontrivial solutions of m-point boundary value problems for singular second-order differential equations with a sign-changing nonlinear term, *J. Comput. Appl. Math.* **224**, pp. 373–382.

Murray J. D (1989). *Mathematical Biology* (Springer-Verlag, Berlin).

Ntouyas S. K., Sficas Y. G. and Tsamatos P. Ch. (1993). An existence principle for boundary value problems for second order functional differential equations, *Nonlinear Anal.* **20**, No. 3, pp. 215–222.

Ntouyas S. K., Sficas Y. G. and Tsamatos P. Ch. (1996). Boundary value problems for functional differential equations, *J. Math. Anal. Appl.* **199**, No. 1, pp. 213–230.

Ntouyas S. K. and Tsamatos P. Ch. (1994). Global existence for functional integro-differential equations of delay and neutral type, *Appl. Anal.* **54**, No. 3–4, pp. 251–262.

Taunton D. and Yin W. K. C. (1997). Existence of solutions of some functional-differential equations, *Comm. Appl. Nonlinear Anal.* **4**, No. 1, pp. 31–43.

Wang H. (2004). Positive periodic solutions of functional differential equations, *J. Differential Equations* **202**, pp. 354–366.

Wazewska-Czyzewska M. and Lasota A. (1988). Mathematical problems of the dynamics of the red blood cell system, *Ann. Polish Math. Soc. Ser. III Appl. Math.* **17**, pp. 23–40.

Chapter 8

Positive Solutions

In this chapter, we focus on positive solutions for certain BVP's for nonlinear ODE's, including positive solutions for nonlinear eigenvalue problems, multiple positive solutions, positive solutions for analogous discrete nonlinear BVP's, symmetric positive solutions, and positive solutions for singular BVP's. For each section fixed point methods are employed using fixed point theorems such as the Guo-Krasnosel'skii cone expansion-compression fixed point theorem, the Leggett and Williams triple fixed point theorem, the Gatica, Oliker and Waltman fixed point theorem for decreasing operators defined on a cone, and the Schauder fixed point theorem. The definitions and basic properties stated in Section 6.1 of Chapter 6 for cones in Banach spaces will be of fundamental importance in this chapter.

8.1 Positive Solutions for Nonlinear Eigenvalue Problems

In this section, we are concerned with determining eigenvalues, λ, for which there exist positive solutions of the BVP,

$$\begin{cases} u'' + \lambda a(t)f(u) = 0, \quad 0 < t < 1, & (8.0_\lambda) \\ u(0) = u(1) = 0, & (8.1) \end{cases}$$

where

(A) $f : [0, \infty) \to [0, \infty)$ is continuous,
(B) $a : [0, 1] \to [0, \infty)$ is continuous and does not vanish identically on any subinterval, and
(C) $f_0 := \lim_{x\to 0^+} \frac{f(x)}{x}$ and $f_\infty := \lim_{x\to\infty} \frac{f(x)}{x}$ exist.

We remark that, if $u(t)$ is a nonnegative solution of (8.0_λ), (8.1), then $u(t)$ is concave on $[0, 1]$.

Our methods involve applying the Guo-Krasnosel'skii fixed point theorem to a completely continuous integral operator whose kernel, $G(t,s)$, is the Green's function for

$$\begin{cases} -y'' = 0, & (8.2) \\ y(0) = y(1) = 0. & (8.3) \end{cases}$$

In particular, from Section 2.3 of Chapter 2, and taking into account the "minus" sign,

$$G(t,s) = \begin{cases} t(1-s), & 0 \le t \le s \le 1, \\ s(1-t), & 0 \le s \le t \le 1, \end{cases} \tag{8.4}$$

from which

$$G(t,s) > 0, \text{ on } (0,1) \times (0,1), \tag{8.5}$$

$$G(t,s) \le G(s,s) = s(1-s), \quad 0 \le t, s \le 1, \tag{8.6}$$

and it is straightforward to show that

$$G(t,s) \ge \frac{1}{4}G(s,s) = \frac{1}{4}s(1-s), \quad \frac{1}{4} \le t \le \frac{3}{4}, \quad 0 \le s \le 1. \tag{8.7}$$

We will apply the following Guo-Krasnosel'skii Fixed Point Theorem to obtain solutions of (8.0_λ), (8.1) for certain λ.

Theorem 8.1. *Let $\mathcal{B}$ be a real Banach space, and let $\mathcal{P} \subset \mathcal{B}$ be a cone in $\mathcal{B}$. Assume Ω_1, Ω_2 are open subsets of $\mathcal{B}$ with $0 \in \Omega_1 \subset \overline{\Omega_1} \subset \Omega_2$, and let*

$$T : \mathcal{P} \cap (\overline{\Omega}_2 \setminus \Omega_1) \to \mathcal{P}$$

be a completely continuous operator such that, either

(i) $\|Tu\| \le \|u\|$, $u \in \mathcal{P} \cap \partial\Omega_1$, *and* $\|Tu\| \ge \|u\|$, $u \in \mathcal{P} \cap \partial\Omega_2$, *or*
(ii) $\|Tu\| \ge \|u\|$, $u \in \mathcal{P} \cap \partial\Omega_1$, *and* $\|Tu\| \le \|u\|$, $u \in \mathcal{P} \cap \partial\Omega_2$.

Then T has a fixed point in $\mathcal{P} \cap (\overline{\Omega}_2 \setminus \Omega_1)$.

8.1.1 *Solutions in the cone*

In this subsection, we will apply Theorem 8.1 to the eigenvalue problem (8.0_λ), (8.1). We note by Theorem 3.1, $u(t)$ is a solution of (8.0_λ), (8.1) if and only if

$$u(t) = \lambda \int_0^1 G(t,s)a(s)f(u(s))\,da, \quad 0 \le t \le 1.$$

For our constructions, let $\mathcal{B} = C[0,1]$, with norm $\|x\| = \sup_{0\le t\le 1}|x(t)|$. Define a cone, $\mathcal{P}$, by

$$\mathcal{P} = \left\{ x \in \mathcal{B} \,|\, x(t) \ge 0 \text{ on } [0,1], \text{ and } \min_{\frac{1}{4}\le t\le \frac{3}{4}} x(t) \ge \frac{1}{4}\|x\| \right\}.$$

Also, let the number $\tau \in [0,1]$ be defined by

$$\int_{\frac{1}{4}}^{\frac{3}{4}} G(\tau,s)a(s)\,ds = \max_{0\le t\le 1} \int_{\frac{1}{4}}^{\frac{3}{4}} G(t,s)a(s)\,ds. \tag{8.8}$$

Theorem 8.2. *Assume that conditions* (A), (B), *and* (C) *are satisfied. Then, for each* λ *satisfying*

$$\frac{4}{\left(\int_{\frac{1}{4}}^{\frac{3}{4}} G(\tau,s)a(s)\,ds\right) f_\infty} < \lambda < \frac{1}{\left(\int_0^1 s(1-s)a(s)\,ds\right) f_0}, \tag{8.9}$$

there exists at least one solution of (8.0_λ), (8.1) *in* $\mathcal{P}$.

Proof. Let λ be given as in (8.9). Now, let $\varepsilon > 0$ be chosen such that

$$\frac{4}{\left(\int_{\frac{1}{4}}^{\frac{3}{4}} G(\tau,s)a(s)\,ds\right)(f_\infty - \varepsilon)} \le \lambda \le \frac{1}{\left(\int_0^1 s(1-s)a(s)\,ds\right)(f_0 + \varepsilon)}.$$

Define an integral operator $T : \mathcal{P} \to \mathcal{B}$ by

$$Tu(t) = \lambda \int_0^1 G(t,s)a(s)f(u(s))\,da, \quad u \in \mathcal{P}. \tag{8.10}$$

We seek a fixed point of T in the cone $\mathcal{P}$.

Notice from (8.5) that, for $u \in \mathcal{P}$, $Tu(t) \ge 0$ on $[0,1]$. Also, for $u \in \mathcal{P}$, we have from (8.6) that

$$Tu(t) = \lambda \int_0^1 G(t,s)a(s)f(u(s))\,ds \le \lambda \int_0^1 s(1-s)a(s)f(u(s))\,ds,$$

so that

$$\|Tu\| \le \lambda \int_0^1 s(1-s)a(s)f(u(s))\,ds. \tag{8.11}$$

And next, if $u \in \mathcal{P}$, we have by (8.7) and (8.11),

$$\begin{aligned}
\min_{\frac{1}{4}\le t\le\frac{3}{4}} Tu(t) &= \min_{\frac{1}{4}\le t\le\frac{3}{4}} \lambda \int_0^1 G(t,s)a(s)f(u(s))\,ds \\
&\ge \frac{\lambda}{4} \int_0^1 s(1-s)a(s)f(u(s))\,ds \\
&\ge \frac{1}{4}\|Tu\|.
\end{aligned}$$

As a consequence, $T : \mathcal{P} \to \mathcal{P}$. In addition, standard arguments show that T is completely continuous.

Now, turning to f_0, there exists an $H_1 > 0$ such that $f(x) \le (f_0 + \varepsilon)x$, for $0 < x \le H_1$. So, choosing $u \in \mathcal{P}$ with $\|u\| = H_1$, we have from (8.6)

$$\begin{aligned} Tu(t) &\le \lambda \int_0^1 s(1-s)a(s)f(u(s))\,ds \\ &\le \lambda \int_0^1 s(1-s)a(s)(f_0+\varepsilon)u(s)\,ds \\ &\le \lambda \int_0^1 s(1-s)a(s)\,ds(f_0+\varepsilon)\|u\| \\ &\le \|u\|. \end{aligned}$$

Consequently, $\|Tu\| \le \|u\|$. So, if we set $\Omega_1 = \{x \in \mathcal{B} \mid \|x\| < H_1\}$, then,

$$\|Tu\| \le \|u\|, \qquad \text{for } u \in \mathcal{P} \cap \partial\Omega_1. \tag{8.12}$$

Next, considering f_∞, there exists an $\overline{H}_2 > 0$ such that $f(x) \ge (f_\infty - \varepsilon)x$, for all $x \ge \overline{H}_2$. Let $H_2 = \max\{2H_1, 4\overline{H}_2\}$ and let $\Omega_2 = \{x \in \mathcal{B} \mid \|x\| < H_2\}$. If $u \in \mathcal{P}$ with $\|u\| = H_2$, then $\min_{\frac{1}{4}\le t\le\frac{3}{4}} u(t) \ge \frac{1}{4}\|u\| \ge \overline{H}_2$, and

$$\begin{aligned} Tu(\tau) &\ge \lambda \int_{\frac{1}{4}}^{\frac{3}{4}} G(\tau,s)a(s)f(u(s))\,ds \\ &\ge \lambda \int_{\frac{1}{4}}^{\frac{3}{4}} G(\tau,s)a(s)(f_\infty-\varepsilon)u(s)\,ds \\ &\ge \frac{\lambda}{4} \int_{\frac{1}{4}}^{\frac{3}{4}} G(\tau,s)a(s)\,ds(f_\infty-\varepsilon)\|u\| \\ &\ge \|u\|. \end{aligned}$$

Thus, $\|Tu\| \ge \|u\|$. Hence

$$\|Tu\| \ge \|u\|, \qquad \text{for } u \in \mathcal{P} \cap \partial\Omega_2. \tag{8.13}$$

Applying (i) of Theorem 8.1 to (8.12) and (8.13) yields that T has a fixed point $u(t) \in \mathcal{P} \cap (\overline{\Omega}_2 \setminus \Omega_1)$. As such $u(t)$ is a desired solution of (8.0_λ), (8.1) for the given λ. The proof is complete. □

Theorem 8.3. *Assume that conditions* (A), (B), (C) *are satisfied. Then, for each* λ *satisfying*

$$\frac{4}{\left(\int_{\frac{1}{4}}^{\frac{3}{4}} G(\tau,s)a(s)\,ds\right) f_0} < \lambda < \frac{1}{\left(\int_0^1 s(1-s)a(s)\,ds\right) f_\infty}, \tag{8.14}$$

there exists at least one solution of (8.0_λ), (8.1) *in* $\mathcal{P}$.

Proof. Let λ be as in (8.14), and choose $\varepsilon > 0$ such that

$$\frac{4}{\left(\int_{\frac{1}{4}}^{\frac{3}{4}} G(\tau,s)a(s)\,ds\right)(f_0-\varepsilon)} \le \lambda \le \frac{1}{\left(\int_0^1 s(1-s)a(s)\,ds\right)(f_\infty+\varepsilon)}.$$

Let T be the cone preserving, completely continuous operator that was defined by (8.10).

Beginning with f_0, there exists an $H_1 > 0$ such that $f(x) \ge (f_0 - \varepsilon)x$, for $0 < x \le H_1$. So, for $u \in \mathcal{P}$ and $\|u\| = H_1$, we have

$$\begin{aligned} Tu(\tau) &= \lambda \int_0^1 G(\tau,s)a(s)f(u(s))\,ds \\ &\ge \lambda \int_{\frac{1}{4}}^{\frac{3}{4}} G(\tau,s)a(s)f(u(s))\,ds \\ &\ge \lambda \int_{\frac{1}{4}}^{\frac{3}{4}} G(\tau,s)a(s)(f_0-\varepsilon)u(s)\,ds \\ &\ge \frac{\lambda}{4} \int_{\frac{1}{4}}^{\frac{3}{4}} G(\tau,s)a(s)(f_0-\varepsilon)\|u\|\,ds \\ &\ge \|u\|. \end{aligned}$$

Thus, $\|Tu\| \ge \|u\|$. So, if we let $\Omega_1 = \{x \in \mathcal{B} \mid \|x\| < H_1\}$, then

$$\|Tu\| \ge \|u\|, \quad \text{for } u \in \mathcal{P} \cap \partial\Omega_1. \tag{8.15}$$

It remains to consider f_∞. There exists an $\overline{H}_2 > 0$ such that $f(x) \le (f_\infty + \varepsilon)x$, for all $x \ge \overline{H}_2$.

There are the two cases: (a) f is bounded, and (b) f is unbounded.

For case (a), suppose $N > 0$ is such that $f(x) \le N$, for all $0 < x < \infty$. Let

$$H_2 = \max\left\{2H_1, N\lambda \int_0^1 s(1-s)a(s)\,ds\right\}.$$

Then, for $u \in \mathcal{P}$ with $\|u\| = H_2$, we have

$$\begin{aligned} Tu(t) &= \lambda \int_0^1 G(t,s)a(s)f(u(s))\,ds \\ &\le \lambda N \int_0^1 s(1-s)a(s)\,ds \\ &\le \|u\|, \end{aligned}$$

so that $\|Tu\| \le \|u\|$. So, if $\Omega_2 = \{x \in \mathcal{B} \mid \|x\| < H_2\}$, then

$$\|Tu\| \le \|u\|, \text{ for } u \in \mathcal{P} \cap \partial\Omega_2. \tag{8.16}$$

For case (b), let $H_2 > \max\{2H_1, \overline{H}_2\}$ be such that $f(x) \le f(H_2)$, for $0 < x \le H_2$. Choosing $u \in \mathcal{P}$ with $\|u\| = H_2$,

$$\begin{aligned} Tu(t) &\le \lambda \int_0^1 s(1-s)a(s)f(u(s))\,ds \\ &\le \lambda \int_0^1 s(1-s)a(s)f(H_2)\,ds \\ &\le \lambda \int_0^1 s(1-s)a(s)\,ds(f_\infty + \varepsilon)H_2 \\ &= \lambda \int_0^1 s(1-s)a(s)\,ds(f_\infty + \varepsilon)\|u\| \\ &\le \|u\|, \end{aligned}$$

and so $\|Tu\| \le \|u\|$. For this case, if we let $\Omega_2 = \{x \in \mathcal{B} \mid \|x\| < H_2\}$, then

$$\|Tu\| \le \|u\|, \text{ for } u \in \mathcal{P} \cap \partial\Omega_2. \tag{8.17}$$

Thus, in either of the cases, an application of part (ii) of Theorem 8.1 yields a solution of (8.0_λ), (8.1) which belongs to $\mathcal{P} \cap (\overline{\Omega}_2 \setminus \Omega_1)$. This completes the proof. $\square$

8.2 Multiplicity of Positive Solutions

In this section, we provide another illustration of how the Guo-Krasnosel'skii fixed point theorem, Theorem 8.1, can be applied to obtain an arbitrary number of positive solutions to the boundary value problem

$$\begin{cases} u'' + a(t)f(u) = 0, \quad 0 \le t \le 1, & (8.18) \\ u(0) = u(1) = 0, & (8.19) \end{cases}$$

with the same hypotheses (A) and (B) of Section 8.1.

In addition, it is assumed that $f_0 := \lim_{x\to 0^+} \frac{f(x)}{x}$ and $f_\infty := \lim_{x\to\infty} \frac{f(x)}{x}$ exist in $[0, +\infty]$.

Define the constants,

$$I := \int_0^1 s(1-s)a(s)\,ds, \quad J := \int_{\frac{1}{4}}^{\frac{3}{4}} s(1-s)a(s)\,ds.$$

The Banach space $\mathcal{B}$, the cone $\mathcal{P}$, and the operator T will be as defined in the last subsection, (Subsection 8.1.1). In addition, for convenience and cleaner notation, define $\eta = \frac{1}{I}$ and $\mu = \frac{4}{J}$. In terms of η and μ, we state the following conditions, which govern the behavior of $f(x)$.

(C_1) There is a $p > 0$ such that $f(x) \leq \eta p$, for $0 \leq x \leq p$.
(C_2) There is a $q > 0$ such that $f(x) \geq \mu q$, for $\frac{q}{4} \leq x \leq q$.

Theorem 8.4. *Suppose there exist distinct $p, q > 0$ such that condition (C_1) holds for p and condition (C_2) holds for q. Then (8.18), (8.19) has a positive solution u such that $\|u\|$ is between p and q.*

Proof. Without loss of generality, we assume $0 < p < q$. If $u \in \mathcal{P}$ and $\|u\| = p$, then

$$\begin{aligned} Tu(t) &= \int_0^1 G(t,s)a(s)f(u(s))\,ds \\ &\leq \int_0^1 s(1-s)a(s)f(u(s))\,ds \\ &\leq \eta p \int_0^1 s(1-s)a(s)\,ds \\ &= \eta p I = p = \|u\|. \end{aligned}$$

Thus, $\|Tu\| \leq \|u\|$ for $\|u\| = p$.

Next, if $u \in \mathcal{P}$ and $\|u\| = q$, then from (8.7) of Section 8.1, we have for $\frac{1}{4} \leq t \leq \frac{3}{4}$,

$$\begin{aligned} Tu(t) &= \int_0^1 G(t,s)a(s)f(u(s))\,ds \\ &\geq \int_{\frac{1}{4}}^{\frac{3}{4}} G(t,s)a(s)f(u(s))\,ds \\ &\geq \mu q \int_{\frac{1}{4}}^{\frac{3}{4}} G(t,s)a(s)\,ds \\ &\geq \mu q \cdot \frac{1}{4} \int_{\frac{1}{4}}^{\frac{3}{4}} s(1-s)a(s)\,ds \\ &= \mu q \cdot \frac{1}{4} \cdot J = q = \|u\|. \end{aligned}$$

Thus, $\|Tu\| \geq \|u\|$ for $\|u\| = q$.

So, if $\Omega_p = \{x \in \mathcal{B} \,|\, \|x\| < p\}$ and $\Omega_q = \{x \in \mathcal{B} \,|\, \|x\| < q\}$, then by Theorem 8.1, the operator T has a fixed point in $\mathcal{P} \cap (\overline{\Omega}_q \setminus \Omega_p)$. As a result, the BVP (8.18), (8.19) has a solution u such that $p < \|u\| < q$. $\square$

Corollary 8.1. *The BVP* (8.18), (8.19) *has a positive solution provided, either*

(C_3) *$f_0 < \eta$ and $f_\infty > 4\mu$, or*
(C_4) *$f_0 > 4\mu$ and $f_\infty < \eta$.*

Proof. Suppose (C_3) holds. Then there is a sufficiently small $p > 0$ and a sufficiently large $q > 0$ such that,

$$\frac{f(x)}{x} \leq \eta, \quad 0 < x \leq p, \qquad \text{and } \frac{f(x)}{x} \geq 4\mu, \quad x \geq \frac{q}{4}.$$

Hence,

$$f(x) \leq \eta x \leq \eta p, \quad 0 \leq x \leq p, \qquad \text{and } f(x) \geq 4\mu x \geq \mu q, \quad \frac{q}{4} \leq x \leq q.$$

In particular, (C_3) implies (C_1) and (C_2).

For the rest of the proof, assume (C_4) holds. Then $f_0 > 4\mu$ and $f_\infty < \eta$. So, there are $0 < p < q$ such that

$$\frac{f(x)}{x} \geq 4\mu, \quad 0 < x < p, \qquad \text{and } \frac{f(x)}{x} \leq \eta, \quad x \geq q. \tag{8.20}$$

Then,

$$f(x) \geq 4\mu x \geq \mu p, \text{ for } \frac{p}{4} \leq x \leq p.$$

Hence condition (C_2) holds for p.

In order to show (C_1) holds, we examine two cases:

Case 1. Suppose $f(x)$ is bounded; that is, $f(x) \leq M$, for all $0 \leq y < \infty$. By (8.20) above, there is a $p^* \geq q$ such that $f(x) \leq M \leq \eta p^*$, for $0 \leq x \leq p^*$ provided $p^* \geq \frac{M}{\eta}$. Hence (C_1) holds for $p = p^*$.

Case 2. Suppose $f(x)$ is unbounded. Then, there is a $p^* \geq q$ such that $f(x) \leq f(p^*)$, for $0 \leq x \leq p^*$. Hence $f(x) \leq f(p^*) \leq \eta p^*$, and ($C_1$) holds for $p = p^*$.

An application of Theorem 8.4 yields the result. □

Now, we provide conditions sufficient for multiple positive solutions of (8.18), (8.19).

Theorem 8.5. *The BVP* (8.18), (8.19) *has at least two positive solutions u_1 and u_2, if* (C_1) *holds for some $p > 0$, and in addition, we have*

$$f_0 > 4\mu \text{ and } f_\infty > 4\mu. \tag{8.21}$$

Moreover, $0 < \|u_1\| < p < \|u_2\|$.

Proof. As in the proof of Corollary 8.1, there exist p_1, p_2 with $0 < p_1 < p < p_2$ satisfying

$$f(x) \geq \mu p_1, \quad \frac{p_1}{4} \leq x \leq p_1, \text{ and } f(x) \leq \mu p_2, \quad \frac{p_2}{4} \leq x \leq p_2.$$

By Theorem 8.4, we obtain the existence of solutions u_1 and u_2 of (8.18), (8.19) with $0 < p_1 < \|u_1\| < p < \|u_2\| < p_2$. □

Similarly, we have the following result.

Theorem 8.6. *The BVP* (8.18), (8.19) *has at least two positive solutions* u_1 *and* u_2, *if* (C_2) *holds for some* $p > 0$, *and in addition, we have*

$$f_0 < \eta \text{ and } f_\infty < \eta. \tag{8.22}$$

Moreover, $0 < \|u_1\| < p < \|u_2\|$.

Criteria for the existence of three (or more) positive solutions may be stated in a similar manner. As examples, we give the following two corollaries.

Corollary 8.2. *Suppose* (C_3) *in Corollary 8.1 holds and suppose there exist* $0 < p_1 < p_2$ *such that* (C_1) *holds for* $p = p_2$ *and* (C_2) *holds for* $p = p_1$. *Then the BVP* (8.18), (8.19) *has at least three positive solutions,* u_1, u_2, u_3, *satisfying* $0 < \|u_1\| < p_1 < \|u_2\| < p_2 < \|u_3\|$.

Corollary 8.3. *Suppose* (C_4) *in Corollary 8.1 holds and suppose there exist* $0 < p_1 < p_2$ *such that* (C_1) *holds for* $p = p_1$ *and* (C_2) *holds for* $p = p_2$. *Then the BVP* (8.18), (8.19) *has at least three positive solutions,* u_1, u_2, u_3, *satisfying* $0 < \|u_1\| < p_1 < \|u_2\| < p_2 < \|u_3\|$.

For final results of this section, we utilize the cone expansion-compression techniques iteratively to obtain various sufficient conditions for the existence of n solutions, for any $n \in \mathbb{N}$. We state two different sufficient conditions for odd n and two different sufficient conditions for even n.

Theorem 8.7 (Any odd number of solutions). *The boundary value problem* (8.18), (8.19) *has at least* n *positive solutions,* $n = 2k + 1$, $k \in \mathbb{N}$, *provided* (C_3) *holds, and there are* $0 < p_1 < p_2 < \cdots < p_{n-1}$ *such that* (C_2) *holds for* p_{2i-1}, $1 \le i \le k$, *while at the same time,* (C_1) *holds for* p_{2i}, $1 \le i \le k$. *Moreover,*

$$0 < \|u_1\| < p_1 < \|u_2\| < p_2 < \cdots < \|u_{n-1}\| < p_{n-1} < \|u_n\|,$$

for the positive solutions $u_1, \ldots, u_n$.

Theorem 8.8 (Any odd number of solutions). *The boundary value problem* (8.18), (8.19) *has at least* n *positive solutions,* $n = 2k + 1$, $k \in \mathbb{N}$, *provided* (C_4) *holds, and there are* $0 < p_1 < p_2 < \cdots < p_{n-1}$ *such that* (C_1) *holds for* p_{2i-1}, $1 \le i \le k$, *while at the same time,* (C_2) *holds for* p_{2i}, $1 \le i \le k$. *Moreover,*

$$0 < \|u_1\| < p_1 < \|u_2\| < p_2 < \cdots < \|u_{n-1}\| < p_{n-1} < \|u_n\|,$$

for the positive solutions $u_1, \ldots, u_n$.

Theorem 8.9 (Any even number of solutions). *The boundary value problem* (8.18), (8.19) *has at least* n *positive solutions,* $n = 2k$, $k \in \mathbb{N}$, *provided* (8.21) *holds, and there are* $0 < p_1 < p_2 < \cdots < p_{n-1}$ *such that* (C_1) *holds for* p_{2i-1}, $1 \le i \le k$, *while at the same time,* (C_2) *holds for* p_{2i}, $1 \le i \le k-1$. *Moreover,*

$$0 < \|u_1\| < p_1 < \|u_2\| < p_2 < \cdots < \|u_{n-1}\| < p_{n-1} < \|u_n\|,$$

for the positive solutions $u_1, \ldots, u_n$.

Theorem 8.10 (Any even number of solutions). *The boundary value problem* (8.18), (8.19) *has at least* n *positive solutions,* $n = 2k$, $k \in \mathbb{N}$, *provided* (8.22) *holds, and there are* $0 < p_1 < p_2 < \cdots < p_{n-1}$ *such that* (C_2) *holds for* p_{2i-1}, $1 \le i \le k$, *while at the same time,* (C_1) *holds for* p_{2i}, $1 \le i \le k-1$. *Moreover,*

$$0 < \|u_1\| < p_1 < \|u_2\| < p_2 < \cdots < \|u_{n-1}\| < p_{n-1} < \|u_n\|,$$

for the positive solutions $u_1, \ldots, u_n$.

8.3 Positive Solutions for Nonlinear Difference Equations

In this section, we are concerned with the existence of positive solutions for the second order nonlinear difference equation,

$$\Delta^2 x(t-1) + a(t)f(x(t)) = 0, \quad t \in [1, T+1], \tag{8.23}$$

satisfying the boundary conditions,

$$x(0) = x(T+2), \tag{8.24}$$

where

(A) $f : \mathbb{R}^+ \to \mathbb{R}^+$ is continuous ($\mathbb{R}^+$ =nonnegative reals),
(B) $a(t)$ is positive valued on $[1, T+1]$.

In this setting, intervals denote *discrete sets of integers.* We will have interest in the *extended real-valued* limits,

$$f_0 := \lim_{u \to 0^+} \frac{f(u)}{u} \text{ and } f_\infty := \lim_{u \to \infty} \frac{f(u)}{u}.$$

In particular, we shall seek positive solutions of (8.23), (8.24) that lie in a cone for the cases when f is *superlinear* (i.e., $f_0 = 0$ and $f_\infty = \infty$), and f is *sublinear* (i.e., $f_0 = \infty$ and $f_\infty = 0$).

Our solutions will be obtained via the Guo-Krasnosel'skii fixed point theorem, (Theorem 8.1), applied to a completely continuous summation operator A whose kernel, $G(t,s)$, is the Green's function for

$$\begin{cases} -\Delta^2 y(t-1) = 0, & t \in [1, T+1], \quad (8.25) \\ y(0) = y(T+2) = 0. & \quad (8.26) \end{cases}$$

In particular,

$$G(t,s) = \frac{1}{T+2} \begin{cases} t(T+2-s), & 0 \le t \le s \le T+1, \\ s(T+2-t), & 1 \le s \le t \le T+2, \end{cases} \tag{8.27}$$

from which

$$G(t,s) > 0 \text{ on } [1, T+1] \times [1, T+1], \tag{8.28}$$

and

$$G(t,s) \le G(s,s) = \frac{s(T+2-s)}{T+2}, \quad 1 \le s \le T+1,\ 0 \le t \le T+2. \tag{8.29}$$

It is not difficult to see that if Y is defined by

$$Y := \left\{ t \in \mathbb{Z} : \frac{T+1}{4} \le t \le \frac{3(T+1)}{4} \right\},$$

and if

$$\sigma = \min \left\{ \frac{\min Y}{T+1}, \frac{T+2-\max Y}{T+1} \right\},$$

then

$$G(t,s) \ge \sigma G(s,s) = \frac{\sigma s(T+2-s)}{T+2}, \quad t \in Y,\ 1 \le s \le T+1. \tag{8.30}$$

Note: $\sigma \ge \frac{1}{4}$.

Remark 8.1. We note at this point that $x(t)$ is a solution of (8.23), (8.24) if and only if

$$x(t) = \sum_{s=1}^{T+1} G(t,s) a(s) f(x(s)), \quad t \in [0, T+2].$$

To apply Theorem 8.1 in obtaining a solution of (8.23), (8.24), we let

$$\mathcal{B} = \{u : [0, T+2] \to \mathbb{R} \,|\, u(0) = u(T+2) = 0\},$$

with norm $\|u\| = \max_{0\le t\le T+2}|u(t)|$. Define a set, $\mathcal{P}$, by

$$\mathcal{P} = \left\{ u \in \mathcal{B} \,|\, u(t) \ge 0 \text{ on } [0, T+2], \text{ and } \min_{t\in Y} u(t) \ge \sigma\|u\| \right\}.$$

Claim: $\mathcal{P}$ is a cone in $\mathcal{B}$.

(i) We first show $\mathcal{P}$ is closed. Let $\{u_k\} \subset \mathcal{P}$ be such that $\|u_k - u_0\| \to 0$, as $k \to \infty$, where $u_0 \in \mathcal{B}$. Then, $u_k(t) \ge 0$ on $[0, T+2]$, and $\min_{t\in Y} u_k(t) \ge \sigma\|u_k\|$, $\forall k$. Thus, given $\varepsilon > 0$, there exists $K \in \mathbb{N}$ such that $-\varepsilon < u_k(t) - u_0(t) < \varepsilon$, for $t \in [0, T+2]$, $k \ge K$, and so $0 \le u_k(t) \le u_0(t) + \varepsilon$, for $t \in [0, T+2]$, $k \ge K$. Therefore, since ε was arbitrary, we have $u_0(t) \ge 0$ on $[0, T+2]$.

Also, $u_k(t) \ge \sigma\|u_k\|$ for $t \in Y$ and all k. Then $\lim_{k\to\infty} u_k(t) \ge \sigma \lim_{k\to\infty}\|u_k\|$ or $u_0(t) \ge \sigma\|u_0\|$ for $t \in Y$. Hence, $u_0 \in \mathcal{P}$, and $\mathcal{P}$ is closed.

(ii) Next, let $u, v \in \mathcal{P}$ and $\alpha, \beta \ge 0$. Then $\alpha u(t) + \beta v(t) \ge 0$, $t \in [0, T+2]$, and

$$\begin{aligned}\min_{t\in Y}\{\alpha u(t) + \beta v(t)\} &\ge \alpha \min_{t\in Y} u(t) + \beta \min_{t\in Y} v(t)\\ &\ge \alpha\|u\| + \beta\|v\|\\ &= \|\alpha u\| + \|\beta v\|\\ &\ge \|\alpha u + \beta v\|.\end{aligned}$$

Hence, $\alpha u + \beta v \in \mathcal{P}$.

(iii) Clearly, if $u, -u \in \mathcal{P}$, then $u(t) = 0$ for all $t \in [0, T+2]$.

We conclude that $\mathcal{P}$ is a cone in $\mathcal{B}$.

Before the main result of the section, let $\tau \in [0, T+2]$ be defined by

$$\sum_{s\in Y} G(\tau, s)a(s) = \max_{t\in[0,T+2]} \sum_{s\in Y} G(t, s)a(s). \tag{8.31}$$

Theorem 8.11. *Assume that conditions (A) and (B) are satisfied. If, either*

(i) $f_0 = 0$ *and* $f_\infty = \infty$ *(i.e., f is superlinear), or*
(ii) $f_0 = \infty$ *and* $f_\infty = 0$ *(i.e., f is sublinear),*

then (8.23), (8.24) *has at least one solution that lies in* $\mathcal{P}$.

Proof. We begin by defining a summation operator $A : \mathcal{P} \to \mathcal{B}$ by

$$Ax(t) = \sum_{s=1}^{T+1} G(t, s)a(s)f(x(s)), \quad x \in \mathcal{P}, \tag{8.32}$$

and we seek a fixed point of A in the cone $\mathcal{P}$.

We note that from (8.28), if $x \in \mathcal{P}$, then $Ax(t) \geq 0$ on $[0, T+2]$. Also from properties of $G(t,s)$, Ax satisfies the boundary conditions (8.24). Next, if $x \in \mathcal{P}$, we have from (8.29),

$$\begin{aligned} Ax(t) &= \sum_{s=1}^{T+1} G(t,s)a(s)f(x(s)) \\ &\leq \sum_{s=1}^{T+1} \frac{s(T+2-s)}{T+2} a(s)f(x(s)), \quad \text{for } t \in [0, T+2], \end{aligned}$$

so that

$$\begin{aligned} \|Ax\| &= \max_{t\in[0,T+2]} |Ax(t)| \\ &\leq \sum_{s=1}^{T+1} \frac{s(T+2-s)}{T+2} a(s)f(x(s)). \end{aligned} \tag{8.33}$$

Hence, if $x \in \mathcal{P}$, then (8.30) and (8.33) imply

$$\begin{aligned} \min_{t\in Y} Ax(t) &= \min_{t\in Y} \sum_{s=1}^{T+1} G(t,s)a(s)f(x(s)) \\ &\geq \sum_{s=1}^{T+1} \min_{t\in Y} G(t,s)a(s)f(x(s)) \\ &\geq \sigma \sum_{s=1}^{T+1} \frac{s(T+2-s)}{T+2} a(s)f(x(s)) \\ &\geq \sigma \|Ax\|. \end{aligned}$$

Thus, $A : \mathcal{P} \to \mathcal{P}$.

Our next argument is in showing A is completely continuous. So, let $x \in \mathcal{P}$ and $\varepsilon > 0$ be given. By the continuity of f, there exists $\delta > 0$ such that for any $u \in [0, \infty)$ with $|x(t) - u| < \delta$, $0 \leq t \leq T+2$, then $|f(x(t)) - f(u)| < \varepsilon$. Hence, let $y \in \mathcal{P}$ with $||x-y|| < \delta$. So, $|x(t)-y(t)| < \delta$, for $0 \leq t \leq T+2$. And we have

$$\begin{aligned} Ax(t) - Ay(t) &\leq \sum_{s=1}^{T+1} \frac{s(T+2-s)}{T+2} a(s)|f(x(s)) - f(y(s))| \\ &< \varepsilon \sum_{s=1}^{T+1} \frac{s(T+2-s)}{T+2} a(s), \quad \text{for } t \in [0, T+2], \end{aligned}$$

Thus, $\|Ax - Ay\| \le \varepsilon \sum_{s=1}^{T+1} \frac{s(T+2-s)}{T+2} a(s)$, and A is continuous.

Now, let $\{x_k\}$ be a bounded sequence in $\mathcal{P}$. Say $\|x_k\| \le M$, for all k. Since f is continuous, there exists $N > 0$ such that $|f(u)| \le N$, for all $u \in \mathbb{R}^+$ with $0 \le u \le M$. Then, for each $t \in [0, T+2]$, and for any k,

$$\begin{aligned} |Ax_k(t)| &\le \sum_{s=1}^{T+1} \frac{s(T+2-s)}{T+2} a(s) |f(x_k(s))| \\ &\le N \sum_{s=1}^{T+1} \frac{s(T+2-s)}{T+2} a(s). \end{aligned}$$

That is, for each $t \in [0, T+2]$, $\{Ax_k(t)\}$ is a bounded sequence of real numbers. By choosing successive subsequences, for each t, there exists a subsequence $\{Ax_{k_j}\}$ which converges (uniformly) for $t \in [0, T+2]$. We conclude that A is completely continuous.

We now examine the cases of the theorem.

(i) Assume first that $f_0 = 0$ and $f_\infty = \infty$. Since $f_0 = 0$, there exist $\varepsilon > 0$ and $H_1 > 0$ such that $\frac{f(u)}{u} \le \varepsilon$, for $0 < u \le H_1$, and $\varepsilon \sum_{s=1}^{T+1} \frac{s(T+2-s)}{T+2} a(s) \le 1$.

Let us choose $x \in \mathcal{P}$ with $\|x\| = H_1$. Then, we have from (8.29),

$$\begin{aligned} Ax(t) &\le \sum_{s=1}^{T+1} \frac{s(T+2-s)}{T+2} a(s) f(x_k(s)) \\ &\le \sum_{s=1}^{T+1} \frac{s(T+2-s)}{T+2} a(s) \varepsilon x(s) \\ &\le \varepsilon \sum_{s=1}^{T+1} \frac{s(T+2-s)}{T+2} a(s) \|x\| \\ &\le \|x\|, \quad \text{for } t \in [0, T+2]. \end{aligned}$$

Therefore, $\|Ax\| \le \|x\|$. Hence, if we set $\Omega_1 = \{u \in \mathcal{B} \mid \|u\| < H_1\}$, then

$$\|Ax\| \le \|x\|, \quad \text{for } x \in \mathcal{P} \cap \partial\Omega_1. \tag{8.34}$$

Since $f_\infty = \infty$, there exist $\lambda > 0$ and $\overline{H}_2 > 0$ such that $f(u) \ge \lambda u$, for $u \ge \overline{H}_2$, and $\lambda\sigma \sum_{s \in Y} G(\tau, s) a(s) \ge 1$.

Set $H_2 = \max\{2H_1, \frac{1}{\sigma}\overline{H}_2\}$, and $\Omega_2 = \{u \in \mathcal{B} \,|\, \|u\| < H_2\}$. If $x \in \mathcal{P} \cap \partial\Omega_2$, so that $\|x\| = H_2$, then $\min_{t\in Y} x(t) \geq \sigma\|x\| \geq \overline{H}_2$. And we have

$$\begin{aligned} Ax(\tau) &= \sum_{s=1}^{T+1} G(\tau,s)a(s)f(x(s)) \\ &\geq \sum_{s\in Y} G(\tau,s)a(s)f(x(s)) \\ &\geq \sum_{s\in Y} G(\tau,s)a(s)\lambda x(s) \\ &\geq \lambda \sum_{s\in Y} G(\tau,s)a(s)\sigma\|x\| \\ &\geq \|x\|. \end{aligned}$$

Thus, $\|Ax\| \geq \|x\|$, and so

$$\|Ax\| \geq \|x\|, \quad \text{for } x \in \mathcal{P} \cap \partial\Omega_2. \tag{8.35}$$

Application of Theorem 8.1 to (8.34) and (8.35) yields a fixed point of A that lies in $\mathcal{P} \cap (\overline{\Omega}_2 \setminus \Omega_1)$. This fixed point is a desired solution of (8.23), (8.24) for the case of f sublinear.

(ii) Assume now that $f_0 = \infty$ and $f_\infty = 0$. Since $f_0 = \infty$, there exist $\overline{\eta} > 0$ and $J_1 > 0$ such that $f(x) \geq \overline{\eta}x$, for $0 < x \leq J_1$, and $\overline{\eta}\sigma \sum_{s\in Y} G(\tau,s)a(s) \geq 1$. In this case, define

$$\Omega_1 = \{u \in \mathcal{B} \,|\, \|u\| < J_1.$$

Then for $x \in \mathcal{P}\cap\partial\Omega_1$, we have $f(x(s)) \geq \overline{\eta}x(s)$, $s \in [1, T+1]$, and moreover, $x(s) \geq \sigma\|x\|$, $s \in Y$. Thus,

$$\begin{aligned} Ax(\tau) &= \sum_{s=1}^{T+1} G(\tau,s)a(s)f(x(s)) \\ &\geq \sum_{s\in Y} G(\tau,s)a(s)f(x(s)) \\ &\geq \sum_{s\in Y} G(\tau,s)a(s)\overline{\eta}x(s) \\ &\geq \overline{\eta} \sum_{s\in Y} G(\tau,s)a(s)\sigma\|x\| \\ &\geq \|x\|, \end{aligned}$$

from which we have

$$\|Ax\| \geq \|x\|, \quad \text{for } x \in \mathcal{P} \cap \partial\Omega_1. \tag{8.36}$$

It remains for us to consider $f_\infty = 0$. In this case, there exist $\overline{\lambda} > 0$ and $\overline{J}_2 > 0$ such that $f(u) \le \overline{\lambda} u$, for $u \ge \overline{J}_2$, and $\overline{\lambda}\sum_{s=1}^{T+1} \frac{s(T+2-s)}{T+2} a(s) \le 1$. There are two cases: (I) f is bounded, and (II) f is unbounded.

(I) Suppose $M > 0$ is such that $f(u) \le M$, for all $0 < u < \infty$. Let $J_2 = \max\{2J_1, M\sum_{s=1}^{T+1} \frac{s(T+2-s)}{T+2} a(s)\}$, and let

$$\Omega_2 = \{u \in \mathcal{B} \,|\, \|u\| < J_2\}.$$

Then, for $x \in \mathcal{P} \cap \partial\Omega_2$, we have

$$\begin{aligned} Ax(t) &= \sum_{s=1}^{T+1} G(\tau, s) a(s) f(x(s)) \\ &\le M \sum_{s=1}^{T+1} \frac{s(T+2-s)}{T+2} a(s) \\ &\le \|x\|, \quad \text{for } t \in [0, T+2], \end{aligned}$$

and so

$$\|Ax\| \le \|x\|, \quad \text{for } x \in \mathcal{P} \cap \partial\Omega_2. \tag{8.37}$$

(II) Let $J_2 > \max\{2J_1, \overline{J}_2\}$ be such that $f(u) \le f(J_2)$, for $0 < u \le J_2$. Let

$$\Omega_2 = \{u \in \mathcal{B} \,|\, \|u\| < J_2\}.$$

Choosing $x \in \mathcal{P} \cap \partial\Omega_2$,

$$\begin{aligned} Ax(t) &\le \sum_{s=1}^{T+1} \frac{s(T+2-s)}{T+2} a(s) f(x(s)) \\ &\le \sum_{s=1}^{T+1} \frac{s(T+2-s)}{T+2} a(s) f(J_2) \\ &\le \sum_{s=1}^{T+1} \frac{s(T+2-s)}{T+2} a(s) \overline{\lambda} f(J_2) \\ &\le J_2 = \|x\|, \quad \text{for } t \in [0, T+2], \end{aligned}$$

and so,

$$\|Ax\| \le \|x\|, \quad \text{for } x \in \mathcal{P} \cap \partial\Omega_2. \tag{8.38}$$

Thus, in each of the subcases, we apply Theorem 8.1 to (8.36), (8.37) or to (8.36), (8.38) to obtain a fixed point of A that lies in $\mathcal{P} \cap (\overline{\Omega}_2 \setminus \Omega_1)$. This fixed point is a solution of (8.23), (8.24). □

8.4 Multiple Symmetric Positive Solutions

This section is devoted to the second order boundary value problem,

$$\begin{cases} y'' + f(y) = 0, \quad 0 \le t \le 1, & (8.39) \\ y(0) = 0 = y(1), & (8.40) \end{cases}$$

where it is assumed throughout that $f : \mathbb{R} \to [0, \infty)$ is continuous. As a consequence of the nonnegativity of f, a solution $y \in C^{(2)}[0,1]$ of (8.39), (8.40) is both nonnegative and concave on $[0,1]$.

Our goal in this section is to impose growth conditions on f which ensure the existence of at least three symmetric positive solutions of (8.39), (8.40). Our ability to obtain symmetric solutions arises from the symmetry of an associated Green's function.

Our techniques make use of a multiple fixed point theorem of a cone preserving operator which is due to Leggett and Williams.

In what follows, let $(\mathcal{B}, \|\cdot\|)$ be a real Banach space and let $\mathcal{P} \subset \mathcal{B}$ be a cone. We recall that $\mathcal{P}$ generates a partial order, $\preceq$, on $\mathcal{B}$ where, for $u, v \in \mathcal{B}$, we say $u \preceq v$ with respect to $\mathcal{P}$, if $v - u \in \mathcal{P}$. So, in some sense, $\preceq$ associates positive elements in $\mathcal{B}$ with the elements of $\mathcal{P}$.

Definition 8.1. A mapping $\alpha : \mathcal{P} \to [0, \infty)$, ($\mathcal{P}$ is a cone), which is continuous, is called a *nonnegative concave functional* on $\mathcal{P}$, if

$$\alpha(tx + (1-t)y) \ge t\alpha(x) + (1-t)\alpha(y),$$

for all $x, y \in \mathcal{P}$ and all $0 \le t \le 1$.

Definition 8.2. For numbers $0 < a < b$ and α a nonnegative concave functional on a cone $\mathcal{P}$, define convex sets $\mathcal{P}_r$ and $\mathcal{P}(\alpha, a, b)$ respectively by

$$\mathcal{P}_r = \{y \in \mathcal{P} \mid \|y\| < r\}$$

and

$$\mathcal{P}(\alpha, a, b) = \{y \in \mathcal{P} \mid a \le \alpha(y), \ \|y\| \le b\}.$$

In obtaining multiple symmetric positive solutions of (8.39), (8.40), the following fixed point theorem of Leggett and Williams will be fundamental.

Theorem 8.12. *Let $A : \overline{\mathcal{P}}_c \to \overline{\mathcal{P}}_c$ be completely continuous and α be a nonnegative continuous concave functional on $\mathcal{P}$ such that $\alpha(y) \le \|y\|$, for all $y \in \overline{\mathcal{P}}_c$. Suppose there exist $0 < a < b < d \le c$ such that*

(C$_1$) $\{y \in \mathcal{P}(\alpha, a, d) \mid \alpha(y) > b\} \neq \emptyset$, *and* $\alpha(Ay) > b$, *for* $y \in \mathcal{P}(\alpha, b, d)$,
(C$_2$) $\|Ay\| < a$, *for* $\|y\| \leq a$, *and*
(C$_3$) $\alpha(Ay) > b$, *for* $y \in \mathcal{P}(\alpha, b, c)$ *with* $\|Ay\| > d$.

Then A has at least three fixed point points y_1, y_2 and y_3 satisfying

$$\|y_1\| < a, \; b < \alpha(y_2), \; \text{and } \|y_3\| > a \text{ with } \alpha(y_3) < b.$$

In applying this fixed point theorem, we will make use of the Green's function, $G(t, s)$, for

$$-y'' = 0 \tag{8.41}$$

satisfying (8.40), which again is given by

$$G(t,s) = \begin{cases} t(1-s), & 0 \leq t \leq s \leq 1, \\ s(1-t), & 0 \leq s \leq t \leq 1. \end{cases} \tag{8.42}$$

Lemma 8.1. *Let $G(t, s)$ be the Green's function for* (8.41), (8.40). *Then*

(a) $0 < G(t, s) \leq G(s, s) = s(1 - s)$, $0 < t, s < 1$. (8.43)
(b) $G(t, s) \geq \frac{1}{4}G(s, s) = \frac{1}{4}s(1 - s)$, *for* $\frac{1}{4} \leq t \leq \frac{3}{4}$, $0 \leq s \leq 1$. (8.44)
(c) $\max_{0 \leq t \leq 1} \int_0^1 G(t, s)\, ds = \frac{1}{8}$. (8.45)
(d) $\min_{\frac{1}{4} \leq t \leq \frac{3}{4}} \int_{\frac{1}{4}}^{\frac{3}{4}} G(t, s)\, ds = \int_{\frac{1}{4}}^{\frac{3}{4}} G(\frac{1}{4}, s)\, ds = \frac{1}{16}$. (8.46)
(e) $\min_{0 \leq r \leq 1} \frac{G(\frac{1}{4}, r)}{G(\frac{1}{2}, r)} = \frac{1}{2}$. (8.47)

Proof. (a) For $t \leq s \leq 1$, $t(1 - s) \leq s(1 - s)$, and for $0 \leq s \leq t$, $s(1 - t) \leq s(1 - s)$. Thus, $G(t, s) \leq s(1 - s) = G(s, s)$.

(b) In the quotient $\frac{t(1-s)}{s(1-s)}$, for $\frac{1}{4} \leq t \leq \frac{3}{4}$ and $0 < s < 1$, the quotient is minimized when $s = 1$ and $t = \frac{1}{4}$. Similarly, the quotient $\frac{s(1-t)}{s(1-s)}$, for $\frac{1}{4} \leq t \leq \frac{3}{4}$ and $0 < s < 1$, is minimized when $t = \frac{3}{4}$ and when $s = 0$. Thus, $\frac{G(t,s)}{G(s,s)} \geq \frac{1}{4}$, $\frac{1}{4} \leq t \leq \frac{3}{4}$, $0 < s < 1$.

(c) See Exercise 26 of Section 2.4.3 for the proof of (c).

Exercise 37. Prove statements (d) and (e). □

Next, let $\mathcal{B} = C[0, 1]$ be endowed with the maximum norm, $\|y\| = \max_{0 \leq t \leq 1} |y(t)|$, and let the cone $\mathcal{P} \subset \mathcal{B}$ be defined by

$$\mathcal{P} = \{y \in \mathcal{B} \mid y \text{ is concave, symmetric and nonnegative valued on } [0, 1]\}.$$

Finally, let the nonnegative, continuous, concave functional $\alpha : \mathcal{P} \to [0, \infty)$ be defined by

$$\alpha(y) = \min_{\frac{1}{4} \le t \le \frac{3}{4}} y(t), \quad y \in \mathcal{P}. \tag{8.48}$$

We observe here that, for each $y \in \mathcal{P}$,

$$\alpha(y) = y\left(\frac{1}{4}\right) \le y\left(\frac{1}{2}\right) = \|y\|, \tag{8.49}$$

and also recall that $y \in \mathcal{B}$ is a solution of (8.39), (8.40) if and only if

$$y(t) = \int_0^1 G(t,s) f(y(s))\, ds, \quad 0 \le t \le 1. \tag{8.50}$$

We now present the main result of this chapter.

Theorem 8.13. *Let* $0 < a < b < \frac{c}{2}$, *and suppose* f *satisfies*

(i) $f(w) < 8a$, *for* $0 \le w \le 1$,
(ii) $f(w) \ge 16b$, *for* $b \le w \le 2b$, *and*
(iii) $f(w) \le 8c$, *for* $0 \le w \le c$.

Then, the BVP (8.39), (8.40) *has three symmetric positive solutions* y_1, y_2 *and* y_3 *satisfying* $\|y_1\| = y_1(\frac{1}{2}) < a$, $b < \alpha(y_2) = y_2(\frac{1}{4})$, *and* $\|y_3\| = y_3(\frac{1}{2}) > a$ *with* $\alpha(y_3) = y_3(\frac{1}{4}) < b$.

Proof. We begin by defining the completely continuous operator $A : \mathcal{B} \to \mathcal{B}$ by

$$Ay(t) = \int_0^1 G(t,s) f(y(s))\, ds,$$

(See the proof of Theorem 3.5 of Section 3.1, for the complete continuity of A). We seek fixed points of A which satisfy the conclusion of the theorem. We note first, if $y \in \mathcal{P}$, then from sign conditions on f and $G(t,s)$,

$$Ay(t) = \int_0^1 G(t,s) f(y(s))\, ds \ge 0, \quad 0 \le t \le 1.$$

In addition, $(Ay)''(t) = -f(y(t)) \le 0$, $0 \le t \le 1$, and so $Ay(t)$ is concave on $[0,1]$. Finally, for $0 \le t \le \frac{1}{2}$,

$$Ay(1-t) = \int_0^1 G(1-t,s) f(y(s))\, ds.$$

Let $v = 1 - s$, then $-dv = ds$, and $G(1-t,s) = G(t,v)$. By the symmetry of $y(s)$,

$$\begin{aligned}\int_0^1 G(1-t,s)f(y(s))\,ds &= -\int_1^0 G(t,v)f(y(-v))\,dv \\ &= \int_0^1 G(t,v)f(y(v))\,dv \\ &= Ay(t).\end{aligned}$$

In particular, $Ay(1-t) = Ay(t)$, $0 \le t \le \frac{1}{2}$, so that Ay is also symmetric. Hence, $A : \mathcal{P} \to \mathcal{P}$.

We now show that the conditions of Theorem 8.12 are satisfied. We noted in (8.49) that $\alpha(y) \le \|y\|$, for all $y \in \mathcal{P}$. Now choose $y \in \overline{\mathcal{P}}_c$. Then $\|y\| \le c$ and by assumption (iii), $f(y(s)) \le 8c$, $0 \le s \le 1$. Thus, from part (c) of Lemma 8.1,

$$\begin{aligned}\|Ay\| &= \max_{0\le t\le 1}\int_0^1 G(t,s)f(y(s))\,ds \\ &\le \max_{0\le t\le 1}\int_0^1 G(t,s)8c\,ds \\ &= c.\end{aligned}$$

Thus, $Ay \in \overline{\mathcal{P}}_c$ so that $A : \overline{\mathcal{P}}_c \to \overline{\mathcal{P}}_c$. In a similar vein, let $y \in \overline{\mathcal{P}}_a$ so that $\|y\| \le a$, and by (i), $f(y(s)) < 8a$, $0 \le s \le 1$. Again by part (c) of Lemma 8.1,

$$\begin{aligned}\|Ay\| &= \max_{0\le t\le 1}\int_0^1 G(t,s)f(y(s))\,ds \\ &\le \max_{0\le t\le 1}\int_0^1 G(t,s)8a\,ds \\ &= a.\end{aligned}$$

Thus, (C_2) of Theorem 8.12 is satisfied. To fulfill (C_1) of Theorem 8.12, we note that $x(t) = 2b$, $0 \le t \le 1$, is a member of $\mathcal{P}(\alpha, b, 2b)$ and $\alpha(x) = \alpha(2b) > b$, and so $\{y \in \mathcal{P}(\alpha, b, 2b) \,|\, \alpha(y) > b\} \ne \emptyset$. In addition, if we choose $y \in \mathcal{P}(\alpha, b, 2b)$, then $\alpha(y) = y(\frac{1}{4}) \ge b$, and so $b \le y(s) \le 2b$, $\frac{1}{4} \le s \le \frac{3}{4}$. Thus, for any $y \in \mathcal{P}(\alpha, b, 2b)$, assumption (ii) yields $f(y(s)) \ge 16b$, $\frac{1}{4} \le s \le \frac{3}{4}$, and hence, from part (d) of Lemma 8.1,

$$\begin{aligned}\alpha(Ay) &= \min_{\frac{1}{4}\le t\le\frac{3}{4}} Ay(t) = Ay\left(\frac{1}{4}\right)\\ &= \int_0^1 G\left(\frac{1}{4},s\right) f(y(s))\,ds > \int_{\frac{1}{4}}^{\frac{3}{4}} G\left(\frac{1}{4},s\right) f(y(s))\,ds\\ &\ge \int_{\frac{1}{4}}^{\frac{3}{4}} G\left(\frac{1}{4},s\right) 16b\,ds = b.\end{aligned}$$

Hence, (C_1) of Theorem 8.12 is satisfied.

We finally exhibit that (C_3) of Theorem 8.12 is also satisfied. (In particular, we show if $y \in \mathcal{P}(\alpha, b, c)$ and $\|Ay\| > 2b$, then $\alpha(Ay) > b$.) Thus, choose $y \in \mathcal{P}(\alpha, b, c)$ such that $\|Ay\| > 2b$. Then, from part (e) of Lemma 8.1,

$$\begin{aligned}\alpha(Ay) &= Ay\left(\frac{1}{4}\right) = \int_0^1 G\left(\frac{1}{4},s\right) f(y(s))\,ds\\ &= \int_0^1 \frac{G\left(\frac{1}{4},s\right)}{G\left(\frac{1}{2},s\right)} G\left(\frac{1}{2},s\right) f(y(s))\,ds\\ &\ge \min_{0\le r\le 1} \frac{G\left(\frac{1}{4},r\right)}{G\left(\frac{1}{2},r\right)} \int_0^1 G\left(\frac{1}{2},s\right) f(y(s))\,ds\\ &= \frac{1}{2} Ay\left(\frac{1}{2}\right) = \frac{1}{2}\|Ay\| > b,\end{aligned}$$

and (C_3) of Theorem 8.12 holds. Hence, Theorem 8.12 yields the conclusions of this theorem. □

8.5 Positive Solutions for a Singular Third Order BVP

Singular nonlinear BVP's appear in a variety of applications and often only positive solutions are important. Our focus will be on a nonlinear problem where the nonlinearity is not defined at zero, even though zero may be prescribed as a boundary value at an endpoint. We also do not require continuity in the independent variable at the endpoints of its interval of definition.

Namely, in this section, we are concerned with positive solutions for the singular third order boundary value problem

$$\begin{cases} u''' = f(x,u), \quad 0 < x < 1, & (8.51)\\ u(0) = u(1) = u''(1) = 0, & (8.52)\end{cases}$$

where $f(t, y)$ is singular at $x = 0, 1$, $y = 0$, and may be singular at $y = \infty$.

We assume the following conditions on f:

(H1) $f(x,y) : (0,1)\times(0,\infty) \to (0,\infty)$ is continuous, and $f(x,y)$ is decreasing in y, for every x,

(H2) $\lim_{y\to 0^+} f(x,y) = +\infty$ and $\lim_{y\to+\infty} f(x,y) = 0$ uniformly on compact subsets of $(0,1)$.

Note: $f(x,y) = \frac{1}{\sqrt[3]{x(1-x)y}}$ satisfies (H1) and (H2).

We will convert the problem (8.51), (8.52) into an integral equation, from which we define a sequence of decreasing integral operators associated with a sequence of perturbed integral equations. Applications of a Gatica, Oliker and Waltman fixed point theorem yield a sequence of fixed points of the integral operators. A solution of (8.51), (8.52) is then obtained from a subsequence of the fixed points.

Let $(\mathcal{B}, \|\cdot\|)$ be a real Banach space, and let $\mathcal{P}$ be a cone in $\mathcal{B}$.

Definition 8.3. For a cone $\mathcal{P} \subset \mathcal{B}$, let $\preceq$ be the partial order induced by $\mathcal{P}$ on $\mathcal{B}$. If $x, y \in \mathcal{B}$ with $x \preceq y$, let $\langle x, y\rangle$ denote the *closed order interval between* x *and* y and be defined by $\langle x,y\rangle := \{z \in B \,|\, x \preceq z \preceq y\}$. $\mathcal{P}$ is *normal* in $\mathcal{B}$ provided there exists $\delta > 0$ such that $\|e_1 + e_2\| \geq \delta$, $\forall e_1, e_2 \in \mathcal{P}$ with $\|e_1\| = \|e_2\| = 1$.

Remark 8.2. If $\mathcal{P}$ is normal in $\mathcal{B}$, then closed order intervals are norm bounded.

We now state the Gatica, Oliker, and Waltman Fixed Point Theorem [Gatica *et al.* (1989)] on which the main result of this section depends.

Theorem 8.14. *Let $\mathcal{B}$ be a Banach space, $\mathcal{P}$ a normal cone, $J \subseteq \mathcal{P}$ such that, if $x, y \in J$, $x \preceq y$, then $\langle x,y\rangle \subseteq J$, and let $T : J \to \mathcal{P}$ be a continuous decreasing mapping which is compact (or completely continuous) on any closed order interval contained in J. Suppose there exists $x_0 \in J$ such that T^2x_0 is defined, and furthermore, Tx_0 and T^2x_0 are order comparable to x_0. Then T has a fixed point in J provided that, either*

(I) *$Tx_0 \preceq x_0$ and $T^2x_0 \preceq x_0$, or $x_0 \preceq Tx_0$ and $x_0 \preceq T^2x_0$, or*

(II) *The complete sequence of iterates $\{T^nx_0\}_{n=0}^{\infty}$ is defined, and there exists $y_0 \in J$ such that $y_0 \preceq T^nx_0$, for every n.*

In setting the stage for application of Theorem 8.14, we let the Banach space $(\mathcal{B}, \|\cdot\|)$ be defined by

$\mathcal{B} := \{u : [0,1] \to \mathbb{R} \,|\, u \text{ is continuous}\}$ with norm $\|u\| := \sup_{0\leq x\leq 1} |u(x)|$.

We define a cone $\mathcal{P} \subset \mathcal{B}$ by $\mathcal{P} := \{u \in \mathcal{B} \,|\, u(x) \geq 0 \text{ on } [0,1]\}$.

We observe that, if $y(x)$ is a solution of (8.51), (8.52), then $y'''(x) \geq 0$, $y(x) \geq 0$ and $y(x)$ is concave.

Next, we define $g(x) : [0,1] \to [0, \frac{3}{4}]$ by $g(x) := \min\{1-x, 3x\}$, and for $\theta > 0$, define $g_\theta(x) := \theta g(x)$. Then

$$\max_{0 \leq x \leq 1} g(x) = \frac{3}{4} \text{ and } \max_{0 \leq x \leq 1} g_\theta(x) = \frac{3\theta}{4}.$$

We will assume hereafter:

(H3) $\int_0^1 f(x, g_\theta(x))\, dx < \infty$, for all $\theta > 0$.

Note: The function $f(x,y) = \frac{1}{\sqrt[3]{x(1-x)y}}$ also satisfies (H3). In particular, for $\theta > 0$,

$$\begin{aligned}\int_0^1 f(x, g_\theta(x))\, dx &= \frac{1}{\sqrt[3]{\theta}} \left[\int_0^{\frac{1}{4}} \frac{1}{\sqrt[3]{3x^2(1-x)}}\, dx + \int_{\frac{1}{4}}^1 \frac{1}{\sqrt[3]{x(1-x)^2}}\, dx \right] \\ &< \frac{\sqrt[3]{3} + \sqrt[3]{3}^4}{\sqrt[3]{\theta}} = 4\sqrt[3]{\frac{3}{\theta}}.\end{aligned}$$

We shall make extensive application of the next theorem due to Graef and Yang [Graef and Yang (2006)].

Theorem 8.15. *Let $u(x) \in C^{(3)}[0,1]$. If $u(x)$ satisfies the boundary conditions (8.52) is such that $u''' \geq 0$ on $[0,1]$, then*

$$u(x) \geq \min\{1-x, 3x\} \sup_{0 \leq x \leq 1} |u(x)|. \tag{8.53}$$

So, from this theorem, for each positive solution $u(x)$ of (8.51), (8.52), there exists $\theta > 0$ such that $g_\theta(x) \leq u(x)$, for $0 \leq x \leq 1$. In particular with $\theta = \sup_{0 \leq x \leq 1} |u(x)|$, then

$$u(x) \geq \min\{1-x, 3x\}\theta = g_\theta(x), \quad 0 \leq x \leq 1.$$

Next, we define $D \subset \mathcal{P}$ by

$$D := \{v \in \mathcal{P} \,|\, \text{there exists } \theta(v) > 0 \text{ such that } g_\theta(x) \leq v(x),\ 0 \leq x \leq 1\}.$$

Observe that, for each $v \in D$ and $\frac{1}{8} \leq x \leq \frac{5}{8}$,

$$u(x) \geq g_\theta(x) = \min\{1-x, 3x\}\theta \geq \frac{3}{8}\theta, \tag{8.54}$$

and for each positive solution $u(x)$ of (8.51), (8.52),

$$u(x) \geq g(x) \sup_{0\leq x\leq 1} |u(x)| = \frac{3}{8} \sup_{0\leq x\leq 1} |u(x)|, \quad \frac{1}{8} \leq x \leq \frac{5}{8}. \tag{8.55}$$

There is a Green's function, $G(x,s)$, for $y''' = 0$ and (8.52), which will play the role of a kernel for certain compact operators meeting the requirements of Theorem 8.14. Directly,

$$G(x,s) = \frac{1}{2}\begin{cases} x(1-x) - x(1-s)^2, & 0 \leq x \leq s \leq 1, \\ x(1-x) - x(1-s)^2 + (x-s)^2, & 0 \leq s \leq x \leq 1, \end{cases}$$

and properties important for us include:

(i) $G(x,s) > 0$ on $(0,1) \times (0,1)$ and continuous on $[0,1] \times [0,1]$.
(ii) $G(0,s) = 0$ for $0 < s \leq 1$, and $G(1,s) = \frac{\partial^2}{\partial x^2}G(1,s) = 0$, $0 \leq s < 1$.
(iii) $\frac{\partial^2}{\partial x^2}G(x,s)$ is continuous as a function of x on $[0,s]$ and on $[s,1]$.
(iv) $\frac{\partial}{\partial x}G(0,s) = \frac{s(2-s)}{2} > 0$ and $\frac{\partial}{\partial x}G(1,s) = -\frac{s^2}{2} < 0$, for $0 < s < 1$.

Now, we define an integral operator $T : D \to \mathcal{P}$ by

$$(Tu)(x) = \int_0^1 G(x,s)f(s,u(s))\,ds, \quad u \in D.$$

We will show that T is well-defined on D and decreasing and that $T : D \to D$. So, first let $v \leq u$ both belong to D. Then, there exists $\theta > 0$ such that $g_\theta(x) \leq v(x)$. By (H1) and (H3), and (i) above,

$$\begin{aligned} 0 &\leq \int_0^1 G(x,s)f(s,u(s))\,ds \\ &\leq \int_0^1 G(x,s)f(s,v(s))\,ds \\ &\leq \int_0^1 G(x,s)f(s,g_\theta(s))\,ds \\ &< \infty. \end{aligned}$$

Next, for $v \in D$, let $w(x) := \int_0^1 G(x,s)f(s,v(s))\,ds \geq 0$, $0 \leq x \leq 1$. From properties of $G(x,s)$, $w'''(x) = f(x,v(s)) > 0$, $0 < x < 1$, and $w(0) = w(1) = w''(1) = 0$, which imply $w''(x) \leq 0$, or that $w(x)$ is concave. Moreover, by Theorem 8.15, $w = Tv \in D$. So $T : D \to D$.

Remark 8.3. It is, of course true that $Tu = u$ if, and only if, u is a solution of (8.51), (8.52). So, we seek solutions of (8.51), (8.52) that belong to D. It follows from (8.54) and (8.55), in the context of our Banach space $\mathcal{B}$, that for each positive solution $u(x)$ of (8.51), (8.52),

$$u(x) \geq g(x)\|u\| \geq \frac{3}{8}\|u\|, \quad \frac{1}{8} \leq x \leq \frac{5}{8}. \tag{8.56}$$

8.5.1 *A priori bounds on norms of solutions*

We now exhibit that solutions of (8.51), (8.52) have positive *a priori* upper and lower bounds on their norms.

Lemma 8.2. *If f satisfies* (H1)-(H3), *then there exists an $S > 0$ such that $\|u\| \le S$, for any solution u of* (8.51), (8.52) *in D.*

Proof. We assume the conclusion is false. Then there exists a sequence $\{u_m\}_{m=1}^{\infty} \subset D$ of solutions of (8.51), (8.52) such that $u_m(x) > 0$ for all $0 < x < 1$, and

$$\|u_m\| \le \|u_{m+1}\| \text{ and } \lim_{m\to\infty} \|u_m\| = \infty.$$

From (8.55) or (8.56),

$$u_m(x) \ge \frac{3}{8}\|u_m\|, \quad \frac{1}{8} \le x \le \frac{5}{8}.$$

So, $\lim_{m\to\infty} u_m(x) = \infty$ uniformly on $[\frac{1}{8}, \frac{5}{8}]$. Now, let

$$M := \max\{G(x,s) \mid (x,s) \in [0,1] \times [0,1]\}.$$

From (H2), there exists $m_0 \in \mathbb{N}$ such that, for all $m \ge m_0$ and $\frac{1}{8} \le x \le \frac{5}{8}$,

$$f(x, u_m(x)) \le \frac{2}{M}.$$

Set $\theta := \|u_{m_0}\|$. Then, for $m \ge m_0$,

$$u_m(x) \ge g_{\|u_m\|}(x) \ge g_{\|u_{m_0}\|}(x) = g_\theta(x), \quad 0 \le x \le 1.$$

And so, for $m \ge m_0$ and $0 \le x \le b$,

$$\begin{aligned} u_m(x) =& Tu_m(x) = \int_0^1 G(x,s) f(s, u_m(s))\, ds \\ =& \left[\int_0^{\frac{1}{8}} + \int_{\frac{5}{8}}^1\right] G(x,s) f(s,u_m(s))\, ds + \int_{\frac{1}{8}}^{\frac{5}{8}} G(x,s) f(s,u_m(s))\, ds \\ \le& \left[\int_0^{\frac{1}{8}} + \int_{\frac{5}{8}}^1\right] G(x,s) f(s,u_m(s))\, ds + \int_{\frac{1}{8}}^{\frac{5}{8}} M \cdot \frac{2}{M}\, ds \\ \le& \left[\int_0^{\frac{1}{8}} + \int_{\frac{5}{8}}^1\right] G(x,s) f(s,u_m(s))\, ds + 1 \\ \le& M \int_0^1 f(s, g_\theta(s))\, ds + 1, \end{aligned}$$

which contradicts $\lim_{m\to\infty} \|u_m\| = \infty$. Therefore, there exists $S > 0$ such that $\|u\| \le S$, for any solution $u \in D$ of (8.51), (8.52). □

Lemma 8.3. *If f satisfies* (H1)-(H3), *then there exists an $R > 0$ such that $\|u\| \geq R$, for any solution u of* (8.51), (8.52) *in D.*

Proof. Again, we assume the conclusion is false. Then, there exists a sequence $\{u_m\}_{m=1}^{\infty} \subset D$ of solutions of (8.51), (8.52) such that $u_m(x) > 0$, for all $0 < x < 1$, and

$$\|u_m\| \geq \|u_{m+1}\| \text{ and } \lim_{m\to\infty} \|u_m\| = 0.$$

In particular, $\lim_{m\to\infty} \|u_m\| = 0$ uniformly on $[0, 1]$.

Now define

$$\overline{m} := \min\left\{G(x,s) \;\middle|\; (x,s) \in \left[\frac{1}{8}, \frac{5}{8}\right] \times \left[\frac{1}{8}, \frac{5}{8}\right]\right\} > 0.$$

From (H2), there exists $\delta > 0$ such that for $\frac{1}{8} \leq x \leq \frac{5}{8}$ and $0 < y < \delta$,

$$f(x,y) > \frac{2}{\overline{m}}.$$

Also, there exists $m_0 \in \mathbb{N}$ such that, for $m \geq m_0$ and $0 < x < 1$,

$$0 < u_m(x) < \frac{\delta}{2}.$$

So, for $m \geq m_0$ and $\frac{1}{8} \leq x \leq \frac{5}{8}$,

$$\begin{aligned} u_m(x) = Tu_m(x) &= \int_0^1 G(x,s) f(s, u_m(s))\, ds \\ &\geq \int_{\frac{1}{8}}^{\frac{5}{8}} G(x,s) f(s, u_m(s))\, ds \\ &\geq \overline{m} \int_{\frac{1}{8}}^{\frac{5}{8}} f(s, u_m(s))\, ds \\ &\geq \overline{m} \int_{\frac{1}{8}}^{\frac{5}{8}} f\left(s, \frac{\delta}{2}\right) ds \\ &\geq \overline{m} \int_{\frac{1}{8}}^{\frac{5}{8}} \frac{2}{\overline{m}}\, ds = 1. \end{aligned}$$

This contradicts $\lim_{m\to\infty} u_m(x) = 0$ uniformly on $[0, 1]$. Thus, there exists an $R > 0$ such that $R \leq \|u\|$ for any solution $u \in D$ of (8.51), (8.52). □

In summary, there exist $0 < R < S$ such that, for each solution $u \in D$ of (8.51), (8.52), we have $R \leq \|u\| \leq S$.

8.5.2 *Existence of positive solutions*

We construct a sequence of operators, $\{T_m\}_{m=1}^{\infty}$, each of which is defined on all of $\mathcal{P}$. Then, via applications of Theorem 8.14, each T_m has a fixed point $\varphi_m \in \mathcal{P}$. Then, we will extract a subsequence from $\{\varphi_m\}$ that converges to a fixed point of T.

Theorem 8.16. *If f satisfies* (H1)-(H3), *then* (8.51), (8.52) *has at least one positive solution $u \in D$.*

Proof. For all $m \in \mathbb{N}$, let

$$u_m(x) := T(m) = \int_0^1 G(x,s)f(s,m)\,ds, \quad 0 \le x \le 1.$$

Since f is decreasing with respect to its second component, we have $0 < u_{m+1}(x) < u_m(x)$, for $0 < x < 1$, and by (H2), $\lim_{m\to\infty} u_m(x) = 0$ uniformly on $[0,1]$.

Next, we define $f_m(x,y) : (0,1) \times [0,\infty) \to (0,\infty)$ by

$$f_m(x,y) := f(x, \max\{y, u_m(x)\}).$$

Then, f_m is continuous and f_m does not have the singularity at $y = 0$ possessed by f. In addition, for $(x,y) \in (0,1) \times (0,\infty)$,

$$f_m(x,y) \le f(x,y) \text{ and } f_m(x,y) \le f(x, u_m(x)).$$

Next, we define a sequence of operators $T_m : \mathcal{P} \to \mathcal{P}$, for $\varphi \in \mathcal{P}$ and $0 \le x \le 1$, by

$$T_m\varphi(x) := \int_0^1 G(x,s)f_m(s,\varphi(s))\,ds.$$

It is standard that each T_m is a compact operator on $\mathcal{P}$. Moreover,

$$\begin{aligned} T_m(0) &= \int_0^1 G(x,s)f_m(s,0)\,ds \\ &= \int_0^1 G(x,s)f(s,\max\{0,u_m(s)\})\,ds \\ &= \int_0^1 G(x,s)f(s,u_m(s))\,ds \ge 0, \end{aligned}$$

and

$$T_m^2(0) = T_m\left(\int_0^1 G(x,s)f_m(s,0)\,ds\right) \ge 0.$$

By Theorem 8.14, with $J = \mathcal{P}$ and $x_0 = 0$, T_m has a fixed point in $\mathcal{P}$, for each m. That is, for each m, there exists $\varphi_m \in \mathcal{P}$ such that $T_m\varphi_m(x) = \varphi_m(x)$, $0 \le x \le 1$. So, for each $m \ge 1$, φ_m satisfies the boundary conditions (8.52), and also

$$\begin{aligned} T_m\varphi_m(x) &= \int_0^1 G(x,s) f_m(s, \varphi_m(x))\, ds \\ &\le \int_0^1 G(x,s) f(s, u_m(s))\, ds = Tu_m(x). \end{aligned}$$

By arguments along the lines of those used in proving Lemma 8.2 and Lemma 8.3, there exist $0 < R < S$ such that $R \le \|\varphi_m\| \le S$, for all m.

Now, let

$$\theta := R.$$

Since $\varphi_m \in \mathcal{P}$ and is a fixed point of T_m, the conditions of Theorem 8.15 hold. So, for all m and $0 \le x \le 1$,

$$\varphi_m(x) \ge g(x)\|\varphi_m\| \ge g(x) \cdot R = g_\theta(x).$$

So, the sequence $\{\varphi_m\}$ is contained in the closed order interval $\langle g_\theta, S\rangle$, and therefore, the sequence belongs to D. Since T is a compact mapping, we may assume $\lim_{m\to\infty} T\varphi_m$ exists; let us say that the limit is φ^*. To complete the proof it suffices to show

$$\lim_{m\to\infty} (T\varphi_m(x) - \varphi_m(x)) = 0 \text{ uniformly on } [0,1].$$

It will follow that $\varphi^* \in \langle g_\theta, S\rangle$.

To that end, let $\varepsilon > 0$ be given, and choose $0 < \delta < 1$ such that

$$\int_0^\delta f(s, g_\theta(s))\, ds + \int_{1-\delta}^1 f(s, g_\theta(s))\, ds < \frac{\varepsilon}{2M},$$

with M as in Lemma 8.2. Then, there exists m_0 such that for all $m \ge m_0$ and for $\delta \le x \le 1-\delta$,

$$u_m(x) \le g_\theta(x) \le \varphi_m(x).$$

We have, for $m \geq m_0$ and $0 \leq x \leq 1$,

$$
\begin{aligned}
|T\varphi_m(x) - \varphi_m(x)| &= |T\varphi_m(x) - T_m\varphi_m(x)| \\
&= \left| \int_0^1 G(x,s)\,[f(s,\varphi_m(s)) - f_m(s,\varphi_m(s))]\,ds \right| \\
&= \left| \int_0^\delta G(x,s)\,[f(s,\varphi_m(s)) - f_m(s,\varphi_m(s))]\,ds \right. \\
&\quad \left. + \int_{1-\delta}^1 G(x,s)\,[f(s,\varphi_m(s)) - f_m(s,\varphi_m(s))]\,ds \right| \\
&\leq M \int_0^\delta [f(s,\varphi_m(s)) + f_m(s,\varphi_m(s))]\,ds \\
&\quad + M \int_{1-\delta}^1 [f(s,\varphi_m(s)) + f_m(s,\varphi_m(s))]\,ds \\
&\leq M \int_0^\delta [f(s,\varphi_m(s)) + f(s,\varphi_m(s))]\,ds \\
&\quad + M \int_{1-\delta}^1 [f(s,\varphi_m(s)) + f(s,\varphi_m(s))]\,ds \\
&\leq 2M \left[\int_0^\delta + \int_{1-\delta}^1 \right] f(s,\varphi_m(s))\,ds \\
&\leq 2M \left[\int_0^\delta + \int_{1-\delta}^1 \right] f(s,g_\theta(s))\,ds \\
&< 2M \cdot \frac{\varepsilon}{2M} = \varepsilon.
\end{aligned}
$$

Thus, for all $m \geq m_0$, $\|T\varphi_m - \varphi_m\| < \varepsilon$. That is, $\lim_{m\to\infty} T\varphi_m(x) - \varphi_m(x) = 0$ uniformly on $[0,1]$. So, for $0 \leq x \leq 1$,

$$
T\varphi^*(x) = T\left(\lim_{m\to\infty} T\varphi_m(x)\right) = T\left(\lim_{m\to\infty} \varphi_m(x)\right) = \lim_{m\to\infty} T\varphi_m(x) = \varphi^*(x),
$$

and φ^* is a desired positive solution of (8.51), (8.52) belonging to D. □

8.6 Positive Solutions for Third Order BVP's with p-Laplacian

In this section, we are concerned with the existence of positive solutions of the BVP consisting of the equation

$$
(\phi_p(u'))'' + a(t)f(u) = 0, \; t \in (0,1), \tag{8.57}
$$

and the nonhomogeneous three-point BC

$$u(0) = \xi u(\eta) + \lambda,\ u'(0) = u'(1) = 0, \tag{8.58}$$

where $\phi_p(x) = |x|^{p-2}x$, $p > 1$, $\xi \in [0,1)$, $\eta \in [0,1]$, $\lambda > 0$ is a parameter, and $a \in C(0,1)$ and $f \in C(\mathbb{R}^+)$ with $\mathbb{R}^+ := [0,\infty)$ are nonnegative functions. Clearly, $\phi_p^{-1}(x) = \phi_q(x)$, where $1/p + 1/q = 1$. By a positive solution of BVP (8.57), (8.58), we mean a function $u \in C^{(1)}[0,1]$ such that $u(t) > 0$ for $t \in [0,1]$, $u(t)$ satisfies Eq. (8.57) a.e. on $(0,1)$, and $u(t)$ satisfies BC (8.58).

We study BVP (8.57), (8.58) and obtain conditions for the existence, nonexistence, and multiplicity of positive solutions in terms of different values of the parameter λ. (See Theorems 8.17–8.20 and Corollaries 8.4–8.7). We also study the uniqueness of positive solutions and dependence of positive solutions on the parameter λ. (See Theorem 8.21).

8.6.1 *Main results*

We need the following assumptions.

(I1) $0 < \int_0^1 s(1-s)a(s)ds < \infty$;

(I2) there exist $0 < \rho < 1$ and $c > 0$ such that

$$f(x) \le \rho L \phi_p(x) \quad \text{for } x \in [0,c], \tag{8.59}$$

where L satisfies

$$0 < L \le \left[\phi_p\left(\frac{1+\xi(\eta-1)}{1-\xi}\right)\int_0^1 s(1-s)a(s)ds\right]^{-1}; \tag{8.60}$$

(I3) there exists $d > 0$ such that

$$f(x) \le M\phi_p(x) \quad \text{for } x \in (d,\infty), \tag{8.61}$$

where M satisfies

$$0 < M < \left[\phi_p\left(\frac{1+\xi(\eta-1)}{1-\xi}2^{q-1}\right)\int_0^1 s(1-s)ds\right]^{-1}; \tag{8.62}$$

(I4) there exist $0 < \tau < 1$ and $e > 0$ such that

$$f(x) \ge N\phi_p(x) \quad \text{for } x \in (e,\infty), \tag{8.63}$$

where N satisfies

$$N > \left[\phi_p\left(c_\tau \int_0^1 \phi_q(r(1-r))dr\right)\int_\tau^1 s(1-s)a(s)ds\right]^{-1} \tag{8.64}$$

with

$$c_\tau = \int_0^\tau \phi_q(r(1-r))dr \in (0,1); \tag{8.65}$$

(I5) $f(x)$ is nondecreasing in x;
(I6) there exists $0 \le \theta < 1$ such that

$$f(\kappa x) \ge (\phi_p(\kappa))^\theta f(x) \quad \text{for any } \kappa \in (0,1) \text{ and } x \in \mathbb{R}^+. \tag{8.66}$$

Remark 8.4. We have the following observations.

(a) Let

$$f_0 = \lim_{x \to 0^+} \frac{f(x)}{\phi_p(x)} \quad \text{and} \quad f_\infty = \lim_{x \to \infty} \frac{f(x)}{\phi_p(x)}.$$

Then, (I2) holds if $f_0 = 0$, (I3) holds if $f_\infty = 0$, and (I4) holds if $f_\infty = \infty$. In particular, for the simple function $f(x) = x^\gamma$ on $\mathbb{R}^+$, (I2) and (I4) hold if $\gamma > p-1$ and (I3) holds and if $\gamma < p-1$.

(b) One class of functions satisfying (I5) and (I6) is given by

$$f(x) = \sum_{i=1}^{m} \delta_i (\phi_p(x))^{\gamma_i},$$

where m is an integer, $\gamma_i \in [0,1)$ and $\delta_i > 0$ for $i = 1, \ldots, m$.

We now state our main results. The first two results give conditions to guarantee that BVP (8.57), (8.58) has a positive solution for small positive λ.

Theorem 8.17. *Assume that* (I1) *and* (I2) *hold. Then BVP* (8.57), (8.58) *has at least one positive solution for* $0 < \lambda \le (1-\xi)(1-\phi_q(\rho))c$.

Corollary 8.4. *Assume that* (I1) *holds and* $f_0 = 0$. *Then BVP* (8.57), (8.58) *has at least one positive solution for sufficiently small* $\lambda > 0$.

The next two results provide conditions for the existence of positive solutions of BVP (8.57), (8.58) for all positive λ.

Theorem 8.18. *Assume that* (I1) *and* (I3) *hold. Then BVP* (8.57), (8.58) *has at least one positive solution for all* $\lambda > 0$.

Corollary 8.5. *Assume that* (I1) *holds and* $f_\infty = 0$. *Then BVP* (8.57), (8.58) *has at least one positive solution for all* $\lambda > 0$.

Theorem 8.19 and its corollary below obtain conditions under which BVP (8.57), (8.58) has no positive solution for large λ.

Theorem 8.19. *Assume that* (I1) *and* (I4) *hold. Then BVP* (8.57), (8.58) *has no positive solution for* $\lambda > (1-\xi)e$.

Corollary 8.6. *Assume that* (I1) *holds and* $f_\infty = \infty$. *Then BVP* (8.57), (8.58) *has no positive solution for sufficiently large* $\lambda > 0$.

The following two results establish the theoretical structure of $(0, \infty)$ in terms of the existence and nonexistence of positive solutions of BVP (8.57), (8.58).

Theorem 8.20. *Assume that* (I1), (I2), (I4), *and* (I5) *hold. Then there exists* $(1-\xi)(1-\phi_q(\rho))c \le \lambda^* \le (1-\xi)e$ *such that BVP* (8.57), (8.58) *has at least two positive solutions for* $0 < \lambda < \lambda^*$, *at least one positive solution for* $\lambda = \lambda^*$, *and no positive solution for* $\lambda > \lambda^*$.

Corollary 8.7. *Assume that* (I1) *and* (I5) *hold,* $f_0 = 0$, *and* $f_\infty = \infty$. *Then there exists* $\lambda^* > 0$ *such that BVP* (8.57), (8.58) *has at least two positive solutions for* $0 < \lambda < \lambda^*$, *at least one positive solution for* $\lambda = \lambda^*$, *and no positive solution for* $\lambda > \lambda^*$.

The last theorem, Theorem 8.21, concerns the uniqueness and dependence of positive solutions of BVP (8.57), (8.58) on the parameter λ. Here, for any $u \in C[0,1]$, we denote $||u|| = \max_{t\in[0,1]} |u(t)|$.

Theorem 8.21. *Assume that* (I1), (I5), *and* (I6) *hold. Then BVP* (8.57), (8.58) *has a unique positive solution* $u_\lambda(t)$ *for any* $\lambda > 0$. *Furthermore, such a solution* $u_\lambda(t)$ *satisfies the following properties:*

(i) $\lim_{\lambda\to\infty} ||u_\lambda|| = \infty$;
(ii) $u_\lambda(t)$ *is strictly increasing in* λ, *that is, if* $\lambda_1 > \lambda_2$, *then* $u_{\lambda_1}(t) > u_{\lambda_2}(t)$ *for* $t \in [0,1]$;
(iii) $u_\lambda(t)$ *is continuous in* λ, *that is, if* $\lambda \to \lambda_0$, *then* $||u_\lambda - u_{\lambda_0}|| \to 0$.

Remark 8.5. By slight modifications of our proofs, results similar to Theorems 8.17–8.21 and Corollaries 8.4–8.7 can also be obtained for the BVP consisting of (8.57) and the BC

$$u(1) = \xi u(\eta) + \lambda,\ u'(0) = u'(1) = 0.$$

We omit the discussions here.

8.6.2 *Proofs of the main results*

Throughout this subsection, we let the Banach space $C[0,1]$ be endowed with the norm $||u|| = \max_{t\in[0,1]} |u(t)|$.

8.6.2.1 *Proofs of Theorems* 8.17–8.19 *and Corollaries* 8.4–8.6

We need the following two lemmas to prove our results.

Lemma 8.4. *Assume* (I1) *holds. Then, for* $h \in C(0,1)$ *with* $h(t) \geq 0$ *on* $(0,1)$ *and* $0 < \int_0^1 s(1-s)h(s)ds < \infty$, *the BVP consisting of the equation*

$$(\phi_p(u'))'' + h(t) = 0, \ t \in (0,1), \tag{8.67}$$

and BC (8.58) *has a unique positive solution* $u(t)$ *satisfying* $u(1) = ||u||$ *and*

$$\begin{aligned} u(t) = &\int_0^t \phi_q\left(\int_0^1 G(r,s)h(s)ds\right)dr \\ &+\frac{\xi}{1-\xi}\int_0^\eta \phi_q\left(\int_0^1 G(r,s)h(s)ds\right)dr + \frac{\lambda}{1-\xi}, \end{aligned} \tag{8.68}$$

where

$$G(t,s) = \begin{cases} t(1-s), & 0 \leq t \leq s \leq 1, \\ s(1-t), & 0 \leq s \leq t \leq 1. \end{cases} \tag{8.69}$$

Proof. Let $u(t)$ satisfy (8.67) and (8.58). From (8.58), $\phi_p(u'(0)) = \phi_p(u'(1)) = 0$. Then, for $t \in [0,1]$, we have

$$\phi_p(u'(t)) = \int_0^1 G(t,s)h(s)ds,$$

and so

$$u'(t) = \phi_q\left(\int_0^1 G(t,s)h(s)ds\right),$$

which in turn implies that

$$u(t) = \int_0^t \phi_q\left(\int_0^1 G(r,s)h(s)ds\right)dr + C,$$

where C is a constant to be determined. Using the assumption that $u(0) = \xi u(\eta) + \lambda$, we find that

$$C = \frac{\xi}{1-\xi}\int_0^\eta \phi_q\left(\int_0^1 G(r,s)h(s)ds\right)dr + \frac{\lambda}{1-\xi}.$$

Thus, (8.68) holds. From (8.68), it is clear that $u(t) > 0$ on $[0,1]$ and $u(1) = ||u||$. This completes the proof of the lemma. □

Lemma 8.5. *Assume* (I1) *holds and let* $\tau \in (0,1)$ *be given in* (I4). *Then the unique solution* $u(t)$ *of BVP* (8.67), (8.58) *satisfies*

$$u(t) \geq c_\tau ||u|| \quad \text{for } t \in [\tau, 1],$$

where c_τ *is defined by* (8.65).

Proof. Note that $G(t,s)$ defined by (8.69) satisfies

$$t(1-t)s(1-s) \leq G(t,s) \leq s(1-s) \quad \text{for } t, s \in [0,1]. \tag{8.70}$$

Then, from (8.68), we have that

$$\begin{aligned} u(t) &\leq t\phi_q\left(\int_0^1 s(1-s)h(s)ds\right) \\ &\quad + \frac{\xi}{1-\xi}\int_0^\eta \phi_q\left(\int_0^1 G(r,s)h(s)ds\right)dr + \frac{\lambda}{1-\xi} \\ &\leq \phi_q\left(\int_0^1 s(1-s)h(s)ds\right) \\ &\quad + \frac{\xi}{1-\xi}\int_0^\eta \phi_q\left(\int_0^1 G(r,s)h(s)ds\right)dr + \frac{\lambda}{1-\xi} \end{aligned}$$

for $t \in [0,1]$, and

$$\begin{aligned} u(t) &\geq \int_0^t \phi_q(r(1-r))dr\phi_q\left(\int_0^1 s(1-s)h(s)ds\right) \\ &\quad + \frac{\xi}{1-\xi}\int_0^\eta \phi_q\left(\int_0^1 G(r,s)h(s)ds\right)dr + \frac{\lambda}{1-\xi} \\ &\geq c_\tau\phi_q\left(\int_0^1 s(1-s)h(s)ds\right) \\ &\quad + \frac{\xi}{1-\xi}\int_0^\eta \phi_q\left(\int_0^1 G(r,s)h(s)ds\right)dr + \frac{\lambda}{1-\xi} \\ &\geq c_\tau\left[\phi_q\left(\int_0^1 s(1-s)h(s)ds\right)\right. \\ &\quad \left.+ \frac{\xi}{1-\xi}\int_0^\eta \phi_q\left(\int_0^1 G(r,s)h(s)ds\right)dr + \frac{\lambda}{1-\xi}\right] \end{aligned}$$

for $t \in [\tau, 1]$. Therefore, $u(t) \geq c_\tau ||u||$ on $[\tau, 1]$. This completes the proof of the lemma. □

Proof of Theorem 8.17. Let $c > 0$ be given in (I2). Define

$$K_1 = \{u \in C[0,1] \ : \ 0 \leq u(t) \leq c \text{ on } [0,1]\}$$

and an operator $T_\lambda : K_1 \to C[0,1]$ by

$$T_\lambda u(t) = \int_0^t \phi_q\left(\int_0^1 G(r,s)a(s)f(u(s))ds\right)dr$$
$$+\frac{\xi}{1-\xi}\int_0^\eta \phi_q\left(\int_0^1 G(r,s)a(s)f(u(s))ds\right)dr + \frac{\lambda}{1-\xi}. \qquad (8.71)$$

Then, K_1 is a closed convex set, and in view of Lemma 8.4, $u(t)$ is a solution of BVP (8.57), (8.58) if and only if u is a fixed point of T_λ. Moreover, a standard argument can be used to show that T_λ is compact. For any $u \in K_1$, from (8.59) and (8.60), we have

$$f(u(t)) \le \rho L\phi_p(u(t)) \le \rho L\phi_p(c) \quad \text{on } [0,1]$$

and

$$\frac{1+\xi(\eta-1)}{1-\xi}\phi_q(L)\phi_q\left(\int_0^1 s(1-s)a(s)ds\right) \le 1.$$

Let $0 < \lambda \le (1-\xi)(1-\phi_q(\rho))c$. Then, from (8.70) and (8.71), it follows that

$$\begin{aligned}
0 &\le T_\lambda u(t) \le T_\lambda u(1)\\
&= \int_0^1 \phi_q\left(\int_0^1 G(r,s)a(s)f(u(s))ds\right)dr\\
&\quad +\frac{\xi}{1-\xi}\int_0^\eta \phi_q\left(\int_0^1 G(r,s)a(s)f(u(s))ds\right)dr + \frac{\lambda}{1-\xi}\\
&\le \phi_q\left(\int_0^1 s(1-s)a(s)f(u(s))ds\right)\\
&\quad +\frac{\xi\eta}{1-\xi}\phi_q\left(\int_0^1 s(1-s)a(s)f(u(s))ds\right) + (1-\phi_q(\rho))c\\
&= \frac{1+\xi(\eta-1)}{1-\xi}\phi_q\left(\int_0^1 s(1-s)a(s)f(u(s))ds\right) + (1-\phi_q(\rho))c\\
&\le \frac{1+\xi(\eta-1)}{1-\xi}\phi_q(L)\phi_q\left(\int_0^1 s(1-s)a(s)ds\right)\phi_q(\rho)c + (1-\phi_q(\rho))c\\
&\le \phi_q(\rho)c + (1-\phi_q(\rho))c = c \quad \text{on } [0,1].
\end{aligned}$$

Thus, $T_\lambda(K_1) \subseteq K_1$. By the Schauder fixed point theorem, T_λ has a fixed point $u \in K_1$, which is clearly a positive solution of BVP (8.57), (8.58). This completes the proof of the theorem. □

Proof of Corollary 8.4. The conclusion is a direct consequence of Theorem 8.17. □

Proof of Theorem 8.18. Let $\lambda > 0$ be fixed and $d > 0$ be given in (I3). Define $D = \max_{x\in[0,d]} f(x)$. Then

$$f(x) \le D \quad \text{for } 0 \le x \le d. \tag{8.72}$$

From (8.62),

$$\frac{1+\xi(\eta-1)}{1-\xi} 2^{q-1}\phi_q(M)\phi_q\left(\int_0^1 s(1-s)a(s)ds\right) < 1.$$

Thus, there exists $d^* > d$ large enough so that

$$\frac{1+\xi(\eta-1)}{1-\xi} 2^{q-1}(\phi_q(D)+\phi_q(M)d^*)\phi_q\left(\int_0^1 s(1-s)a(s)ds\right)$$
$$+\frac{\lambda}{1-\xi} \le d^*. \tag{8.73}$$

Let

$$K_2 = \{u \in C[0,1] \;:\; 0 \le u(t) \le d^* \text{ on } [0,1]\}.$$

For $u \in K_2$, define

$$I_1^u = \{t \in [0,1] \;:\; 0 \le u(t) \le d\}$$

and

$$I_2^u = \{t \in [0,1] \;:\; d < u(t) \le d^*\}.$$

Then, $I_1^u \cup I_2^u = [0,1]$, $I_1^u \cap I_2^u = \emptyset$, and (8.61) implies that

$$f(u(t)) \le M\phi_p(u(t)) \le M\phi_p(d^*) \quad \text{for } t \in I_2^u. \tag{8.74}$$

Let the compact operator T_λ be defined by (8.71). Then, from (8.70), (8.72) and (8.74), we have that

$$\begin{aligned}
0 \le T_\lambda u(t) &\le \phi_q\left(\int_0^1 s(1-s)a(s)f(u(s))ds\right)\\
&\quad + \frac{\xi\eta}{1-\xi}\phi_q\left(\int_0^1 s(1-s)a(s)f(u(s))ds\right) + \frac{\lambda}{1-\xi}\\
&= \frac{1+\xi(\eta-1)}{1-\xi}\\
&\quad \times\phi_q\left(\int_{I_1^u} s(1-s)a(s)f(u(s))ds + \int_{I_2^u} s(1-s)a(s)f(u(s))ds\right) + \frac{\lambda}{1-\xi}\\
&\le \frac{1+\xi(\eta-1)}{1-\xi}\phi_q\left(D\int_{I_1^u} s(1-s)a(s)ds + M\phi_p(d^*)\int_{I_2^u} s(1-s)a(s)ds\right)\\
&\quad +\frac{\lambda}{1-\xi}\\
&\le \frac{1+\xi(\eta-1)}{1-\xi}\phi_q(D+M\phi_p(d^*))\phi_q\left(\int_0^1 s(1-s)a(s)ds\right) + \frac{\lambda}{1-\xi}.
\end{aligned}$$

Using the inequality $(a+b)^r \leq 2^r(a^r+b^r)$ for any $a,b,r>0$ (see, for example, [Hardy *et al.* (1988)]) and from (8.73), we obtain that

$$\begin{aligned}
0 &\leq T_\lambda u(t) \\
&\leq \frac{1+\xi(\eta-1)}{1-\xi}2^{q-1}(\phi_q(D)+\phi_q(M)d^*)\phi_q\left(\int_0^1 s(1-s)a(s)ds\right) \\
&\quad+\frac{\lambda}{1-\xi} \leq d^*,
\end{aligned}$$

which means that $T_\lambda : K_2 \to K_2$. Hence, by the Schauder fixed point theorem, there exists a fixed point $u \in K_2$ of T_λ, which is clearly a positive solution of BVP (8.57), (8.58). This completes the proof of the theorem. □

Proof of Corollary 8.5. The conclusion is a direct consequence of Theorem 8.18. □

Proof of Theorem 8.19. Assume, to the contrary, that BVP (8.57), (8.58) has a positive solution $u(t)$ for $\lambda > (1-\xi)e$. Then, by Lemma 8.4,

$$\begin{aligned}
u(t) &= \int_0^t \phi_q\left(\int_0^1 G(r,s)a(s)f(u(s))ds\right)dr \\
&\quad+\frac{\xi}{1-\xi}\int_0^\eta \phi_q\left(\int_0^1 G(r,s)a(s)f(u(s))ds\right)dr+\frac{\lambda}{1-\xi}.
\end{aligned}$$

Hence, $u(t) > e$ on $[0,1]$. From (8.63) and (8.64), we have that

$$f(u(t)) \geq N\phi_p(u(t)) \quad \text{on } [0,1]$$

and

$$c_\tau\phi_q(N)\phi_q\left(\int_\tau^1 s(1-s)a(s)ds\right)\int_0^1 \phi_q(r(1-r))dr > 1.$$

Then, by (8.70) and Lemma 8.5, we reach that

$$\begin{aligned}
||u|| = u(1) &> \int_0^1 \phi_q\left(\int_0^1 G(r,s)a(s)f(u(s))ds\right)dr \\
&\geq \phi_q\left(\int_0^1 s(1-s)a(s)f(u(s))ds\right)\int_0^1 \phi_q(r(1-r))dr \\
&\geq \phi_q(N)\phi_q\left(\int_\tau^1 s(1-s)a(s)\phi_p(u(s))ds\right)\int_0^1 \phi_q(r(1-r))dr \\
&\geq ||u||c_\tau\phi_q(N)\phi_q\left(\int_\tau^1 s(1-s)a(s)ds\right)\int_0^1 \phi_q(r(1-r)) \\
&> ||u||,
\end{aligned}$$

which is a contradiction and thus completes the proof of the theorem. □

Proof of Corollary 8.6. The conclusion is a direct consequence of Theorem 8.19. □

8.6.2.2 *Proofs of Theorem* 8.20 *and Corollary* 8.7

We will use the lower and upper solution method to prove our results. We first present the following definition and several lemmas.

Definition 8.4. A function $\alpha \in C^{(1)}[0,1]$ is said to be a *lower solution* of BVP (8.57), (8.58) if

$$(\phi_p(\alpha'(t)))'' + a(t)f(\alpha(t)) \geq 0 \quad \text{a.e. on } (0,1),$$

$$\alpha(0) \leq \xi\alpha(\eta) + \lambda,\ \alpha'(0) \leq 0,\ \alpha'(1) \leq 0.$$

A function $\beta \in C^1[0,1]$ is said to be an *upper solution* of BVP (8.57), (8.58) if

$$(\phi_p(\beta'(t)))'' + a(t)f(\beta(t)) \leq 0 \quad \text{a.e. on } (0,1), \tag{8.75}$$

$$\beta(0) \geq \xi\beta(\eta) + \lambda,\ \beta'(0) \geq 0,\ \beta'(1) \geq 0. \tag{8.76}$$

Lemma 8.6. *Assume that* $\xi \in [0,1)$ *and* $u_1, u_2 \in C^1[0,1]$ *satisfy*

$$(\phi_p(u_1'(t)))'' - (\phi_p(u_2'(t)))'' \leq 0 \quad \textit{a.e. on } (0,1),$$
$$u_1(0) - u_2(0) \geq \xi(u_1(\eta) - u_2(\eta)),$$
$$u_1'(0) - u_2'(0) \geq 0,\ u_1'(1) - u_2'(1) \geq 0.$$

Then, $u_1(t) \geq u_2(t)$ *for* $t \in [0,1]$.

Proof. Let $w(t) = \phi_p(u_1'(t)) - \phi_p(u_2'(t))$, $t \in [0,1]$. Then, $w''(t) \leq 0$ a.e. on $(0,1)$, $w(0) \geq 0$, and $w(1) \geq 0$. Thus, $w(t) \geq 0$ on $[0,1]$, which implies that $u_1'(t) - u_2'(t) \geq 0$ for $t \in [0,1]$. Hence, $u_1(t) - u_2(t)$ is nondecreasing on $[0,1]$. Then,

$$u_1(0) - u_2(0) \geq \xi(u_1(\eta) - u_2(\eta)) \geq \xi(u_1(0) - u_2(0)).$$

If $u_1(0) - u_2(0) < 0$, then we have $\xi \geq 1$, which contradicts the assumption that $\xi \in [0,1)$. Thus, $u_1(0) - u_2(0) \geq 0$. Therefore, $u_1(t) \geq u_2(t)$ on $[0,1]$. This completes the proof of the lemma. □

Lemma 8.7. *Assume that* (I1) *and* (I5) *hold and BVP* (8.57), (8.58) *has a lower solution* $\alpha(t)$ *and an upper solution* $\beta(t)$ *satisfying*

$$\alpha(t) \leq \beta(t) \quad \textit{for } t \in [0,1]. \tag{8.77}$$

Then BVP (8.57), (8.58) *has at least one solution* $u(t)$ *satisfying*

$$\alpha(t) \leq u(t) \leq \beta(t) \quad \textit{for } t \in [0,1]. \tag{8.78}$$

Proof. For $u \in C[0,1]$ and $t \in [0,1]$, define

$$\tilde{u}(t) = \max\{\alpha(t), \min\{u(t), \beta(t)\}\} \tag{8.79}$$

and an operator

$$\tilde{T}_\lambda u(t) = \int_0^t \phi_q\left(\int_0^1 G(r,s)a(s)f(\tilde{u}(s))ds\right)dr$$
$$+\frac{\xi}{1-\xi}\int_0^\eta \phi_q\left(\int_0^1 G(r,s)a(s)f(\tilde{u}(s))ds\right)dr + \frac{\lambda}{1-\xi}.$$

Then, it can be shown that $\tilde{T}_\lambda(C[0,1])$ is relatively compact and a fixed point of $\tilde{T}_\lambda$ is equivalent to a solution of the BVP consisting of the equation

$$(\phi_p(u'))'' + a(t)f(\tilde{u}) = 0, \ t \in (0,1), \tag{8.80}$$

and BC (8.58). From the Schauder fixed point theorem, $\tilde{T}_\lambda$ has at least one fixed point $u \in C[0,1]$. Now, to complete the proof, it suffices to show that $u(t)$ satisfies (8.78). From (I5), (8.75), and (8.76), we have

$$(\phi_p(\beta'(t)))'' - (\phi_p(u'(t)))'' \le -a(t)f(\beta(t)) + a(t)f(\tilde{u}(t)) \le 0 \quad \text{a.e. on } (0,1),$$
$$\beta(0) - u(0) \ge \xi(\beta(\eta) - u(\eta)),$$
$$\beta'(0) - u'(0) \ge 0, \ \beta'(1) - u'(1) \ge 0.$$

Then, Lemma 8.6 implies that $\beta(t) \ge u(t)$ on $[0,1]$. By a similar argument, we can show that $u(t) \ge \alpha(t)$ for $t \in [0,1]$. Hence, (8.78) holds. This completes the proof of the lemma. □

Lemma 8.8. *Assume that* (I1) *and* (I5) *hold and BVP* (8.57), (8.58) *with* $\lambda = \bar{\lambda} > 0$ *has a positive solution* $\bar{u}(t)$. *Then, for any* $0 < \lambda < \bar{\lambda}$, *BVP* (8.57), (8.58) *has a positive solution* $u(t)$ *satisfying* $u(t) < \bar{u}(t)$ *for* $t \in [0,1]$.

Proof. Let $\alpha(t) = 0$ and $\beta(t) = \bar{u}(t)$ for $t \in [0,1]$. Then, for $0 < \lambda < \bar{\lambda}$, it is easy to see that $\alpha(t)$ and $\beta(t)$ are lower and upper solutions of BVP (8.57), (8.58), respectively, satisfying (8.77). By Lemma 8.7, BVP (8.57), (8.58) has a solution $u(t)$ satisfying $0 \le u(t) \le \bar{u}(t)$ on $[0,1]$. Moreover, since

$$u(t) = \int_0^t \phi_q\left(\int_0^1 G(r,s)a(s)f(u(s))ds\right)dr$$
$$+\frac{\xi}{1-\xi}\int_0^\eta \phi_q\left(\int_0^1 G(r,s)a(s)f(u(s))ds\right)dr + \frac{\lambda}{1-\xi}$$

and

$$\bar{u}(t) = \int_0^t \phi_q\left(\int_0^1 G(r,s)a(s)f(\bar{u}(s))ds\right)dr$$
$$+\frac{\xi}{1-\xi}\int_0^\eta \phi_q\left(\int_0^1 G(r,s)a(s)f(\bar{u}(s))ds\right)dr + \frac{\bar{\lambda}}{1-\xi},$$

from (I5), we see that $0 < u(t) < \bar{u}(t)$ for $t \in [0,1]$. This completes the proof of the lemma. □

Lemma 8.9. *Assume that* (I1) *and* (I4) *hold. Let* $\Omega \subseteq \mathbb{R}$ *be a bounded set. Then all possible positive solutions* $u(t)$ *of BVP* (8.57), (8.58) *with* $\lambda \in \Omega$ *satisfy* $u(t) < \Lambda$ *for* $t \in [0,1]$, *where* $\Lambda = \Lambda(\Omega)$ *is a positive constant depending on* Ω *only.*

Proof. Assume, to the contrary, that there exists a sequence $\{u_k(t)\}_{k=1}^\infty$ of positive solutions of BVP (8.57), (8.58) at $\lambda_k \in \Omega$ such that $||u_k|| \to \infty$ as $k \to \infty$. By Lemma 8.5, $u_k(t) \geq c_\tau ||u_k||$ for $t \in [\tau, 1]$. Then, $u_k(t) \to \infty$ as $k \to \infty$ on $[\tau, 1]$. Thus, from (8.63), we have

$$f(u_k(t)) \geq N\phi_p(u_k(t)) \quad \text{for } t \in [\tau, 1]$$

for k large enough, where N satisfies (8.64). Now, as in the proof of Theorem 8.19, we can reach a contradiction and thus complete the proof of the lemma. □

Proof of Theorem 8.20. Let

$$E = \{\lambda > 0 \ : \ \text{BVP (8.57), (8.58) has a positive solution}\}.$$

By Theorem 8.17, $E \neq \emptyset$. Define $\lambda^* = \sup E$. Then, by Theorems 8.17, 8.19, and Lemma 8.8, $(1-\xi)(1-\phi_q(\rho))c \leq \lambda^* \leq (1-\xi)e$, BVP (8.57), (8.58) has at least one positive solution for $0 < \lambda < \lambda^*$, and no positive solution for $\lambda > \lambda^*$.

We now show that BVP (8.57), (8.58) has at least one positive solution for $\lambda = \lambda^*$. Clearly, there exists a sequence $\{\lambda_k\}_{k=1}^\infty \subset (0, \lambda^*)$ such that $\lambda_k \to \lambda^*$ as $k \to \infty$. Let $u_k(t)$ be a positive solution of BVP (8.57), (8.58) with $\lambda = \lambda_k$. By Lemma 8.9, $\{u_k(t)\}_{k=1}^\infty$ is uniformly bounded, i.e., there exists $B > 0$ such that $||u_k|| < B$ for $k \in \mathbb{N}$. From (8.70) and the fact that

$$u_k'(t) = \phi_q\left(\int_0^1 G(t,s)a(s)f(u_k(s))ds\right),$$

it follows that

$$0 \leq u_k'(t) \leq \max_{x\in[0,B]} \phi_q(f(x))\phi_q\left(\int_0^1 G(t,s)a(s)ds\right) < \infty \quad \text{for } k \in \mathbb{N}.$$

Thus, $\{u_k(t)\}_{k=1}^{\infty}$ is equicontinuous on $[0,1]$. Therefore, by the Arzelà-Ascoli Theorem, there exists a subsequence $\{u_{k_l}(t)\}_{l=1}^{\infty}$ of $\{u_k(t)\}_{k=1}^{\infty}$ that converges uniformly in $C[0,1]$ to a solution $u(t)$ of BVP (8.57), (8.58) at $\lambda = \lambda^*$.

In the remainder of the proof, let $0 < \lambda < \lambda^*$ be fixed, we show the existence of a second solution of BVP (8.57), (8.58). Clearly, there exist $\hat{\lambda}, \check{\lambda} > 0$ such that

$$0 < \hat{\lambda} < \lambda < \check{\lambda} < \lambda^*.$$

Let $\check{u}(t)$ be a positive solution of BVP (8.57), (8.58) with λ replaced by $\check{\lambda}$. Then, from Lemma 8.8, BVP (8.57), (8.58) has a positive solution $u_1(t)$ satisfying $u_1(t) < \check{u}(t)$ on $[0,1]$. Again by Lemma 8.8, BVP (8.57), (8.58) with λ replaced by $\hat{\lambda}$ has a positive solution $\hat{u}(t)$ satisfying $\hat{u}(t) < u_1(t)$ on $[0,1]$. Thus, we have

$$0 < \hat{u}(t) < u_1(t) < \check{u}(t) \quad \text{for } t \in [0,1]. \tag{8.81}$$

Let

$$\alpha(t) = \hat{u}(t) \quad \text{and} \quad \beta(t) = \check{u}(t). \tag{8.82}$$

Then, it is easy to check that $\alpha(t)$ and $\beta(t)$ are lower and upper solutions of BVP (8.57), (8.58), respectively. Define a set $\mathcal{O}$ in $C[0,1]$ by

$$\mathcal{O} = \{u \in C[0,1] \ : \ \alpha(t) < u(t) < \beta(t) \text{ on } [0,1]\}.$$

Then, $\mathcal{O}$ is a bounded open subset of $C[0,1]$, and from (8.81), $u_1 \in \mathcal{O}$.

Let $\bar{\lambda} > \lambda^*$ be fixed. Define compact operators $T_1, T_2 : [\lambda, \bar{\lambda}] \times C[0,1] \to C[0,1]$ by

$$\begin{aligned} T_1(\mu, u)(t) = &\int_0^t \phi_q\left(\int_0^1 G(r,s)a(s)f(u(s))ds\right) dr \\ &+\frac{\xi}{1-\xi}\int_0^{\eta} \phi_q\left(\int_0^1 G(r,s)a(s)f(u(s))ds\right) dr + \frac{\mu}{1-\xi} \end{aligned}$$

and

$$\begin{aligned} T_2(\mu, u)(t) = &\int_0^t \phi_q\left(\int_0^1 G(r,s)a(s)f(\tilde{u}(s))ds\right) dr \\ &+\frac{\xi}{1-\xi}\int_0^{\eta} \phi_q\left(\int_0^1 G(r,s)a(s)f(\tilde{u}(s))ds\right) dr + \frac{\mu}{1-\xi}, \end{aligned}$$

where $\tilde{u}$ is defined by (8.79) with $\alpha(t)$ and $\beta(t)$ given in (8.82). Clearly, $u(t)$ is a solution of BVP (8.57), (8.58) if and only if $u(t) = T_1(\lambda, u(t))$.

Let $\Lambda = \Lambda(\Omega)$ be given by Lemma 8.9 with $\Omega = (0, \lambda^*]$ and define a cone P by

$$P = \{u \in C[0,1] \ : \ u(t) \geq 0 \text{ on } [0,1]\}. \tag{8.83}$$

Since $T_2(\lambda, u(t))$ is bounded on P, there exists $R > \Lambda$ such that $||T_2(\lambda, u)|| < R$ for all $u \in P$. Let $B(\mathbf{0}, R)$ be the open ball centered at $\mathbf{0}$ with radius R in $C[0,1]$, where $\mathbf{0}$ is the zero element of $C[0,1]$. Then $\mathcal{O} \subseteq B(\mathbf{0}, R)$ and by Lemma 6.16, we have

$$i(T_2(\lambda, \cdot), P \cap B(\mathbf{0}, R), P) = 1. \tag{8.84}$$

In the following, we first prove the claim:

Claim: $T_2(\lambda, \cdot)$ has no fixed point on $P \cap (\overline{B(\mathbf{0}, R)} \setminus \mathcal{O})$.

In fact, if $T_2(\lambda, \cdot)$ has a fixed point u_2 on $P \cap (\overline{B(\mathbf{0}, R)} \setminus \mathcal{O})$, then $u_2(t)$ is a solution of BVP (8.80), (8.58). Using Lemma 8.6, as in the proof of Lemma 8.7, it can be shown that

$$\alpha(t) \leq u_2(t) \leq \beta(t) \quad \text{for } t \in [0,1]. \tag{8.85}$$

Thus, $u_2(t)$ is a solution of BVP (8.57), (8.58). In view of (8.82), we have

$$\begin{aligned}\alpha(t) = &\int_0^t \phi_q\left(\int_0^1 G(r,s)a(s)f(\alpha(s))ds\right)dr \\ &+\frac{\xi}{1-\xi}\int_0^\eta \phi_q\left(\int_0^1 G(r,s)a(s)f(\alpha(s))ds\right)dr + \frac{\hat{\lambda}}{1-\xi},\end{aligned}$$

$$\begin{aligned}u_2(t) = &\int_0^t \phi_q\left(\int_0^1 G(r,s)a(s)f(u_2(s))ds\right)dr \\ &+\frac{\xi}{1-\xi}\int_0^\eta \phi_q\left(\int_0^1 G(r,s)a(s)f(u_2(s))ds\right)dr + \frac{\lambda}{1-\xi},\end{aligned}$$

and

$$\begin{aligned}\beta(t) = &\int_0^t \phi_q\left(\int_0^1 G(r,s)a(s)f(\beta(s))ds\right)dr \\ &+\frac{\xi}{1-\xi}\int_0^\eta \phi_q\left(\int_0^1 G(r,s)a(s)f(\beta(s))ds\right)dr + \frac{\check{\lambda}}{1-\xi}.\end{aligned}$$

From (I5), (8.85), and the fact that $\hat{\lambda} < \lambda < \check{\lambda}$, we see that

$$\alpha(t) < u_2(t) < \beta(t) \quad \text{for } t \in [0,1].$$

Then, $u_2 \in \mathcal{O}$. This contradicts the assumption that $u_2 \in P \cap (\overline{B(\mathbf{0}, R)} \setminus \mathcal{O})$. Hence, the claim is true.

Now, note that $T_1(\lambda,\cdot) = T_2(\lambda,\cdot)$ on $\mathcal{O}$, from the above claim, the excision property of the fixed point index, and (8.84), we obtain

$$\begin{aligned} i(T_1(\lambda,\cdot), P\cap \mathcal{O}, P) &= i(T_2(\lambda,\cdot), P\cap \mathcal{O}, P) \\ &= i(T_2(\lambda,\cdot), P\cap B(\mathbf{0},R), P) = 1. \end{aligned} \tag{8.86}$$

Since $\bar{\lambda} > \lambda^*$, we have $T_1(\bar{\lambda}, u) \neq u$ for any $u \in P$. Thus,

$$i(T_1(\bar{\lambda},\cdot), P\cap B(\mathbf{0},R), P) = 0. \tag{8.87}$$

Let the compact operator $H : [0,1]\times P\cap(\overline{B(\mathbf{0},R)}) \to C[0,1]$ be defined by

$$H(\tau,\cdot) = T_1((1-\tau)\lambda + \tau\bar{\lambda},\cdot).$$

We claim that $H(\tau,u) \neq u$ for $(\tau,u) \in [0,1]\times(P\cap(\partial B(\mathbf{0},R))$. Otherwise, there exists $(\tau^*,u^*) \in [0,1]\times \partial B(\mathbf{0},R)$ such that $H(\tau^*,u^*) = u^*$. This implies that $u^*(t)$ is a solution of BVP (8.57), (8.58) with λ replaced by $\tilde{\mu} = (1-\tau^*)\lambda + \tau^*\bar{\lambda}$. Then, $0 \leq \tilde{\mu} \leq \lambda^*$. Thus, by Lemma 8.9, $||u^*|| < \Lambda$. But this contradicts the assumptions that $u^* \in \partial B(\mathbf{0},R)$ and $R > \Lambda$. From the homotopy invariance of the Leray-Schauder degree and (8.87), it follows that

$$\begin{aligned} i(T_1(\lambda,\cdot), P\cap B(\mathbf{0},R), P) &= i(H(0,\cdot), P\cap B(\mathbf{0},R), P) \\ &= i(H(1,\cdot), P\cap B(\mathbf{0},R), P) \\ &= i(T_1(\bar{\lambda},\cdot), P\cap B(\mathbf{0},R), P) = 0. \end{aligned} \tag{8.88}$$

In view of $R > \Lambda$ and by Lemma 8.9, $T_1(\lambda,\cdot)$ has no fixed point on $P\cap \partial B(\mathbf{0},R)$. Moreover, from the above claim, $T_2(\lambda,\cdot)$ has no fixed point on $\partial\mathcal{O}$. Then, since $T_1(\lambda,\cdot) = T_2(\lambda,\cdot)$ on $\partial\mathcal{O}$, $T_1(\lambda,\cdot)$ has no fixed point on $\partial\mathcal{O}$. By the additivity property of the fixed point index, (8.86), and (8.88), we have

$$i(T_1(\lambda,\cdot), P\cap(B(\mathbf{0},R)\setminus\overline{\mathcal{O}}), P) = -1.$$

Hence, $T_1(\lambda,\cdot)$ has a fixed point u_2 in $P\cap(B(\mathbf{0},R)\setminus\overline{\mathcal{O}})$, which is a positive solution of BVP (8.57), (8.58). Since $u_1 \in \mathcal{O}$, $u_2(t) \not\equiv u_1(t)$ on $[0,1]$, i.e., $u_2(t)$ is a second solution of BVP (8.57), (8.58). This completes the proof of the theorem. □

Proof of Corollary 8.7. The conclusion is a direct consequence of Theorem 8.20. □

8.6.2.3 *Proof of Theorem* 8.21

We first recall the following two definitions.

Definition 8.5. A cone P in a real Banach space X is called *solid* if P° is not empty, where P° is the interior of P.

Definition 8.6. Let P be a solid cone in a real Banach space X, $T : P^\circ \to P^\circ$ be an operator, and $0 \le \theta < 1$. Then, T is called a θ–concave operator if

$$T(\kappa u) \ge \kappa^\theta Tu \quad \text{for any } \kappa \in (0,1) \text{ and } u \in P^\circ.$$

We refer the reader to [Guo and Lakshmikantham (1988), Theorem 2.2.6] for the following lemma and its proof.

Lemma 8.10. *Assume that P is a normal solid cone in a real Banach space X, $0 \le \theta < 1$, and $T : P^\circ \to P^\circ$ is a θ-concave increasing operator. Then T has only one fixed point in P°.*

Proof of Theorem 8.21. Let the cone P be defined by (8.83). Then, P is a normal solid cone in $C[0,1]$ with

$$P^\circ = \{u \in C[0,1] \ : \ u(t) > 0 \text{ on } [0,1]\}.$$

For any fixed $\lambda > 0$, let $T_\lambda : P \to C[0,1]$ be defined by (8.71). Define $A : P \to C[0,1]$ by

$$\begin{aligned} Au(t) &= \int_0^t \phi_q\left(\int_0^1 G(r,s)a(s)f(u(s))ds\right) dr \\ &\quad + \frac{\xi}{1-\xi}\int_0^\eta \phi_q\left(\int_0^1 G(r,s)a(s)f(u(s))ds\right) dr. \end{aligned}$$

Then, from (I5), it is easy to see that A is increasing in $u \in P$ and

$$T_\lambda u(t) = Au(t) + \frac{\lambda}{1-\xi}.$$

Clearly, $T_\lambda : P^\circ \to P^\circ$. We now show that T_λ is a θ–concave increasing operator. In fact, for $u_1, u_2 \in P$ with $u_1(t) \ge u_2(t)$ on $[0,1]$, we have

$$\begin{aligned} T_\lambda u_1(t) &= Au_1(t) + \frac{\lambda}{1-\xi} \\ &\ge Au_2(t) + \frac{\lambda}{1-\xi} = T_\lambda u_2(t), \end{aligned} \tag{8.89}$$

i.e., T_λ is increasing. Moreover, (I6) implies that

$$\begin{aligned}
T_\lambda(\kappa u)(t) &= A(\kappa u)(t) + \frac{\lambda}{1-\xi} \\
&\geq \kappa^\theta \int_0^t \phi_q\left(\int_0^1 G(r,s)a(s)f(u(s))ds\right) dr \\
&\quad + \kappa^\theta \frac{\xi}{1-\xi}\int_0^\eta \phi_q\left(\int_0^1 G(r,s)a(s)f(u(s))ds\right) dr + \frac{\lambda}{1-\xi} \\
&= \kappa^\theta Au(t) + \frac{\lambda}{1-\xi} \\
&\geq \kappa^\theta\left(Au(t) + \frac{\lambda}{1-\xi}\right) = \kappa^\theta T_\lambda u(t),
\end{aligned} \tag{8.90}$$

i.e., T_λ is θ-concave. By Lemma 8.10, T_λ has a unique fixed point u_λ in P°, which is the unique positive solution of BVP (8.57), (8.58). The first part of the theorem is proved.

In the remainder of the proof, we show the "furthermore" part of the theorem. Note that

$$\begin{aligned}
u_\lambda(t) &= \int_0^t \phi_q\left(\int_0^1 G(r,s)a(s)f(u_\lambda(s))ds\right) dr \\
&\quad + \frac{\xi}{1-\xi}\int_0^\eta \phi_q\left(\int_0^1 G(r,s)a(s)f(u_\lambda(s))ds\right) dr + \frac{\lambda}{1-\xi}.
\end{aligned}$$

Then, part (i) is obvious.

Next, we show part (ii). Assume $\lambda_1 > \lambda_2 > 0$. Then we first claim that $u_{\lambda_1}(t) \geq u_{\lambda_2}(t)$ on $[0,1]$. In fact, let

$$\bar\gamma = \sup\{\gamma \ : \ u_{\lambda_1}(t) \geq \gamma u_{\lambda_2}(t) \text{ on } [0,1]\}.$$

We assert that $\bar\gamma \geq 1$. For otherwise, we have $0 < \bar\gamma < 1$. Then, noting (8.89) and (8.90), it follows that

$$\begin{aligned}
u_{\lambda_1}(t) = T_{\lambda_1}u_{\lambda_1}(t) &\geq T_{\lambda_1}(\bar\gamma u_{\lambda_2})(t) \\
&> T_{\lambda_2}(\bar\gamma u_{\lambda_2})(t) \\
&\geq (\bar\gamma)^\theta T_{\lambda_2}u_{\lambda_2}(t) \\
&= (\bar\gamma)^\theta u_{\lambda_2}(t) > \bar\gamma u_{\lambda_2}(t) \quad \text{on } [0,1].
\end{aligned}$$

This contradicts the definition of $\bar\gamma$. Thus, $u_{\lambda_1}(t) \geq u_{\lambda_2}(t)$ for $t \in [0,1]$. Therefore, we have

$$\begin{aligned}
u_{\lambda_1}(t) &= Au_{\lambda_1}(t) + \frac{\lambda_1}{1-\xi} \\
&\geq Au_{\lambda_2}(t) + \frac{\lambda_1}{1-\xi} \\
&> Au_{\lambda_2}(t) + \frac{\lambda_2}{1-\xi} = u_{\lambda_2}(t) \quad \text{on } [0,1].
\end{aligned}$$

Thus, $u_\lambda(t)$ is strictly increasing in λ.

Finally, we show part (iii). For any given $\lambda_0 > 0$, from part (i), we have

$$u_\lambda(t) < u_{\lambda_0}(t) \quad \text{on } [0,1] \text{ for any } 0 < \lambda < \lambda_0. \tag{8.91}$$

Let

$$\bar{\delta} = \sup\{\delta > 0 \ : \ u_\lambda(t) \geq \delta u_{\lambda_0}(t), \quad t \in [0,1], \ \lambda_0/2 < \lambda < \lambda_0\}.$$

Then, $0 < \bar{\delta} < 1$, and for $\lambda_0/2 < \lambda < \lambda_0$, $u_\lambda(t) \geq \bar{\delta} u_{\lambda_0}(t)$ on $[0,1]$. Moreover, in virtue of (8.89) and (8.90), we see that

$$\begin{aligned} u_\lambda(t) = T_\lambda u_\lambda(t) &\geq T_\lambda(\bar{\delta} u_{\lambda_0})(t) \\ &= A(\bar{\delta} u_{\lambda_0})(t) + \frac{\lambda}{1-\xi} \\ &> \frac{\lambda}{\lambda_0}\left(A(\bar{\delta} u_{\lambda_0})(t) + \frac{\lambda_0}{1-\xi}\right) \\ &= \frac{\lambda}{\lambda_0} T_{\lambda_0}(\bar{\delta} u_{\lambda_0})(t) \\ &\geq \frac{\lambda}{\lambda_0}(\bar{\delta})^\theta T_{\lambda_0} u_{\lambda_0}(t) = \frac{\lambda}{\lambda_0}(\bar{\delta})^\theta u_{\lambda_0}(t) \quad \text{on } [0,1]. \end{aligned}$$

Hence, from the definition of $\bar{\delta}$, we see that

$$\frac{\lambda}{\lambda_0}(\bar{\delta})^\theta \leq \bar{\delta},$$

i.e.,

$$\bar{\delta} \geq \left(\frac{\lambda}{\lambda_0}\right)^{1/(1-\theta)}.$$

Then, for $t \in [0,1]$ and $\lambda_0/2 < \lambda < \lambda_0$, we have

$$u_\lambda(t) \geq \bar{\delta} u_{\lambda_0}(t) \geq \left(\frac{\lambda}{\lambda_0}\right)^{1/(1-\theta)} u_{\lambda_0}(t). \tag{8.92}$$

From (8.91) and (8.92), we obtain that

$$||u_{\lambda_0} - u_\lambda|| \leq \left(1 - \left(\frac{\lambda}{\lambda_0}\right)^{1/(1-\theta)}\right) ||u_{\lambda_0}|| \quad \text{for } \lambda_0/2 < \lambda < \lambda_0.$$

Thus, we have

$$||u_{\lambda_0} - u_\lambda|| \to 0 \quad \text{as } \lambda \to \lambda_0^-.$$

Similarly, we can show that

$$||u_{\lambda_0} - u_\lambda|| \to 0 \quad \text{as } \lambda \to \lambda_0^+.$$

Hence, part (iii) holds. This completes the proof of the theorem. □

Bibliography

Agarwal R. P., O'Regan D. and Wong P. J. Y. (1999). *Positive Solutions of Differential, Difference and Integral Equations* (Kluwer Academic Publishers, Dordrecht).

Anderson D. R. and Avery R. I. (2002). Fixed point theorem of cone expansion and compression of functional type, *J. Difference Equ. Appl.* **8**, No. 11, pp. 1073–1083.

Anderson D. R. and Avery R. I. (2011). Existence of a periodic solution for continuous and discrete periodic second-order equations with variable potentials, *J. Appl. Math. Comput.* **37**, No. 1–2, pp. 297–312.

Anderson D. R., Avery R. I., Henderson J. and Liu X. Y. (2011). Existence of positive solutions of a second order right focal boundary value problem, *Commun. Appl. Nonlinear Anal.* **18**, pp. 41–52.

Anderson D. R., Avery R. I., Henderson J., Liu X. Y. and Lyons J. W. (2011). Existence of a positive solution for a right focal discrete boundary value problem, *J. Difference Equ. Appl* **17**, No. 11, pp. 1635–1642.

Anderson D. R., Avery R. I. and Peterson A. C. (1998). Three positive solutions to a discrete focal boundary value problem, *J. Comput. Appl. Math.* **88**, No. 1, pp. 103–118.

Avery R. I. (1998). Existence of multiple positive solutions to a conjugate boundary value problem, *Math. Sci. Res. Hot-Line* **2**, pp. 1–6.

Avery R. I. (1999). A generalization of the Leggett-Williams fixed point theorem, *Math. Sci. Res. Hot-Line* **3**, pp. 9–14.

Avery R. I., Anderson D. R. and Henderson J. (2016). An extension of the compression-expansion fixed point theorem of functional type, *Electron. J. Differenital Equations* No. 253, pp. 1–12.

Avery R. I., Davis J. M. and Henderson J. (2000). Three symmetric positive solutions for Lidstone problems by a generalization of the Leggett-Williams theorem, *Electron. J. Differential Equations* **2000**, No. 40, pp. 1–15.

Avery R. I., Graef J. R. and Liu X. (2015). Compression fixed point theorems of operator type, *J. Fixed Point Theory Appl.* **17**, No. 1, pp. 83–97.

Hardy G. H., Littlewood J. E. and Pólya, G (1899). *Inequalities*, Reprint of the 1952 Edition (Cambridge University Press, Cambridge).

Avery R. I. and Henderson J. (2000). Three symmetric positive solutions for a econd-order boundary value problem, *Appl. Math. Lett.* **13**, No. 3, pp. 1–7.

Avery R. I., Henderson J. and Liu X. (2015). Omitted ray fixed point theorem, *J. Fixed Point Theory Appl.* **17**, No. 2, pp. 313–330.

Avery R. I. and Peterson A. C. (2001). Three positive fixed points of nonlinear operators on ordered Banach spaces, Advances in Difference Equations III, *Comput. Math. Appl.* **42**, No. 3–5, pp. 313–322.

DaCunha J. J., Davis J. M. and Singh P. (2004). Existence results for singular three point boundary value problems on time scales, *J. Math. Anal. Appl.* **295**, No. 2, pp. 378–391.

Davis J. M., Eloe P. W. and Henderson J. (1999). Triple positive solutions and dependence on higher order derivatives, *J. Math. Anal. Appl.* **237**, No. 2, pp. 710–720.

Davis J. M., Erbe L. H. and Henderson J. (2001). Multiplicity of positive solutions for higher order Sturm-Liouville problems, *Rocky Mountain J. Math.* **31**, No. 1, pp. 169–184.

Dix J. G., Padhi S. and Pati S. (2010). Multiple positive periodic solutions for a nonlinear first order functional difference equation, *J. Difference Equ. Appl.* **16**, No. 9, pp. 1037–1046.

Davis J. M. and Henderson J. (1999). Triple positive symmetric solutions for a Lidstone boundary value problem, *Differential Equations Dynam. Systems* **7**, pp. 321–330.

Eloe P. W. (2000). Nonlinear eigenvalue problems for higher order Lidstone boundary value problems, *Electron. J. Qual. Theory Differ. Equ.* **2000**, No. 2, pp. 1–8.

Eloe P. W. and Henderson J. (1991). Singular nonlinear boundary value problems for higher order ordinary differential equations, *Nonlinear Anal.* **17**, No. 1, pp. 1–10.

Eloe P. W. and Henderson J. (1997). Positive solutions for $(n-1, n)$ conjugate boundary value problems, *Nonlinear Anal.* **28**, No. 10, pp. 1669–1680.

Eloe P. W. and Henderson J. (1997). Singular $(n-1, 1)$ conjugate boundary value problems, *Georgian Math. J.* **4**, No. 5, pp. 401–412.

Eloe P. W. and Henderson J. (1997). Singular nonlinear $(k, n-k)$ conjugate boundary value problems, *J. Differential Equations* **133**, No. 1, pp. 136–151.

Eloe P. W. and Henderson J. (1998). Singular nonlinear multipoint conjugate boundary value problems, *Commun. Appl. Anal.* **2**, No. 4, pp. 497–511.

Erbe L. H., S. Hu and Wang H. (1994). Multiple positive solutions of some boundary value problems, *J. Math. Anal. Appl.* **184**, No. 3, pp. 640–648.

Erbe L. H. and Wang H. (1994). On the existence of positive solutions of ordinary differential equations, *Proc. Amer. Math. Soc.* **120**, No. 3, pp. 743–748.

Gatica J. A., Oliker V. and Waltman P. E. (1989). Singular boundary value problems for second-order ordinary differential equations, *J. Differential Equations* **79**, No. 1, pp. 62–78.

Graef J. R., Henderson J. and Yang B. (2008). Positive solutions to a singular third order nonlocal boundary value problem, *Indian J. Math.* **50**, No. 2, pp. 317–330.

Graef J. R., Henderson J. and Yang B. (2009). Existence of positive solutions of a higher order nonlocal singular boundary value problem, *Dyn. Contin. Discrete Impuls. Syst. Ser. A Math. Anal.* **16**, **Differential Equations and Dynamical Systems, Suppl. S1**, pp. 147–152.

Graef J. R., Kong L. and Kong Q. (2010) Ambrosetti–Prodi–type results for a third order multi–point boundary value problem, *Nonlinear Stud.* **17**, 121–130.

Graef J. R., Kong L. and Yang B. (2010). Positive solutions for third order multi-point singular boundary value problems, *Czechoslovak Math. J.* **60**, 173–182.

Graef J. R., Kong L. and Wang H. (2008). Existence, multiplicity and dependence on a parameter for a periodic boundary value problem, *J. Differential Equations* **245**, No. 5, pp. 1185–1197.

Graef J. R., Qian C. and Yang B. (2003). Multiple symmetric positive solutions of a class of boundary value problems for higher order ordinary differential equations, *Proc. Amer. Math. Soc.* **131**, No. 2, pp. 577–585.

Graef J. and Liu X. Y. (2014). Existence of positive solutions of a fractional boundary value problem involving with bounded linear operators, *J. Nonlinear Funct. Anal.* **2014**, Article ID 13.

Graef J. R. and Yang B. (2006). Positive solutions to a multi-point higher order boundary value problem, *J. Math. Anal. Appl.* **316**, No. 2, pp. 409–421.

Guo D. and Lakshmikantham V. (1988). *Nonlinear Problems in Abstract Cones* (Academic Press, New York).

Henderson J. and Liu X. Y. (2013), Positive solutions of fractional differential equations with bounded linear functional conditions, *Nonlinear Stud.* **20**, No. 4, pp. 559–570.

Henderson J., Liu X. Y., Sutherland S. and Tian Y. (2013). Positive solutions for a second order impulsive BVP with bounded linear operator conditions, *Dynam. Cont. Dis. Ser. A*, **20**, No. 5, pp. 571–592.

Henderson J. and Luca R. (2013). Existence and multiplicity for positive solutions of a system of higher-order multi-point boundary value problems, *Nonlinear Differential Equations Appl.* **20**, No. 3, pp. 1035–1054.

Henderson J. and Luca R. (2014). On a second-order nonlinear discrete multi-point eigenvalue problem, *J. Difference Equ. Appl.* **20**, No. 7, pp. 1005–1018.

Henderson J., Luca R., Nelms C. and Yang A. (2015). Positive solutions for a singular third order boundary value problem, *Differ. Equ. Appl.* **7**, No. 4, pp. 437–447.

Henderson J., Ntouyas S. K. and Purnaras I. K. (2008). Positive solutions for systems of three-point nonlinear discrete boundary value problems, *Neural Parallel Sci. Comput.* **16**, pp. 209–224.

Henderson J., Ntouyas S. K. and Purnaras I. K. (2009). Positive solutions for systems of nonlinear discrete boundary value problems, *J. Difference Equ. Appl.* **15**, No. 10, pp. 895–912.

Henderson J. and Singh P. (2004). An nth order singular three-point boundary value problem, *Comm. Appl. Nonlinear Anal.* **11** No. 1, pp. 39–50.

Henderson J. and Wang H. (1997). Positive solutions for nonlinear eigenvalue problems, *J. Math. Anal. Appl.* **208**, No. 1, pp. 252–259.

Henderson J. and Wang H. (2005). Nonlinear eigenvalue problems for quasilinear systems, *Comput. Math. Appl.* **49**, No. 11–12, pp. 1941–1949.

Henderson J. and Yin W. K. C. (1991). Singular boundary value problems, *Bull. Inst. Math. Acad. Sinica* **19**, No. 3, pp. 229–242.

Henderson J. and Yin W. K. C. (1998). Singular $(k, n-k)$ boundary value problems between conjugate and right focal, *J. Comput. Appl. Math.* **88**, No. 1, pp. 57–69.

Kaufmann E. R. and Kosmatov N. (2007). Elastic beam problem with higher order derivatives, *Nonlinear Anal. Real World Appl.* **8**, No. 3, pp. 811–821.

Kong L., Piao D. and Wang L. (2009). Positive solutions for third order boundary value problems with p-Laplacian, *Result. Math.* **55**, 111–128.

Krasnosel'skii M. A. (1964). *Topological Methods in the Theory of Nonlinear Integral Equations*, (translated from the Russian by A. H. Armstrong; translation edited by J. Burlak), A Pergamon Press Book (The Macmillan Co., New York).

Leggett R. W. and Williams L. R. (1979). Multiple positive fixed points of nonlinear operators on ordered Banach spaces, *Indiana Univ. Math. J.* **28**, No. 4, pp. 673–688.

Luca R. (2012). Existence of positive solutions for a second-order $m+1$-point discrete boundary value problem, *J. Difference Equ. Appl.* **18**, No. 5, pp. 865–877.

Lyons J. W. (2015). An application of an Avery type fixed point theorem to a second order antiperiodic boundary value problem, *Discrete Contin. Syn. Syst.*, **2015, Dynamical Systems, Differential Equations and Applications. 10th AIMS Conference, Suppl.**, pp. 615–620.

Lyons J. W. and Neugebauer J. T. (2015). Existence of an antisymmetric solution of a boundary value problem with antisymmetric boundary conditions, *Electron. J. Qual. Theory Differ. Equ.* **2015**, No. 72, pp. 1–15.

Maroun M. (2005). Positive solutions to a third-order right focal boundary value problem, *Comm. Appl. Nonlinear Anal.* **12**, No. 3, pp. 71–82.

Maroun M. (2007). Positive solutions to an nth order right focal boundary value problem, *Electron. J. Qual. Theory Differ. Equ.* **2007**, No. 4, pp. 1–17.

O'Regan D. (1994). *Theory of Singular Boundary Value Problems* (World Scientific Publishing Co., Inc., River Edge, NJ).

Singh P. (2004). A second-order singular three-point boundary value problem, *Appl. Math. Lett.* **17**, No. 8, pp. 969–976.

Wong P. J. Y. (1998). Two-point right focal eigenvalue problems for difference equations, *Dynam. Systems Appl.* **7**, No. 3, pp. 345–364.

Tian Y., Henderson J., Liu X. Y. and Sutherland S. (2013), Existence and uniqueness of positive solutions for even order singular impulsive boundary value problem, *Dyn. Contin. Discrete Impuls. Syst. Ser. A Math. Anal.* **20**, No. 1, pp. 107–120.

Zeidler E. (1986). *Nonlinear Functional Analysis and Its Applications I: Fixed-Point Theorems* (Springer, New York).

Chapter 9

Boundary Data Smoothness

As an application of the Brouwer Invariance of Domain Theorem (see Theorem 4.3), we saw, in the presence of continuity of the nonlinearity and uniqueness conditions on boundary value problems, that solutions of boundary value problems depend continuously on the boundary conditions (see Lemma 4.4). In this chapter, we assume the nonlinearity is a $C^{(1)}$-function, and we obtain results about differentiability of solutions of boundary value problems with respect to the boundary conditions.

9.1 Differentiation of Solutions with Respect to Boundary Conditions

We will be concerned with differentiating solutions of boundary value problems with respect to boundary data for the second order ODE,

$$y'' = f(x, y, y'), \quad a < x < b, \tag{9.1}$$

satisfying

$$y(x_1) = y_1, \quad \int_{x_1}^{x_2} y(x)\, dx = y_2, \tag{9.2}$$

where $a < x_1 < x_2 < b$, and $y_1, y_2 \in \mathbb{R}$, and where we assume:

(i) $f(x, u_1, u_2) : (a, b) \times \mathbb{R}^2 \to \mathbb{R}$ is continuous,
(ii) $\frac{\partial f}{\partial u_i}(x, u_1, u_2) : (a, b) \times \mathbb{R}^2 \to \mathbb{R}$ is continuous, for $i = 1, 2$, and
(iii) Solutions of IVP's for (9.1) extend to (a, b).

Condition (iii) is not necessary for the spirit of this chapter's results, however, by assuming (iii), we avoid continually making statements in terms of solutions' maximal intervals of existence.

In addition to (i)–(iii), under uniqueness assumptions on solutions of (9.1), (9.2), we will establish analogues of a Peano result concerning differentiation of solutions of (9.1) with respect to initial conditions. For our results on differentiation with respect to boundary conditions, given a solution $y(x)$ of (9.1), we will give much attention to the *variational equation for* (9.1) *along* $y(x)$, which is defined by

$$z'' = \frac{\partial f}{\partial u_1}(x, y(x), y'(x))z + \frac{\partial f}{\partial u_2}(x, y(x), y'(x))z'. \tag{9.3}$$

The theorem for which we seek an analog can be stated in the context of (9.1) as follows:

Theorem 9.1 (Peano). *Assume that with respect to* (9.1), *conditions* (i)–(iii) *are satisfied. Let* $x_0 \in (a, b)$ *and* $y(x) := y(x, x_0, c_1, c_2)$ *denote the solution of* (9.1) *satisfying the initial conditions,* $y(x_0) = c_1$, $y'(x_0) = c_2$. *Then,*

(a) $\frac{\partial y}{\partial c_1}$ *and* $\frac{\partial y}{\partial c_2}$ *exist on* (a, b), *and* $\alpha_i := \frac{\partial y}{\partial c_i}$, $i = 1, 2$, *are solutions of* (9.3) *along* $y(x)$ *and satisfy the respective initial conditions,*

$$\alpha_1(x_0) = 1, \quad \alpha_1'(x_0) = 0, \quad \alpha_2(x_0) = 0, \quad \alpha_2'(x_0) = 1.$$

(b) $\frac{\partial y}{\partial x_0}$ *exists on* (a, b), *and* $\beta := \frac{\partial y}{\partial x_0}$ *is the solution of* (9.3) *along* $y(x)$ *and satisfies the initial conditions*

$$\beta(x_0) = -y'(x_0), \quad \beta'(x_0) = -y''(x_0).$$

(c) $\frac{\partial y}{\partial x_0} = -y'(x_0)\frac{\partial y}{\partial c_1} - y''(x_0)\frac{\partial y}{\partial c_2}$.

Our analog of Theorem 9.1 will require uniqueness of solutions of (9.1), (9.2), a condition we list as an assumption:

(iv) Given $a < x_1 < x_2 < b$, if $y(x_1) = z(x_1)$ and $\int_{x_1}^{x_2} y(x)\,dx = \int_{x_1}^{x_2} z(x)\,dx$, where $y(x)$ and $z(x)$ are solutions of (9.1), then $y(x) \equiv z(x)$.

We will also make extensive use of a similar uniqueness condition on (9.3) along solution a $y(x)$ of (9.1):

(v) Given $a < x_1 < x_2 < b$ and a solution $y(x)$ of (9.1), if $u(x_1) = 0$ and $\int_{x_1}^{x_2} u(x)\,dx = 0$, where $u(x)$ is a solution of (9.3) along $y(x)$, then $u(x) \equiv 0$.

As stated at the beginning of this chapter, in the presence of uniqueness, solutions of boundary value problems depend continuously upon boundary conditions. For the BVP's of this chapter, we state such a continuous dependence result. The proof, which we omit, involves a straightforward application of Theorem 4.3.

Theorem 9.2. *Assume* (i)–(iv) *are satisfied with respect to* (9.1). *Let* $u(x)$ *be a solution of* (9.1) *on* (a,b), *and let* $a < c < x_1 < x_2 < d < b$ *be given. Then, there exists a* $\delta > 0$ *such that for* $|x_i - t_i| < \delta$, $i = 1,2$, $|u(x_1) - y_1| < \delta$, *and* $|\int_{x_1}^{x_2} u(x)\,dx - y_2| < \delta$, *there exists a unique solution* $u_\delta(x)$ *of* (9.1) *such that* $u_\delta(t_1) = y_1$, $\int_{t_1}^{t_2} u_\delta(x)\,dx = y_2$, *and* $\{u_\delta^{(j)}(x)\}$ *converges uniformly to* $u^{(j)}(x)$, *as* $\delta \to 0$, *on* $[c,d]$, *for* $j = 0,1$.

We now present the main result of this chapter.

Theorem 9.3. *Assume conditions* (i)–(v) *are satisfied. Let* $u(x)$ *be a solution of* (9.1) *on* (a,b). *Let* $a < x_1 < x_2 < b$ *be given, so that* $u(x) = u(x, x_1, x_2, u_1, u_2)$, *where* $u(x_1) = u_1$ *and* $\int_{x_1}^{x_2} u(x)\,dx = u_2$. *Then*

(a) $\frac{\partial u}{\partial u_1}$ *and* $\frac{\partial u}{\partial u_2}$ *exist on* (a,b), *and* $r_i := \frac{\partial u}{\partial u_i}$, $i = 1,2$, *are solutions of* (9.3) *along* $u(x)$ *and satisfy the respective boundary conditions,*

$$r_1(x_1) = 1, \quad \int_{x_1}^{x_2} r_1(x)\,dx = 0,$$

$$r_2(x_1) = 0, \quad \int_{x_1}^{x_2} r_2(x)\,dx = 1.$$

(b) $\frac{\partial u}{\partial x_1}$ *and* $\frac{\partial u}{\partial x_2}$ *exist on* (a,b), *and* $z_i := \frac{\partial u}{\partial x_i}$, $i = 1,2$, *are solutions of* (9.3) *along* $u(x)$ *and satisfy the respective boundary conditions,*

$$z_1(x_1) = -u'(x_1), \quad \int_{x_1}^{x_2} z_1(x)\,dx = u(x_1),$$

$$r_2(x_1) = 0, \quad \int_{x_1}^{x_2} z_2(x)\,dx = -u(x_2).$$

(c) $\frac{\partial u}{\partial x_1} = -u'(x_1)\frac{\partial u}{\partial u_1} + u(x_1)\frac{\partial u}{\partial u_2}$, $\frac{\partial u}{\partial x_2} = -u(x_2)\frac{\partial u}{\partial u_2}$.

Proof. For part (a), we begin with $\frac{\partial u}{\partial u_1}$. Let $\delta > 0$ be as in Theorem 9.2. Let $0 < |h| < \delta$ be given and define

$$r_{1h}(x) = \frac{1}{h}[u(x, x_1, x_2, u_1 + h, u_2) - u(x, x_1, x_2, u_1, u_2)].$$

Note that $u(x_1, x_1, x_2, u_1 + h, u_2) = u_1 + h$, and $u(x_1, x_1, x_2, u_1, u_2) = u_1$, so that for every $h \neq 0$,

$$r_{1h}(x_1) = \frac{1}{h}[u_1 + h - u_1] = 1,$$

and

$$\int_{x_1}^{x_2} r_{1h}(x)\,dx = \frac{1}{h}\int_{x_1}^{x_2} [u(x,x_1,x_2,u_1+h,u_2) - u(x,x_1,x_2,u_1,u_2)]\,dx$$
$$= \frac{1}{h}[u_1 - u_2] = 0.$$

Let

$$\beta_2 = u'(x_1,x_1,x_2,u_1,u_2),$$

and

$$\varepsilon_2 = \varepsilon_2(h) = u'(x,x_1,x_2,u_1+h,u_2) - \beta_2.$$

By Theorem 9.2, $\varepsilon_2 = \varepsilon_2(h) \to 0$, as $h \to 0$. Using the notation of Theorem 9.1 for solutions of IVP's for (9.1) and viewing solutions u as solutions y of IVP's, we have

$$r_{1h}(x) = \frac{1}{h}[y(x,x_1,u_1+h,\beta_2+\varepsilon_2) - y(x,x_1,u_1,\beta_2)].$$

Then, using a telescoping sum,

$$r_{1h}(x) = \frac{1}{h}\{[y(x,x_1,u_1+h,\beta_2+\varepsilon_2) - y(x,x_1,u_1,\beta_2+\varepsilon_2)]$$
$$+ [y(x,x_1,u_1,\beta_2+\varepsilon_2) - y(x,x_1,u_1,\beta_2)]\}.$$

By Theorem 9.1 and the Mean Value Theorem, we obtain,

$$\begin{aligned} r_{1h}(x) =& \frac{1}{h}\alpha_1(x,y(x,x_1,u_1+\overline{h},\beta_2+\varepsilon_2))(u_1+h-u_1) \\ &+ \frac{1}{h}\alpha_2(x,y(x,x_1,u_1,\beta_2+\overline{\varepsilon}_2))(\beta_2+\varepsilon_2-\beta_2) \\ =& \alpha_1(x,y(x,x_1,u_1+\overline{h},\beta_2+\varepsilon_2)) \\ &+ \frac{\varepsilon_2}{h}\alpha_2(x,y(x,x_1,u_1,\beta_2+\overline{\varepsilon}_2)), \end{aligned}$$

where $\alpha_i(x,y(\cdot))$, $i=1,2$, is the solution of (9.3) along $y(\cdot)$ and satisfies in each case

$$\alpha_1(x_1,y(\cdot)) = 1, \quad \alpha_1'(x_1,y(\cdot)) = 0,$$
$$\alpha_2(x_1,y(\cdot)) = 0, \quad \alpha_2'(x_1,y(\cdot)) = 1,$$

and where $u_1+\overline{h}$ is between u_1 and u_1+h, and $\beta_2+\overline{\varepsilon}_2$ is between β_2 and $\beta_2+\varepsilon_2$. To show $\lim_{h\to 0} r_{1h}(x)$ exists, it suffices to show $\lim_{h\to 0}\frac{\varepsilon_2}{h}$ exists.

Since $\alpha_2(x, y(\cdot))$ is a nontrivial solution of (9.3) along $y(\cdot)$ and since $\alpha_2(x_1, y(\cdot)) = 0$, it follows from (v) that

$$\int_{x_1}^{x_2} \alpha_2(x, y(\cdot))\, dx \neq 0.$$

Above, we observed that $\int_{x_1}^{x_2} r_{1h}(x)\, dx = 0$, and so we have

$$\frac{\varepsilon_2}{h} = \frac{-\int_{x_1}^{x_2} \alpha_1(x, y(x, x_1, u_1 + \overline{h}, \beta_2 + \varepsilon_2))\, dx}{\int_{x_1}^{x_2} \alpha_2(x, y(x, x_1, u_1, \beta_2 + \overline{\varepsilon}_2))\, dx}.$$

By continuous dependence on initial conditions, we can let $h \to 0$, so that

$$\begin{aligned}\lim_{h\to 0} \frac{\varepsilon_2}{h} &= \frac{-\int_{x_1}^{x_2} \alpha_1(x, y(x, x_1, u_1, \beta_2))\, dx}{\int_{x_1}^{x_2} \alpha_2(x, y(x, x_1, u_1, \beta_2))\, dx} \\ &= \frac{-\int_{x_1}^{x_2} \alpha_1(x, u(\cdot))\, dx}{\int_{x_1}^{x_2} \alpha_2(x, u(\cdot))\, dx} \\ &:= D.\end{aligned}$$

Let $r_1(x) := \lim_{h\to 0} r_{1h}(x)$. Then

$$r_1(x) = \frac{\partial u}{\partial u_1}(x, x_1, x_2, u_1, u_2),$$

and moreover,

$$\begin{aligned} r_1(x) &= \lim_{h\to 0} r_{1h}(x) \\ &= \alpha_1(x, y(x, x_1, u_1, \beta_2)) + D\alpha_2(x, y(x, x_1, u_1, \beta_2)) \\ &= \alpha_1(x, u(x, x_1, x_2, u_1, u_2)) + D\alpha_2(x, u(x, x_1, x_2, u_1, u_2)),\end{aligned}$$

which is a solution of (9.3) along $u(x)$. Also, from the boundary conditions satisfied by $r_{1h}(x)$,

$$r_1(x_1) = \lim_{h\to 0} (x_1) = 1, \text{ and}$$

$$\int_{x_1}^{x_2} r_1(x)\, dx = \lim_{h\to 0} \int_{x_1}^{x_2} r_{1h}(x)\, dx = 0.$$

The argument is complete for $\frac{\partial u}{\partial u_1}$.

We will also include some of the details for $\frac{\partial u}{\partial u_2}$. Again, let $\delta > 0$ be as in Theorem 9.2. Let $0 < |h| < \delta$ be given and define

$$r_{2h} = \frac{1}{h}[u(x, x_1, x_2, u_1, u_2 + h) - u(x, x_1, x_2, u_1, u_2)].$$

This time, for $h \neq 0$,

$$r_{2h}(x_1) = \frac{1}{h}[u_1 - u_1] = 0,$$

and

$$\begin{aligned}\int_{x_1}^{x_2} r_{2h}(x)\,dx &= \frac{1}{h}\int_{x_1}^{x_2} [u(x, x_1, x_2, u_1, u_2 + h) - u(x, x_1, x_2, u_1, u_2)]\,dx \\ &= \frac{1}{h}[u_2 + h - u_2] \\ &= 1.\end{aligned}$$

Let

$$\begin{aligned}&\beta_2 = u'(x_1, x_1, x_2, u_1, u_2), \text{ and} \\ &\varepsilon_2 = \varepsilon_2(h) = u'(x_1, x_1, x_2, u_1, u_2 + h) - \beta_2.\end{aligned}$$

Again, $\varepsilon_2 \to 0$, as $h \to 0$. As before, using the notation of solutions of IVP's for (9.1) as in Theorem 9.1, we have

$$r_{2h}(x) = \frac{1}{h}[y(x, x_1, u_1, \beta_2 + \varepsilon_2) - y(x, x_1, u_1, \beta_2)].$$

By Theorem 9.1 and the Mean Value Theorem,

$$r_{2h}(x) = \frac{\varepsilon_2}{h}\alpha_2(x, y(x, x_1, u_1, \beta_2 + \bar{\varepsilon}_2)),$$

where $\beta_2 + \bar{\varepsilon}_2$ is between β_2 and $\beta_2 + \varepsilon_2$, and $\alpha_2(x, y(\cdot))$ is the solution of (9.3) along $y(\cdot)$ and satisfies

$$\alpha_2(x_1, y(\cdot)) = 0, \quad \alpha_2'(x_1, y(\cdot)) = 1.$$

To show $\lim_{h\to 0} r_{2h}(x)$ exists, it suffices to show $\lim_{h\to 0} \frac{\varepsilon_2}{h}$ exists. By assumption (v),

$$\int_{x_1}^{x_2} \alpha_2(x, y(\cdot))\,dx \neq 0,$$

and above, we have $\int_{x_1}^{x_2} r_{2h}(x)\,dx = 1$. Hence,

$$\frac{\varepsilon_2}{h} = \frac{1}{\int_{x_1}^{x_2} \alpha_2(x, y(x, x_1, u_1, \beta_2 + \bar{\varepsilon}_2))\,dx},$$

from which,

$$\begin{aligned}\lim_{h\to 0}\frac{\varepsilon_2}{h} &= \frac{1}{\int_{x_1}^{x_2} \alpha_2(x, y(x, x_1, u_1, \beta_2))\,dx}, \\ &= \frac{1}{\int_{x_1}^{x_2} \alpha_2(x, u(\cdot))\,dx}, \\ &:= E.\end{aligned}$$

If we set $r_2(x) = \lim_{h\to 0} r_{2h}(x)$, then by construction,

$$r_2(x) = \frac{\partial u}{\partial u_2}(x, x_1, x_2, u_1, u_2),$$

and moreover,

$$r_2(x) = \lim_{h\to 0} r_{2h}(x) = E\alpha_2(x, u(x, x_1, x_2, u_1, u_2)),$$

which is a solution of (9.3) along $u(x)$. And from the boundary conditions satisfied by $r_{2h}(x)$, we have

$$r_2(x) = 0 \text{ and } \int_{x_1}^{x_2} r_2(x)\,dx = 1.$$

The argument is complete for $\frac{\partial u}{\partial u_2}$.

For part (b), we produce the details for $\frac{\partial u}{\partial x_1}$, with the arguments for $\frac{\partial u}{\partial x_2}$ being somewhat along the same lines. Again, we let $\delta > 0$ be as in Theorem 9.2 and let $0 < |h| < \delta$ be given. Define

$$z_{1h}(x) = \frac{1}{h}[u(x, x_1 + h, x_2, u_1, u_2) - u(x, x_1, x_2, u_1, u_2)].$$

First, we consider boundary conditions.

By employing the Mean Value Theorem for Integrals, we have for $h \neq 0$,

$$\begin{aligned}
\int_{x_1}^{x_2} z_{1h}(x)\,dx &= \frac{1}{h}\int_{x_1}^{x_2} [u(x, x_1 + h, x_2, u_1, u_2) - u(x, x_1, x_2, u_1, u_2)]\,dx \\
&= \frac{1}{h}\bigg\{\int_{x_1}^{x_1+h} u(x, x_1 + h, x_2, u_1, u_2)\,dx \\
&\qquad + \int_{x_1+h}^{x_1} u(x, x_1 + h, x_2, u_1, u_2)\,dx \\
&\qquad - \int_{x_1}^{x_2} u(x, x_1, x_2, u_1, u_2)\,dx\bigg\} \\
&= \frac{1}{h}\bigg\{\int_{x_1}^{x_1+h} u(x, x_1 + h, x_2, u_1, u_2)\,dx + u_2 - u_2\bigg\} \\
&= \frac{1}{h}u(c_h, x_1 + h, x_2, u_1, u_2)h \\
&= u(c_h, x_1 + h, x_2, u_1, u_2),
\end{aligned}$$

for some c_h inclusively between x_1 and $x_1 + h$. By Theorem 9.2, we can compute the limit,

$$\lim_{h\to 0}\int_{x_1}^{x_2} z_{1h}(x)\,dx = u(x_1, x_1, x_2, u_1, u_2) = u(x_1).$$

Next, we apply the Mean Value Theorem in looking at

$$\begin{aligned} z_{1h}(x_1) &= \frac{1}{h}[u(x_1, x_1+h, x_2, u_1, u_2) - u(x_1, x_1, x_2, u_1.u_2)] \\ &= \frac{1}{h}[u(x_1, x_1+h, x_2, u_1, u_2) - u(x_1+h, x_1+h, x_2, u_1.u_2)] \\ &= -\frac{1}{h}u'(\xi_h, x_1+h, x_2, u_1, u_2)h \\ &= -u'(\xi_h, x_1+h, x_2, u_1, u_2), \end{aligned}$$

where ξ_h is between x_1 and x_1+h. So, in passing to the limit, we have

$$\lim_{h\to 0} z_{1h}(x_1) = -u'(x_1, x_1, x_2, u_1, u_2) = -u'(x_1).$$

Finally, we deal with $\lim_{h\to 0} z_{1h}(x)$. This time, let

$$\begin{aligned} &\beta_2 = u'(x_1, x_1, x_2, u_1, u_2) \text{ and} \\ &\varepsilon_2 = \varepsilon_2(h) = u'(x_1+h, x_1+h, x_2, u_1, u_2) - \beta_2. \end{aligned}$$

As before, $\varepsilon_2 \to 0$, as $h \to 0$. As in part (a), we view solutions u as solutions of initial value problems y, and we have

$$\begin{aligned} z_{1h}(x) &= \frac{1}{h}[u(x, x_1+h, x_2, u_1, u_2) - u(x, x_1, x_2, u_1.u_2)] \\ &= \frac{1}{h}\big[y(x, x_1+h, u_1, \beta_2+\varepsilon_2) - y(x, x_1, u_1, \beta_2)\big] \\ &= \frac{1}{h}\big[y(x, x_1+h, u_1, \beta_2+\varepsilon_2) - y(x, x_1, u_1, \beta_2+\varepsilon_2) \\ &\qquad + y(x, x_1, u_1, \beta_2+\varepsilon_2) - y(x, x_1, u_1, \beta_2)\big] \\ &= \frac{1}{h}\big\{\beta(x, y(x, x_1+h, u_1, \beta_2+\varepsilon_2))h \\ &\qquad + \alpha_2(x, y(x, x_1, u_1, \beta_2+\bar{\varepsilon}_2))\varepsilon_2\big\} \\ &= \beta(x, y(x, x_1+\bar{h}, u_1, \beta_2+\varepsilon_2)) \\ &\quad + \frac{\alpha_2}{h}(x, y(x, x_1, u_1, \beta_2+\bar{\varepsilon}_2)), \end{aligned}$$

where $\beta(x, y(\cdot))$ is the solution of (9.3) along $y(\cdot)$ and satisfies,

$$\begin{aligned} &\beta(x_1, y(\cdot)) = -y'(x_1) = -u'(x_1), \text{ and} \\ &\beta'(x_1, y(\cdot)) = -y''(x_1) = -u''(x_1), \end{aligned}$$

and $\alpha_2(x, y(\cdot))$ is the solution of (9.3) along $y(\cdot)$ and satisfies,

$$\alpha_2(x_1, y(\cdot)) = 0, \text{ and } \alpha_2'(x_1, y(\cdot)) = 1,$$

and moreover, $\beta_2 + \overline{\varepsilon_2}$ is between β_2 and $\beta_2 + \varepsilon_2$, and $x_1 + \overline{h}$ is between x_1 and $x_1 + h$. To show $\lim_{h\to 0} z_{1h}(x)$ exists, it suffices to show $\lim_{h\to 0} \frac{\varepsilon_2}{h}$ exists.

As argued before, it follows from (v) that

$$\int_{x_1}^{x_2} \alpha_2(x, y(\cdot))\, dx \neq 0.$$

So

$$\frac{\varepsilon_2}{h} = \frac{\int_{x_1}^{x_2} z_{1h}\, dx - \int_{x_1}^{x_2} \beta(x, y(x, x_1 + \overline{h}, u_1, \beta_2 + \varepsilon_2))\, dx}{\int_{x_1}^{x_2} \alpha_2(x, y(x, x_1, u_1, \beta_2 + \overline{\varepsilon}_2))\, dx}.$$

Passing to the limit and using an above boundary condition,

$$\lim_{h\to 0} \frac{\varepsilon_2}{h} = \frac{u(x_1) - \int_{x_1}^{x_2} \beta(x, u(\cdot))\, dx}{\int_{x_1}^{x_2} \alpha_2(x, u(\cdot))\, dx} := H.$$

From above,

$$z_{1h}(x) = \beta(x, y(x, x_1 + \overline{h}, u_1, \beta_2 + \varepsilon_2)) + \frac{\varepsilon_2}{h}\alpha_2(x, y(x, x_1, u_1, \beta_2 + \overline{\varepsilon}_2))$$

and so we can let $h \to 0$. If we let $z_1(x) := \lim_{h\to 0} z_{1h}(x)$, we have $z_1(x) = \frac{\partial u(x)}{\partial x_1}$, and

$$z_1(x) = \frac{\partial u(x)}{\partial x_1} = \lim_{h\to 0} z_{1h}(x) = \beta(x, u(x)) + H\alpha_2(x, u(x)),$$

which is a solution of (9.3) along $u(x)$. From above computations,

$$z_1(x_1) = \lim_{h\to 0} z_{1h}(x_1) = -u'(x_1),$$

$$\int_{x_1}^{x_2} z_1(x)\, dx = \lim_{h\to 0} \int_{x_1}^{x_2} z_{1h}(x)\, dx = u(x_1).$$

This completes the proof of part (b).

Part (c) is immediate by verifying that both sides of the respective equations satisfy the same boundary conditions, and then assumption (v) establishes the equalities. □

Bibliography

Benchohra M., Hamani S., Henderson J., Ntouyas S. K. and Ouahab A. (2007). Differentiation and differences for solutions of nonlocal boundary value problems for second order difference equations, *Internat. J. Difference Eq.* **2**, No. 1, pp. 37–47.

Datta A. and Henderson J. (1992). Differentiation of solutions of difference equations with respect to right focal boundary values, *Panamer. Math. J.* **2**, pp. 1–16.

Davis J. M. (2000). Differentiation of solutions of Lidstone boundary value problems with respect to the boundary data, *Math. Comput. Modelling* **32**, No. 5–6, pp. 675–685.

Ehme J. (1993). Differentiation of solutions of boundary value problems with respect to nonlinear boundary conditions, *J. Differential Equations* **101**, pp. 139–147.

Ehme J. and Henderson J. (1996). Functional boundary value problems and smoothness of solutions, *Nonlinear Anal.* **26**, pp. 139–148.

Ehrke J. E., Henderson J., Kunkel C. J. and Sheng Q. (2007). Boundary data smoothness for solutions of nonlocal boundary value problems for second order differential equations, *J. Math. Anal. Appl.* **333**, pp. 191–203.

Hartman P. (1964). *Ordinary Differential Equations* (Wiley, New York).

Henderson J. (1984). Right focal point boundary value problems for ordinary differential equations and variational equations, *J. Math. Anal. Appl.* **48**, pp. 363–377.

Henderson J. (1987). Disconjugacy, disfocalty and differentiation with respect to boundary conditions, *J. Math. Anal. Appl.* **121**, pp. 1–9.

Henderson J. and Lyons J. W. (2009). Characterization of partial derivatives with respect to boundary conditions for nonlocal boundary value problems for N-th order differential equations, *Int. J. Pure Appl. Math.* **56**, No. 2, pp. 235–257.

Henderson J. and Tisdell C. C. (2004). Boundary data smoothness for solutions of three point boundary value problems for second order ordinary differential equations, *Z. Anal. Anwendungen* **23**, No. 3, 631-640.

Lyons J. W. (2011). Differentiation of solutions of nonlocal boundary value problems with respect to boundary data, *Electron. J. Qual. Theory Differ. Equ.* **2011**, No. 51, pp. 1–11.

Lyons J. W. (2014). Disconjugacy, differences and differentiation for solutions of non-local boundary value problems for nth order difference equations, *J. Difference Equ. Appl.* **20**, No. 2, pp. 296–311.

Peterson A. C. (1976). An expression for the first conjugate point for an nth order nonlinear differential equation, *Proc. Amer. Math. Soc.* **61**, No. 2, pp. 300–304.

Peterson A. C. (1978). Existence-uniqueness for ordinary differential equations, *J. Math. Anal. Appl.* **64**, No. 1, pp. 166–172.

Peterson A. C. (1981). Existence-uniqueness for focal-point boundary value problems, *SIAM J. Math. Anal.* **12**, No. 2, pp. 173–185.

Peterson A. C. (1982). A disfocality function for a nonlinear ordinary differential equation, *Rocky Mountain J. Math.* **12**, No. 4, pp. 741–752.

Spencer J. (1975). Relations between boundary value functions for a nonlinear differential equation and its variational equation, *Canad. Math. Bull.* **18**, pp. 269–276.

Chapter 10

Nodal Solutions of BVP's for ODE's

In this chapter, we first study nodal solutions for BVP's with separated two-point BC's and then study nodal solutions for BVP's with nonlocal BC's. Our proofs in this chapter are mainly based on a bifurcation theorem by Rabinowitz, the shooting method, and some suitable energy functions.

10.1 Nodal Solutions of Nonlinear BVP's with Separated BC's

In this section, we are concerned with the existence and nonexistence of nodal solutions of the BVP consisting of the equation

$$-(p(t)y')' + q(t)y = w(t)f(y),\ t \in (a, b), \tag{10.1}$$

and the separated BC

$$\begin{aligned} &\cos\alpha\ y(a) - \sin\alpha\ (py')(a) = 0, \quad \alpha \in [0, \pi), \\ &\cos\beta\ y(b) - \sin\beta\ (py')(b) = 0, \quad \beta \in (0, \pi], \end{aligned} \tag{10.2}$$

where $a, b \in \mathbb{R}$ with $a < b$. We assume throughout, and without further mention, that the following conditions hold:

(J1) $p, q, w \in C[a, b]$, $p > 0$ on $[a, b]$ and $w > 0$ a.e. on $[a, b]$;
(J2) f is continuous on $\mathbb{R}$ and $f(y)y > 0$ for $y \neq 0$;
(J3) there exist extended real numbers f_0, f_∞ such that $0 \le f_0, f_\infty \le \infty$ and

$$\lim_{y\to 0} \frac{f(y)}{y} = f_0 \quad \text{and} \quad \lim_{|y|\to\infty} \frac{f(y)}{y} = f_\infty. \tag{10.3}$$

By a solution of BVP (10.1), (10.2), we mean a function $y \in C[a, b]$ such that $py' \in C^{(1)}[a, b]$, and y satisfies Eq. (10.1) and BC (10.2). Moreover, if $y(t)$ changes sign in (a, b), then y is said to be a *nodal solution*. A simple zero of a nodal solution $y(t)$ is a point $t_0 \in (a,b)$ such that $y(t_0) = 0$ and $(py')(t_0) \neq 0$.

We utilize the Rabinowitz bifurcation method and the Sturm comparison theorem to study the existence as well as the nonexistence of nodal solutions of BVP (10.1), (10.2) under the general conditions (J1)–(J3). More specifically, let $\lambda_n, n = 0,1,2,\dots$, be the nth eigenvalues of the corresponding linear Sturm-Liouville problem. We show that BVP (10.1), (10.2) has a pair of solutions with exactly n zeros in (a,b) if λ_n is in the interior of the range of $f(y)/y$; and does not have any solution with exactly n zeros in (a,b) if λ_n is outside the range of $f(y)/y$. These conditions become necessary and sufficient when $f(y)/y$ is strictly between f_0 and f_∞. The existence of multiple and even an infinite number of nodal solutions are derived as consequences. In light of the work in [Kong and Zettl (1996); Kong *et al.* (2008, 1999)] on the dependence of eigenvalues of SLPs on the problem, we also discuss the changes of the number of different types of nodal solutions as the equation or the BC changes.

10.1.1 *Main results*

It is well known (see, for example, [Weidmann (1987)]) that the linear SLP consisting of the equation

$$-(p(t)y')' + q(t)y = \lambda w(t)y \tag{10.4}$$

and BC (10.2) has a countable number of eigenvalues λ_i, $i = 0,1,2,\dots$, which are bounded below and unbounded above, and can be ordered to satisfy

$$-\infty < \lambda_0 < \lambda_1 < \cdots < \lambda_i < \cdots, \quad \text{and } \lambda_i \to \infty \text{ as } i \to \infty. \tag{10.5}$$

Furthermore, any eigenfunction associated with λ_i has i simple zeros in (a,b).

In the sequel, we let $\mathbb{N}_k = \{k, k+1, k+2, \dots\}$ for $k = 0,1,2,\dots$, and let λ_m be the first positive eigenvalue of SLP (10.4), (10.2). The first theorem is about the existence of nodal solutions of BVP (10.1), (10.2).

Theorem 10.1. *Assume for some $k \in \mathbb{N}_m$ we have*

$$\textit{either} \quad 0 \le f_0 < \lambda_k < f_\infty \le \infty \quad \textit{or} \quad 0 \le f_\infty < \lambda_k < f_0 \le \infty.$$

Then BVP (10.1), (10.2) has two solutions $u_{k,+}$ and $u_{k,-}$ such that they both have exactly k simple zeros in (a,b), $u_{k,+}$ is positive near a, and $u_{k,-}$ is negative near a.

The following is an immediate consequence of Theorem 10.1.

Corollary 10.1. *Assume either* (i) $f_0 = 0$ *and* $f_\infty = \infty$, *or* (ii) $f_0 = \infty$ *and* $f_\infty = 0$. *Then, for each* $k \in \mathbb{N}_m$, *BVP* (10.1), (10.2) *has two nodal solutions* $u_{k,+}$ *and* $u_{k,-}$ *such that they both have exactly* k *simple zeros in* (a, b), $u_{k,+}$ *is positive near* a, *and* $u_{k,-}$ *is negative near* a.

Now we present a result on the nonexistence of nodal solutions of BVP (10.1), (10.2).

Theorem 10.2. (i) *Assume* $f(y)/y < \lambda_n$ *for some* $n \in \mathbb{N}_0$ *and all* $y \neq 0$. *Then BVP (10.1), (10.2) has no solution with exactly* i *zeros in* (a, b) *for any* $i \geq n$.

(ii) *Assume* $\lambda_n < f(y)/y$ *for some* $n \in \mathbb{N}_0$ *and all* $y \neq 0$. *Then BVP (10.1), (10.2) has no solution with exactly* i *zeros in* (a, b) *for any* $i \leq n$.

(iii) *Assume either* $f_0 < \infty$ *or* $f_\infty < \infty$ *and* $f(y)/y \neq \lambda_n$ *for any* $n \in \mathbb{N}_0$ *and all* $y \neq 0$. *Then BVP (10.1), (10.2) has no non-trivial solution.*

The combination of Theorems 10.1 and 10.2 leads to the following corollary.

Corollary 10.2. (i) *Assume* $f_0 < f(y)/y < f_\infty$ *for all* $y \neq 0$. *Then, for* $n \in \mathbb{N}_m$, *BVP (10.1), (10.2) has two solutions which have exactly* n *simple zeros in* (a, b) *and have opposite signs near* a *if and only if* $f_0 < \lambda_n < f_\infty$.

(ii) *Assume* $f_\infty < f(y)/y < f_0$ *for all* $y \neq 0$. *Then, for* $n \in \mathbb{N}_m$, *BVP (10.1), (10.2) has two solutions which have exactly* n *simple zeros in* (a, b) *and have opposite signs near* a *if and only if* $f_\infty < \lambda_n < f_0$.

In the following, we present results on the changes of the number of different types of nodal solutions to BVP (10.1), (10.2) as the problem changes. The first one is about the changes as the interval $[a, b]$ shrinks, more specifically, as $b \to a+$. We discuss both the cases when one of f_0 and f_∞ is infinite and both of them are finite.

Theorem 10.3. (i) *Assume either* $f_0 < \infty$ *and* $f_\infty = \infty$, *or* $f_\infty < \infty$ *and* $f_0 = \infty$.

(a) *For any* $n \in \mathbb{N}_1$, *there exists* $b_n > a$ *such that for any* $b \in (a, b_n)$ *and for any* $i \geq n$, *BVP (10.1), (10.2) has two solutions which have exactly* i *simple zeros in* (a, b) *and have opposite signs near* a.

(b) *Let* $\alpha \in [0, \pi)$ *and* $\beta \in (0, \pi]$. *Then, for* $\alpha < \beta$, *there exists* $b_0 > a$ *such that for any* $b \in (a, b_0)$, *BVP (10.1), (10.2) has a positive solution* *and a negative solution; and for $\beta < \alpha$, there exists $b_0 > a$ such that for any $b \in (a, b_0)$, BVP (10.1), (10.2) has no positive or negative solution.*

(ii) *Assume $f_0 < \infty$ and $f_\infty < \infty$. Let $\alpha \in [0, \pi)$ and $\beta \in (0, \pi]$ with $\alpha \neq \beta$. Then there exists $b_* > a$ such that for any $b \in (a, b_*)$, BVP (10.1), (10.2) has no nontrivial solution.*

Remark 10.1. (i) For the Dirichlet BC $y(a) = y(b) = 0$, we know that $\alpha = 0$ and $\beta = \pi$.

(a) If either $f_0 < \infty$ and $f_\infty = \infty$, or $f_\infty < \infty$ and $f_0 = \infty$, then there exists $b_0 > a$ such that for any $b \in (a, b_0)$ and for any $i \geq 0$, BVP (10.1), (10.2) has two solutions which have exactly i simple zeros in (a, b) and have opposite signs near a.

(b) If $f_0 < \infty$ and $f_\infty < \infty$, then there exists $b_* > a$ such that for any $b \in (a, b_*)$, BVP (10.1), (10.2) has no nontrivial solution.

(ii) The case when $\alpha = \beta \in (0, \pi)$ is much more complicated. Further results can be obtained for the existence of solutions with no zero in (a, b) based on the work in [Kong *et al.* (2008)]. We omit the details.

The theorem below is about the nonexistence of positive and negative solutions and solutions with one zero in (a, b) for some values of α and β in BC (10.2).

Theorem 10.4. *Let Eq. (10.1) be fixed.*

(i) *For each $\beta \in (0, \pi]$, there exists $\alpha_* \in [0, \pi)$ such that BVP (10.1), (10.2) has no positive or negative solution for any $\alpha \in (\alpha_*, \pi)$.*

(ii) *For each $\alpha \in [0, \pi)$, there exists $\beta_* \in (0, \pi]$ such that BVP (10.1), (10.2) has no positive or negative solution for any $\beta \in (0, \beta_*)$.*

(iii) *There exist $\alpha_* \in [0, \pi)$ and $\beta_* \in (0, \pi]$ such that BVP (10.1), (10.2) has no positive or negative solution nor solution with one zero in (a, b) for any $\alpha \in (\alpha_*, \pi)$ and $\beta \in (0, \beta_*)$.*

We then state results on the changes of the number of different types of nodal solutions to BVP (10.1), (10.2) as the functions $1/p$, q and w change in a given direction. In Theorems 10.5–10.7 we assume $w(t) > 0$ on $[a, b]$.

Let $s \in \mathbb{R}$ and $h \in C[a, b]$ such that $h > 0$ on $[a, b]$, and consider the equation

$$-(p(t)y')' + [q(t) + sh(t)]y = w(t)f(y) \quad \text{on } [a, b]. \tag{10.6}$$

Theorem 10.5. (i) *For any* $n \in \mathbb{N}_0$, *there exists* $s_n \leq 0$ *such that for any* $s < s_n$ *and for any* $i \leq n$, *BVP (10.6), (10.2) has no solution with exactly* i *zeros in* (a, b).

(ii) *Assume either* $f_0 < \infty$ *and* $f_\infty = \infty$, *or* $f_\infty < \infty$ *and* $f_0 = \infty$. *Then for any* $n \in \mathbb{N}_0$, *there exists* $s_n \geq 0$ *such that for any* $s > s_n$ *and for any* $i \geq n$, *BVP (10.6), (10.2) has two solutions which have exactly* i *simple zeros in* (a, b) *and have opposite signs near* a.

(iii) *Assume* $f_0 < \infty$ *and* $f_\infty < \infty$. *Then there exists* $s_* \geq 0$ *such that for any* $s > s_*$, *BVP (10.6), (10.2) has no nontrivial solution.*

Let $s \geq 0$ and $h \in C[a, b]$ such that $h > 0$ on $[a, b]$, and consider the equation

$$-(p(t)y')' + q(t)y = [w(t) + sh(t)]f(y) \quad \text{on } [a, b]. \tag{10.7}$$

Theorem 10.6. *Assume* $f(y)/y \geq f_* > 0$ *for all* $y \neq 0$. *Then for any* $n \in \mathbb{N}_0$, *there exists* $s_n \geq 0$ *such that for any* $s > s_n$ *and for any* $i \leq n$, *BVP (10.7), (10.2) has no solution with exactly* i *zeros in* (a, b).

Let $s \geq 0$ and $h \in C[a, b]$ such that $h > 0$ on $[a, b]$, and consider the equation

$$-\left[\frac{1}{1/p(t) + sh(t)}y'\right]' + q(t)y = w(t)f(y) \quad \text{on } [a, b]. \tag{10.8}$$

Theorem 10.7. *Let* $q^* = \max\{q(t)/w(t) : t \in [a, b]\}$ *and assume* $f(y)/y \geq f_* > q^*$ *for all* $y \neq 0$. *Then for any* $n \in \mathbb{N}_0$, *there exists* $s_n \geq 0$ *such that for any* $s > s_n$ *and for any* $i \leq n$, *BVP (10.8), (10.2) has no solution with exactly* i *zeros in* (a, b).

Remark 10.2. Theorems 10.3–10.7 show that we can "create" or "eliminate" certain types of nodal solutions by changing the interval $[a, b]$, the BC (10.2), or the coefficient functions $1/p$, q, w. Since the eigenvalues of SLP (10.4), (10.2) can be easily computed using computer software, we are able to construct specific BVPs (10.1), (10.2) which have or do not have nodal solutions with prescribed zeros in $[a, b]$.

10.1.2 *Proofs of the main results*

The proof of Theorem 10.1, is based on a well-known bifurcation theorem by Rabinowitz [Rabinowitz (1971)] on nonlinear BVPs. To present Rabinowitz's result, we first introduce some notations below.

Let

$$E = \{y \in C[a,b] : \ py' \in C[a,b] \text{ and } y \text{ satisfies BC (10.2)}\}.$$

Then E with the norm

$$||y||_1 = \max_{a \le t \le b} (\|y\| + \|py'\|),$$

where $\|y\| = \max_{a \le t \le b} |y(t)|$, is a Banach space. For $k \in \mathbb{N}_0$, let S_k^+ be the set of functions in E which have exactly k simple zeros on (a,b) and are positive near a, and let $S_k^- = -S_k^+$ and $S_k = S_k^+ \cup S_k^-$. We also define $\Phi_k^\pm = \mathbb{R} \times S_k^\pm$ and $\Phi_k = \mathbb{R} \times S_k$. Let $\mathcal{P} = \mathbb{R} \times E$ under the product topology. As in [Rabinowitz (1971)], we say that a continuum C of $\mathcal{P}$ meets ∞ if C is unbounded.

Consider the equation

$$-(p(t)y')' + q(t)y = \lambda w(t)y + w(t)g(y), \tag{10.9}$$

where $g \in C(\mathbb{R}, \mathbb{R})$ such that $g(y)y > 0$ for $y \neq 0$, and $g(y) = o(y)$ as $y \to 0$. The following lemma, known as the Rabinowitz Bifurcation Theorem appears as Theorem 2.3 in [Rabinowitz (1971)].

Lemma 10.1. *For $i \in \mathbb{N}_0$ and $\nu \in \{+,-\}$, there exists a continuum C_i^ν of solutions $(\lambda, y(t))$ of Eq. (10.9) in $\Phi_i^\nu \cup \{(\lambda_i, 0)\}$ that meets $(\lambda_i, 0)$ and ∞ in $\mathcal{P}$, where λ_i is the ith eigenvalue of SLP (10.4), (10.2).*

From the remark after Theorem 2.3 in [Rabinowitz (1971)], we have the following remark.

Remark 10.3. Lemma 10.1 can be extended to the case where the function g in Eq. (10.9) is replaced by a continuous operator $G : C[a,b] \to C[a,b]$ satisfying the following condition:

(i) $G(y(t)) = o(\|y\|)$ as $\|y\| \to 0$ uniformly for $t \in [a,b]$.

(ii) If (λ, y) is a solution of the modified equation (10.9) with y having a double zero in $[a,b]$, then $y \equiv 0$.

In addition to Lemma 10.1 and Remark 10.3, we need the following lemma in the proof of Theorem 10.1.

Lemma 10.2. *Let $h \in C(\mathbb{R}, \mathbb{R})$ such that $h(y)y > 0$ for $y \neq 0$, and $q_n \in C[a,b]$ such that $q_n(t) \to q_\infty(t) \in C([a,b], \mathbb{R})$ as $n \to \infty$ uniformly on $[a,b]$. Let $\{(c_n, d_n)\}_{n=1}^\infty$ be a sequence of subintervals of (a,b) such that $(c_n, d_n) \to (c,d)$ as $n \to \infty$. For $n \in \mathbb{N}$, assume $y_n(t)$ is a solution of the BVP consisting of the equation*

$$-(p(t)y')' + q_n(t)y = w(t)h(y) \tag{10.10}$$

and the BC

$$\begin{aligned} \cos\bar{\alpha}\ y(c_n) - \sin\bar{\alpha}\ (py')(c_n) = 0, \quad \bar{\alpha} \in [0,\pi), \\ \cos\bar{\beta}\ y(d_n) - \sin\bar{\beta}\ (py')(d_n) = 0, \quad \bar{\beta} \in (0,\pi], \end{aligned} \tag{10.11}$$

such that on (c_n, d_n)

$$y_n(t) > 0 \text{ for all } n \in \mathbb{N} \quad \text{or} \quad y_n(t) < 0 \text{ for all } n \in \mathbb{N}.$$

Then, for any closed interval $I \subset (c,d)$*, there exists* $r > 0$ *such that for sufficiently large* $n \in \mathbb{N}$

$$|y_n(t)| \geq r \max_{s\in I} |y_n(s)| \quad \text{for } t \in I. \tag{10.12}$$

Proof. Without loss of generality we only prove the case when $y_n(t) > 0$ on (c_n, d_n), $n \in \mathbb{N}$. For $n \in \mathbb{N}$ let $u_n(t)$ and $v_n(t)$ be the solutions of the equation

$$-(p(t)y')' + q_n(t)y = 0 \tag{10.13}$$

satisfying the initial conditions

$$u_n(c_n) = \sin\bar{\alpha}, \ (pu_n')(c_n) = \cos\bar{\alpha}, \tag{10.14}$$

and

$$v_n(d_n) = \sin\bar{\beta}, \ (pv_n')(d_n) = \cos\bar{\beta}, \tag{10.15}$$

and let θ_n, ϕ_n, ψ_n be the Prüfer angles of y_n, u_n, v_n, respectively, satisfying

$$\theta_n(c_n) = \phi_n(c_n) = \bar{\alpha} \quad \text{and} \quad \theta_n(d_n) = \psi_n(d_n) = \bar{\beta}.$$

Noting that $w(t)h(y_n(t))/y_n(t) > 0$ for $t \in (c_n, d_n)$, by the Sturm comparison theorem for Prüfer angles (see, for example, [Weidmann (1987)]), we have

$$0 < \phi_n(t) < \theta_n(t) \leq \pi \quad \text{for } t \in (c_n, d_n]. \tag{10.16}$$

In particular,

$$0 < \phi_n(d_n) < \theta_n(d_n) = \psi_n(d_n) \leq \pi.$$

This shows that u_n and v_n are linearly independent on $[c_n, d_n]$. Moreover, from (10.16) we see that $0 < \phi_n(t) < \pi$ for $t \in (c_n, d_n]$ and hence $u_n(t) > 0$ on $(c_n, d_n]$. With a similar argument we find that $v_n(t) > 0$ on $[c_n, d_n)$.

Let $\rho_n := p(t)[u_n'v_n - u_nv_n'](t)$. Then ρ_n is a nonzero constant on $[c_n, d_n]$. Define

$$H_n(t,s) = \frac{1}{\rho_n}\begin{cases} u_n(t)v_n(s), & c_n \leq t \leq s \leq d_n, \\ u_n(s)v_n(t), & c_n \leq s \leq t \leq d_n. \end{cases}$$

Then $H_n(t,s)$ is the Green's function for BVP (10.13), (10.11), $H_n(t,s) > 0$ for $t, s \in (c_n, d_n)$, and

$$\frac{H_n(t,s)}{H_n(s,s)} = \begin{cases} \dfrac{u_n(t)}{u_n(s)}, & c_n \le t \le s \le d_n, \\ \dfrac{v_n(t)}{v_n(s)}, & c_n \le s \le t \le d_n. \end{cases}$$

For $t \in [c_n, d_n]$ denote

$$\bar{l}_n(t) = \max\left\{ \sup_{t<s\le d_n} \frac{u_n(t)}{u_n(s)}, \sup_{c_n \le s<t} \frac{v_n(t)}{v_n(s)} \right\} \tag{10.17}$$

and

$$\underline{l}_n(t) = \min\left\{ \inf_{t<s\le d_n} \frac{u_n(t)}{u_n(s)}, \inf_{c_n \le s<t} \frac{v_n(t)}{v_n(s)} \right\}. \tag{10.18}$$

It is easy to see that $\bar{l}_n$ and $\underline{l}_n$ are well defined and for $t \in I$

$$0 < \min_{s\in I} \underline{l}_n(s) \le \underline{l}_n(t) \le \bar{l}_n(t) \le \max_{s\in I} \bar{l}_n(s) < \infty. \tag{10.19}$$

Moreover,

$$\underline{l}_n(t) H_n(s,s) \le H_n(t,s) \le \bar{l}_n(t) H_n(s,s) \quad \text{for } t, s \in [c_n, d_n]. \tag{10.20}$$

Since $y_n(t)$ is a solution of BVP (10.10), (10.11),

$$y_n(t) = \int_{c_n}^{d_n} H_n(t,s) w(s) h(y_n(s)) ds.$$

From (10.20) we have that for $t \in I$

$$\max_{t\in I} y_n(t) \le \max_{t\in I} \bar{l}_n(t) \int_{c_n}^{d_n} H_n(s,s) w(s) h(y_n(s)) ds$$

and

$$y_n(t) \ge \min_{t\in I} \underline{l}_n(t) \int_{c_n}^{d_n} H_n(s,s) w(s) h(y_n(s)) ds.$$

Therefore, for $t \in I$

$$y_n(t) \ge \left(\min_{t\in I} \underline{l}_n(t) / \max_{t\in I} \bar{l}_n(t) \right) \max_{t\in I} y_n(t). \tag{10.21}$$

Now we show that there exists $r > 0$ such that

$$\min_{t\in I} \underline{l}_n(t) / \max_{t\in I} \bar{l}_n(t) \ge r$$

for all sufficiently large $n \in \mathbb{N}$. Without loss of generality we may assume that $\lambda = 0$ is not an eigenvalue of the SLPs consisting of the equation

$$-(p(t)y')' + q_\infty(t) y = \lambda w(t) y \tag{10.22}$$

and one of the BCs

$$\cos\bar{\alpha}\ y(c) - \sin\bar{\alpha}\ (py')(c) = 0, \quad y(d) = 0 \tag{10.23}$$

and

$$y(c) = 0, \quad \cos\bar{\beta}\ y(d) - \sin\bar{\beta}\ (py')(d) = 0.$$

For otherwise, we can replace $q_n(t)$ and $h(y)$ by $q_n(t)+\epsilon w(t)$ and $h(y)+\epsilon y$, respectively in (10.10), for small $\epsilon > 0$. Then for the modified problem, $\lambda = 0$ is no longer an eigenvalue, and the above argument applies exactly the same way.

Let $u(t)$ and $v(t)$ be the solutions of Eq. (10.13) satisfying (10.14) and (10.15) with q_n, c_n, d_n replaced by q_∞, c, d, respectively. Then for $t \in (c, d)$

$$\lim_{n\to\infty} u_n(t) = u(t) \ \text{ and } \ \lim_{n\to\infty} v_n(t) = v(t).$$

We claim that $u(t) > 0$ on $(c, d]$ and $v(t) > 0$ on $[c, d)$. In fact, $u_n(t) > 0$ on $(c_n, d_n]$ for $n \in \mathbb{N}$ implies that $u(t) \geq 0$ on $(c, d]$. If $u(t^*) = 0$ for some $t^* \in (c, d)$, Then t^* is a double zero and hence $u(t) \equiv 0$ on $[c, d]$ by the uniqueness of solutions of the linear initial value problems; if $u(d) = 0$, then $\lambda = 0$ is an eigenvalue of SLP (10.22), (10.23) with associated eigenfunction $u(t)$, contradicting the assumption. Similarly for $v(t)$.

We define $\bar{l}(t)$ and $\underline{l}(t)$ in the same way as $\bar{l}_n(t)$ and $\underline{l}_n(t)$ in (10.17) and (10.18), where u_n, v_n, c_n, d_n are replaced by u, v, c, d, respectively. Then, with the same argument as for $\bar{l}_n(t)$ and $\underline{l}_n(t)$, we have

$$0 < \min_{s\in I} \underline{l}(s) \leq \underline{l}(t) \leq \bar{l}(t) \leq \max_{s\in I} \bar{l}(s) < \infty, \ t \in I.$$

Since

$$\lim_{n\to\infty} \bar{l}_n(t) = \bar{l}(t) \ \text{ and } \ \lim_{n\to\infty} \underline{l}_n(t) = \underline{l}(t)$$

uniformly on I, we see that

$$\lim_{n\to\infty} \frac{\min_{t\in I} \underline{l}_n(t)}{\max_{t\in I} \bar{l}_n(t)} = \frac{\min_{t\in I} \underline{l}(t)}{\max_{t\in I} \bar{l}(t)}.$$

Let $r = \frac{1}{2}(\min_{t\in I} \underline{l}(t)/\max_{t\in I} \bar{l}(t))$. The for sufficiently large $n \in \mathbb{N}$, (10.12) follows from (10.21). $\square$

Proof of Theorem 10.1. We use the notation $Ly := -(p(t)y')' + q(t)y$ throughout the proof.

(I) We first prove the case where $0 \leq f_0 < \lambda_k < f_\infty \leq \infty$. Let $g(y) := f(y) - f_0 y$. Then Eq. (10.1) is equivalent to

$$Ly = f_0 w(t)y + w(t)g(y), \tag{10.24}$$

and $g(y) = o(y)$ as $y \to 0$. We consider the equation

$$Ly = \lambda w(t)y + w(t)g(y). \tag{10.25}$$

Then, by Lemma 10.1, for $\nu \in \{+, -\}$, there exists a continuum C_k^ν of solutions of (10.25) in $\Phi_k^\nu \cup \{(\lambda_k, 0)\}$ that meets $(\lambda_k, 0)$ and ∞ in $\mathcal{P}$. It is easily seen that any solution of (10.25) in Φ_k^ν of the form (f_0, y) yields a solution y of BVP (10.1), (10.2) with k simple zeros. In the following we show that C_k^ν crosses the hyperplane $\{f_0\} \times E$ in $\mathcal{P}$. Let $\{\mu_n, u_n\}_{n=1}^\infty \subset C_k^\nu$ satisfy

$$\mu_n + ||y_n||_1 \to \infty \quad \text{as } n \to \infty. \tag{10.26}$$

Our purpose is to show that $\mu_n \le f_0$ for some $n \in \mathbb{N}$, and hence we reach the conclusion by the connectedness of C_k^ν and the fact that $\lambda_k > f_0$.

Assume the contrary, i.e., $\mu_n > f_0$ for all $n \in \mathbb{N}$. For each $n \in \mathbb{N}$, let $\tau(0, n) = a$ and $\tau(k+1, n) = b$, and let

$$\tau(1, n) < \tau(2, n) < \cdots < \tau(k, n)$$

be the zeros of y_n. Then, after taking a subsequence if necessary,

$$\lim_{n\to\infty} \tau(i, n) := \tau(i), \quad i = 0, \ldots, k+1.$$

Since

$$b - a = \tau(k+1, n) - \tau(0, n) = \sum_{i=0}^{k} (\tau(i+1, n) - \tau(i, n)),$$

we have

$$b - a = \sum_{i=0}^{k} (\tau(i+1) - \tau(i)).$$

Then, there exists $i_0 \in \{0, 1, \ldots, k\}$ such that $\tau(i_0) < \tau(i_0 + 1)$. Let I be a closed subinterval of $(\tau(i_0), \tau(i_0+1))$ with positive length. Since

$$(-1)^{i_0} \nu y_n(t) > 0 \quad \text{for all } n \in \mathbb{N} \text{ and } t \in (\tau(i_0, n), \tau(i_0+1, n)),$$

there exists a positive integer N_0 such that

$$(-1)^{i_0} \nu y_n(t) > 0 \quad \text{for all } n \ge N_0 \text{ and } t \in I. \tag{10.27}$$

Now we show that $\{\mu_n\}_{n=1}^\infty$ is bounded above, i.e., there exists $K > 0$ such that

$$f_0 < \mu_n \le K \quad \text{for all } n \in \mathbb{N}. \tag{10.28}$$

For otherwise, by taking a subsequence if necessary, we have $\mu_n \to \infty$. Note that, for $t \in I$, Eq. (10.25) can be written as

$$Ly = (\lambda + g(y)/y)w(t)y \tag{10.29}$$

with $g(y)/y = f(y)/y - f_0$ bounded below, and every solution of Eq. (10.4) with sufficiently large λ has at least two zeros in I. By the Sturm Comparison theorem, every solution of Eq. (10.29), and hence of Eq. (10.25), changes sign in I for $\lambda = \mu_n$ with sufficiently large $n \in \mathbb{N}$. This contradicts (10.27) and hence proves (10.28). Combining (10.26) and (10.28) we see that

$$\|y_n\|_1 \to \infty \quad \text{as } n \to \infty. \tag{10.30}$$

(a) We assume $f_\infty < \infty$. We observe that the function g in (10.25) can be written as

$$g(y) = f(y) - f_0 y = (f_\infty - f_0)y + h(y),$$

where

$$h(y) = f(y) - f_\infty y = o(|y|) \ \text{ as } \ |y| \to \infty.$$

Thus, for $n \in \mathbb{N}$, $y_n(t)$ is a solution of the BVP consisting of the equation

$$Ly = (\mu_n + f_\infty - f_0)w(t)y + w(t)h(y)$$

and BC (10.2). With the same reasoning as in the proof of Lemma 10.2 we may assume that $\lambda = 0$ is not an eigenvalue of SLP (10.4), (10.2). Therefore, there is a Green's function $H(t, s)$ for the BVP consisting of the equation $Ly = 0$ and BC (10.2). Then,

$$y_n(t) = \int_a^b H(t,s)\left[(\mu_n + f_\infty - f_0)w(s)y_n(s) + w(s)h(y_n(s))\right] ds,$$

and hence

$$\begin{aligned} \frac{y_n(t)}{\|y_n\|_1} &= \int_a^b H(t,s) \\ &\quad \times \left[(\mu_n + f_\infty - f_0)w(s)\frac{y_n(s)}{\|y_n\|_1} + w(s)\frac{h(y_n(s))}{\|y_n\|_1}\right] ds. \end{aligned} \tag{10.31}$$

It is easy to see that the sequence $\{y_n(t)/\|y_n\|_1\}_{n=1}^{\infty}$ is uniformly bounded and equicontinuous on $[a, b]$. By the Arzelà-Ascoli Theorem, we may assume, after taking a subsequence if necessary, that

$$\lim_{n\to\infty} \frac{y_n(t)}{\|y_n\|_1} = y_*(t) \quad \text{for some } y_* \in C[a,b] \text{ with } \|y_*\|_1 = 1.$$

Note that $\{\mu_n\}_{n=1}^{\infty}$ is bounded, we may also assume that $\lim_{n\to\infty}\mu_n = \mu_* \in \mathbb{R}$. Denote $h^*(y) = \max_{-y\le z\le y}|h(z)|$. Then from $h(y) = o(|y|)$ as $|y| \to \infty$, we have $h^*(y) = o(|y|)$ as $|y| \to \infty$. Hence for $t \in (a,b)$

$$\frac{|h(y_n(t))|}{||y_n||_1} \le \frac{h^*(||y_n||)}{||y_n||_1} \le \frac{h^*(||y_n||_1)}{||y_n||_1}.$$

This, together with (10.30), shows that

$$\lim_{n\to\infty}|h(y_n(t))|/||y_n||_1 = 0 \quad \text{for all } t \in (a,b).$$

Taking limits as $n \to \infty$ in (10.31) and using the Lebesgue dominated convergence theorem, we obtain

$$y_*(t) = (\mu_* + f_\infty - f_0)\int_a^b H(t,s)w(s)y_*(s)ds,$$

i.e., y_* is a nontrivial solution of the equation

$$Ly = (\mu_* + f_\infty - f_0)w(t)y$$

satisfying BC (10.2). In other words, y_* is an eigenfunction of SLP (10.4), (10.2) associated with the eigenvalue $\mu_* + f_\infty - f_0$. Since $y_n, n = 1,2,\ldots,$ has k simple zeros in (a,b) and y_* has only simple zeros in $[a,b]$, we conclude that y_* has exactly k simple zeros in (a,b). As a result, $\mu_* + f_\infty - f_0 = \lambda_k$, or $\mu_* = \lambda_k - f_\infty + f_0 < f_0$. This contradicts the assumption that $\mu_n > f_0$ for all $n \in \mathbb{N}$ and thus completes the proof.

(b) We assume $f_\infty = \infty$. We show that in this case, by taking a subsequence if necessary, for the closed interval I defined earlier, we have

$$\max_{t\in I}|y_n(t)| \to \infty \qquad \text{as } n \to \infty. \tag{10.32}$$

Assume for the contrary, there exists $M_1 > 0$ such that $|y_n(t)| \le M_1$ for $t \in I$ and $n \in \mathbb{N}$. Then we claim that for $n \in \mathbb{N}$, there exists $t_n \in I$ such that $|(py_n')(t_n)| \le M_2 := M_1(\int_I 1/p(t)dt)^{-1}$. For otherwise, $|(py')(t)| > M_2$ for all $t \in I$ and hence $y_n(t)$ and $(py_n')(t)$ do not change sign on I. Let $I = [c_0, d_0]$. Then

$$\max\{|y_n(c_0)|, |y_n(d_0)|\} > |y_n(d_0) - y_n(c_0)| = \int_I \frac{1}{p(t)}|(py')(t)|dt \ge M_1,$$

contradicting the boundedness assumption of y_n.

For $t \in [a,b]$, by integrating (10.25) with $\lambda = \mu_n$ and $y = y_n(t)$ from t_n to t we obtain

$$\begin{aligned}(py_n')(t) &= (py_n')(t_n)\\ &+ \int_{t_n}^t \{[q(\tau) - (\mu_n - f_0)w(\tau)]y_n(\tau) - w(\tau)f(y_n(\tau))\}\,d\tau\end{aligned} \tag{10.33}$$

and

$$y_n(t) = y_n(t_n) + (py_n')(t_n) \int_{t_n}^{t} \frac{1}{p(s)} ds + \int_{t_n}^{t} \left(\int_{\tau}^{t} \frac{1}{p(s)} ds \right) \times \{[q(\tau) - (\mu_n - f_0)w(\tau)]y_n(\tau) - w(\tau)f(y_n(\tau))\} \, d\tau. \quad (10.34)$$

Recall that for sufficiently large $n \in \mathbb{N}$, $I \subset (\tau(i_0, n), \tau(i_0 + 1, n))$ and $f(y_n(t))$ has the same sign as $y_n(t)$. Then, from (10.34), for $t \in [t_n, \tau(i_0 + 1, n)]$

$$\begin{aligned}|y_n(t)| &= |y_n(t_n)| + (py_n')(t_n)\mathrm{sgn}(y_n(t)) \int_{t_n}^{t} \frac{1}{p(s)} ds \\ &+ \int_{t_n}^{t} \left(\int_{\tau}^{t} \frac{1}{p(s)} ds \right) \{[q(\tau) - (\mu_n - f_0)w(\tau)]|y_n(\tau)| - w(\tau)|f(y_n(\tau))|\} \, d\tau \\ &\le |y_n(t_n)| + |(py_n')(t_n)| \int_{a}^{b} \frac{1}{p(s)} ds \\ &+ \int_{t_n}^{t} \left(\int_{a}^{b} \frac{1}{p(s)} ds \right) \{[|q(\tau)| + (\mu_n - f_0)w(\tau)]|y_n(\tau)|\} \, d\tau \\ &\le M_1 + M_2 \int_{a}^{b} \frac{1}{p(s)} ds \\ &+ \int_{a}^{b} \frac{1}{p(s)} ds \int_{t_n}^{t} \{[|q(\tau)| + (K - f_0)w(\tau)]|y_n(\tau)|\} \, d\tau,\end{aligned}$$

where K is defined by (10.28). By the Gronwall inequality, we have that for $t \in [t_n, \tau(i_0 + 1, n)]$,

$$|y_n(t)| \le \left(M_1 + M_2 \int_{a}^{b} \frac{1}{p(s)} ds \right) \times \exp\left\{ \int_{a}^{b} \frac{1}{p(s)} ds \int_{a}^{b} [|q(\tau)| + (K - f_0)w(\tau)] \, d\tau \right\}.$$

This means that $\{y_n(t)\}_{n=1}^{\infty}$ is uniformly founded on $[t_n, \tau(i_0+1, n)]$. Then by (10.33), $\{(py_n')(t)\}_{n=1}^{\infty}$ is uniformly bounded on $[t_n, \tau(i_0+1, n)]$. In particular, $\{y_n(\tau(i_0 + 1, n))\}_{n=1}^{\infty}$ and $\{(py_n')(\tau(i_0 + 1, n))\}_{n=1}^{\infty}$ are bounded. By the same process, we see that $\{y_n(t)\}_{n=1}^{\infty}$ and $\{(py_n')(t)\}_{n=1}^{\infty}$ are uniformly bounded on $[\tau(i_0 + 1, n), \tau(i_0 + 2, n)]$ and thereafter on intervals $[\tau(i, n), \tau(i + 1, n)]$ for all $i = i_0 + 1, \ldots, k$. With the backward Gronwall inequality we also see that $\{y_n(t)\}_{n=1}^{\infty}$ and $\{(py_n')(t)\}_{n=1}^{\infty}$ are uniformly bounded on $[\tau(i,n), \tau(i+1,n)]$ for all $i = 0, \ldots, i_0 - 1$. This contradicts (10.30) and thus proves (10.32).

We observe that for all $n \in \mathbb{N}$, $y_n(t)$ satisfy the same BC (10.11) with $c_n = \tau(i_0, n)$, $d_n = \tau(i_0 + 1, n)$,

$$\bar{\alpha} = \begin{cases} \alpha & \text{if } i_0 = 0, \\ 0 & \text{if } i_0 > 0, \end{cases} \quad \text{and} \quad \bar{\beta} = \begin{cases} \beta & \text{if } i_0 = k, \\ \pi & \text{if } i_0 < k. \end{cases}$$

By Lemma 10.2 with $q_n(t) = q(t) - \mu_n w(t)$, there exists $r > 0$ such that (10.12) holds, and hence $|y_n(t)| \to \infty$ uniformly on I as $n \to \infty$. Since

$$\frac{g(y)}{y} = \frac{f(y)}{y} - f_0 \to \infty \quad \text{as } |y| \to \infty,$$

it follows from (10.25) and (10.28) that for any $\lambda^* > 0$ there exists $N \in \mathbb{N}$ such that for $n \geq N$

$$Ly_n(t) = \left(\mu_n + \frac{g(y_n(t))}{y_n(t)}\right) w(t)y_n(t) \geq \lambda^* w(t)y_n(t).$$

Note that every solution of Eq. (10.4) with $\lambda = \lambda^*$ sufficiently large has at least two zeros in I. By the Sturm Comparison Theorem, every solution of Eq. (10.25) changes sign in I for sufficiently large $n \in \mathbb{N}$. This contradicts the definition of the interval I and thus completes the proof.

(II) Finally, we prove the case where $0 \leq f_\infty < \lambda_k < f_0 \leq \infty$. Let $g(y) := f(y) - f_\infty y$. Then Eq. (10.1) becomes

$$Ly = f_\infty w(t)y + w(t)g(y), \tag{10.35}$$

and $g(y) = o(y)$ as $|y| \to \infty$. We consider the equation

$$Ly = \lambda w(t)y + w(t)g(y). \tag{10.36}$$

By the standard transformation $z = y/\|y\|^2$ for $y \neq 0$ (see for example [Ambrosetti and Hess (1980); Ma and Thompson (2004)]), Eq. (10.36) becomes

$$Lz = \lambda w(t)z + w(t)G(z), \tag{10.37}$$

where $G : C[a,b] \to C[a,b]$ is defined by

$$G(z) = \begin{cases} ||z||^2 g\left(\frac{z}{||z||^2}\right), & z \neq 0, \\ 0, & z = 0. \end{cases}$$

It is easy to verify that G is continuous at $z = 0$, $z \neq 0$ is a solution of BVP (10.37), (10.2) if and only if $y = z/||z||^2$ is a solution of BVP (10.36), (10.2), and $\|z\| = 1/\|y\|$.

Now we show that

$$||G(z)|| = o(||z||) \quad \text{as } ||z|| \to 0. \tag{10.38}$$

To do this, let $g^* : [0,\infty) \to [0,\infty)$ be given by

$$g^*(y) = \max_{x\in[-y,y]} |g(x)|.$$

Then, g^* is nondecreasing and from the condition $g(y) = o(y)$ we see that

$$\lim_{y\to\infty} \frac{g^*(y)}{y} = 0. \tag{10.39}$$

Thus for $z \neq 0$ and $t \in [a,b]$

$$\begin{aligned} |G(z(t))| &= ||z||^2 g\left(\frac{z(t)}{||z||^2}\right) \le ||z||^2 g^*\left(\frac{|z(t)|}{||z||^2}\right) \\ &\le ||z||^2 g^*\left(\frac{||z||}{||z||^2}\right) \le ||z|| \frac{g^*(1/||z||)}{1/||z||}. \end{aligned}$$

This, together with (10.39), implies (10.38).

We then show that the only solution of Eq. (10.37) with a double zero in $[a,b]$ is the trivial solution. In fact, if $t_0 \in [a,b]$ is a double zero of a solution $z(t)$, then $z(t_0) = (pz')(t_0) = 0$. Without loss of generality we only show that $z(t) \equiv 0$ on $[t_0,b]$. For any $t \in [t_0,b]$ and $s \in [a,b]$ let

$$z(t;s) = \begin{cases} z(t), & t < s \le b, \\ z(s), & t_0 \le s \le t, \\ 0, & a \le s \le t_0. \end{cases}$$

Then $z(t;\cdot) \in C[a,b]$. Since $z(t;s) \to 0$ as $t \to t_0+$ uniformly for $s \in [a,b]$, by (10.38)

$$||G(z(t;\cdot)|| = o(||z(t;\cdot)||) \quad \text{as } t \to t_0.$$

This implies that there exists $M > 0$ such that for $t \ge t_0$ sufficiently close to t_0 and $0 \le s \le t$ we have

$$|G(z(t;s))| \le M||z(t;\cdot)||.$$

Let $z^*(t) = ||z(t;\cdot)|| = \max_{s\in[a,b]} |z(t;s)|$. Then there exists $t^* \in [0,t]$ such that $z^*(t) = |z(t^*)|$. For $t \ge t_0$ sufficiently close to t_0, by integrating

Eq. (10.37) twice from t_0 to t^* we obtain

$$\begin{aligned}
z^*(t) &= |z(t^*)| \\
&= \left| \int_{t_0}^{t^*} \left(\int_{\tau}^{t^*} 1/p(s)ds \right) [(q(\tau) - \lambda w(\tau))z(\tau) - w(\tau)G(z(\tau))]\, d\tau \right| \\
&\le \left(\int_a^b 1/p(s)ds \right) \int_{t_0}^{t} [|q(\tau) - \lambda w(\tau)||z(\tau;\tau)| + w(\tau)|G(z(\tau;\tau))|] d\tau \\
&\le \left(\int_a^b 1/p(s)ds \right) \int_{t_0}^{t} [|q(\tau)| + (\lambda + M)w(\tau)]\|z(\tau;\cdot)\| d\tau \\
&\le \left(\int_a^b 1/p(s)ds \right) \int_{t_0}^{t} [|q(\tau)| + (\lambda + M)w(\tau)]z^*(\tau)d\tau.
\end{aligned}$$

Hence the Gronwall inequality implies that $z^*(t) \equiv 0$ and so $z(t) \equiv 0$ for $t \ge t_0$ sufficiently close to t_0. The conclusion on the whole interval $[t_0, b]$ is then obtained by extension.

Now, all the conditions of Lemma 10.1 and Remark 10.3 are satisfied. Therefore, for $\nu \in \{+,-\}$, there exists a continuum C_k^ν of solutions of (10.37) in $\Phi_k^\nu \cup \{(\tilde{\lambda}_k, 0)\}$ that meets $(\lambda_k, 0)$ and ∞ in $\mathcal{P}$. It is clear that every solution of Eq. (10.37) of the form (f_∞, z) in C_k^ν with $z \ne 0$ yields a solution $y = z/\|z\|$ of BVP (10.35), (10.2) with k simple zeros.

The rest of the proof is essentially in the same way as Part (I) except that f_0 and f_∞ interchange their roles. We omit the details. □

Proof of Theorem 10.2. The proof for Parts (i) and (ii) is based on the comparison theorem for Prüfer angles. Although the proof of Theorem 10.1 (i) and (ii) in [Kong (2007)] is for the special case with $p \equiv 1$ and other stronger assumptions, it works here for the general problem. We omit the details.

To Prove part (iii), we only need to show that under our assumptions, no nontrivial solution $y(t)$ of BVP (10.1), (10.2) can have an infinite number of zeros, and hence the conclusion follows from Parts (i) and (ii). For otherwise, there exist a nontrivial solution $y(t)$ and $t_* \in [a, b]$ such that $y(t_*) = (py')(t_*) = 0$.

It is easy to see that $y(t) \equiv 0$ on $[a, b]$.

Assume $f_\infty < \infty$. If $y(t) \not\equiv 0$ on $[a, b]$, let $z = y/\|y\|^2$. Then $z(t)$ is a nontrivial solution of Eq. (10.37) satisfying $z(t_0) = (pz')(t_0) = 0$. Then by Part (II) of the proof of Theorem 10.1, we have $z(t) \equiv 0$ on $[a, b]$. We have reached a contradiction and this completes the proof. □

Proof of Theorem 10.3. This theorem is a generalization of [Kong (2007), Theorem 2.4] to our case with weaker assumptions. The proof is essentially the same and hence is omitted. □

Proof of Theorem 10.4. This theorem is a generalization of [Kong (2007), Theorem 2.6] to our case with weaker assumptions. The proof is essentially the same and hence is omitted. □

Proof of Theorem 10.5. For $s \in \mathbb{R}$ and $i \in \mathbb{N}_0$, we denote by $\lambda_i(s)$ the ith eigenvalue of the SLP consisting of the equation

$$-(p(t)y')' + [q(t) + sh(t)]y = \lambda w(t)y \quad \text{on } [a,b]$$

and BC (10.2). Let $h_* = \min\{h(t)/w(t) : t \in [a,b]\}$, and denote by $\mu_i(s)$ the ith eigenvalue of the SLP consisting of the equation

$$-(p(t)y')' + [q(t) + sh_*w(t)]y = \mu w(t)y \quad \text{on } [a,b] \tag{10.40}$$

and BC (10.2).

(i) Since for $s \leq 0$

$$q(t) + sh(t) \leq q(t) + sh_*w(t),$$

by Theorem 4.2, (6) in [Kong and Zettl (1996)], $\lambda_i(s) \leq \mu_i(s)$ for all $s \leq 0$ and $i \geq 0$. Note that Eq. (10.40) is the same as the equation

$$-(p(t)y')' + q(t)y = (\mu - sh_*)w(t)y \quad \text{on } [a,b].$$

Thus, for $s \leq 0$ and $i \geq 0$, $\mu_i(s) - sh_* = \mu_i(0)$ which implies that

$$\mu_i(s) = \mu_i(0) + sh_* \to -\infty \ \text{ as } s \to -\infty,$$

and hence

$$\lambda_i(s) \to -\infty \ \text{ as } s \to -\infty, \ i \geq 0.$$

Then, for any $n \in \mathbb{N}$ there exists $s_n \leq 0$ such that $\lambda_i < 0$ for all $i \leq n$ and $s < s_n$. Therefore, the conclusion follows from Theorem 10.2.

(ii) Without loss of generality assume $f_0 < \infty$ and $f_\infty = \infty$. Similar to the argument in (i) we have

$$\lambda_i(s) \to \infty \ \text{ as } s \to \infty, \ i \geq 0. \tag{10.41}$$

Then for any $n \in \mathbb{N}_0$ there exists $s_n \geq 0$ such that for any $s > s_n$ we have $f_0 < \lambda_n(s) < f_\infty$ and hence $f_0 < \lambda_i(s) < f_\infty$ for all $i \geq n$. Therefore, the conclusion follows from Theorem 10.1.

(iii) From (10.41), there exists $s_* \geq 0$ such that for any $s > s_*$ we have that $\lambda_n(s) > f^* := \sup\{f(y)/y : y \in (0,\infty)\}$. Therefore, the conclusion follows from Theorem 10.2. □

Proof of Theorem 10.6. For $s \geq 0$ and $i \in \mathbb{N}_0$, we denote by $\lambda_i(s)$ the ith eigenvalue of the SLP consisting of the equation

$$-(p(t)y')' + q(t)y = \lambda[w(t) + sh(t)]y \quad \text{on } [a, b]$$

and BC (10.2). Let $h_* = \min\{h(t)/w(t) : t \in [a, b]\}$, and denote by $\mu_i(s)$ the ith eigenvalue of the SLP consisting of the equation

$$-(p(t)y')' + q(t)y = \mu(1 + sh_*)w(t)y \quad \text{on } [a, b]$$

and BC (10.2). Since for $s \geq 0$

$$w(t) + sh(t) \geq (1 + sh_*)w(t),$$

by Theorem 4.2, (7) in [Kong and Zettl (1996)],

$$\lambda_i(s) \leq \mu_i(s) \ \text{ for all } s \geq 0 \text{ and } i \geq 0, \text{ whenever } \lambda_i(s) \geq 0. \tag{10.42}$$

Note that for $i \geq 0$, $\mu_i(s)(1 + sh_*) = \mu_i(0)$, we have

$$\mu_i(s) = \frac{\mu_i(0)}{1 + sh_*} \to 0 \ \text{ as } s \to \infty.$$

This together with (10.42) implies that $\lambda_i(s) < f_*$ as $s \to \infty$. Then, for any $n \in \mathbb{N}$ there exists $s_n \geq 0$ such that $\lambda_i < f_*$ for all $s > s_n$ and $i \leq n$. Therefore, the conclusion follows from Theorem 10.2. □

Proof of Theorem 10.7. For $s \geq 0$ and $i \in \mathbb{N}_0$, we denote by $\lambda_i(s)$ the ith eigenvalue of the SLP consisting of the equation

$$-\left[\frac{1}{1/p(t) + sh(t)} y'\right]' + q(t)y = \lambda w(t)y$$

and BC (10.2) with an eigenfunction $u_i(t, s)$. Let $\theta_i(t, s)$ be the Prüfer angle of $u_i(t, s)$ satisfying $\theta_i(a, s) = \alpha$. Then

$$\theta_i'(t, s) = \left[\frac{1}{p(t)} + sh(t)\right] \cos^2 \theta_i(t, s) + [\lambda_i w(t) - q(t)] \sin^2 \theta_i(t, s). \tag{10.43}$$

By Theorem 4.2, (5) in [Kong and Zettl (1996)], $\lambda_i(s)$ is decreasing and hence

$$\lim_{s \to \infty} \lambda_i(s) = \lambda_i^* \in [-\infty, \infty).$$

We show that $\lambda_i^* < f_*$ and then the conclusion follows from Theorem 10.2.

Assume the contrary, i.e., $\lambda_i^* \geq f_*$. Let $w_* = \min\{w(t) : t \in [a, b]\}$. By (10.43)

$$\begin{aligned}\theta_i'(t, s) &\geq \left[\frac{1}{p(t)} + sh(t)\right] \cos^2 \theta_i(t, s) + [\lambda_i^* w(t) - q(t)] \sin^2 \theta_i(t, s) \\ &\geq \left[\frac{1}{p(t)} + sh(t)\right] \cos^2 \theta_i(t, s) + (f_* - q^*)w_* \sin^2 \theta_i(t, s).\end{aligned}$$

Let $\phi(t,s)$ be the solution of the equation

$$\phi'(t,s) = \left[\frac{1}{p(t)} + sh(t)\right]\cos^2\phi(t,s) + (f_* - q^*)w_* \sin^2\phi(t,s) \tag{10.44}$$

satisfying $\phi(a,s) = \alpha$. By the theory of differential inequalities we have $\phi(t,s) \le \theta_i(t,s)$. In particular,

$$\phi(b,s) \le \theta_i(b,s) = i\pi + \beta. \tag{10.45}$$

We observe from (10.44) and the Sturm comparison theorem that $\phi(t,s)$ is strictly increasing in t and s, and $0 < \phi(t,s) < (i+1)\pi$ for $t \in [a,b]$ and $s \ge 0$. Let $\phi^*(t) = \lim_{s\to\infty}\phi(t,s)$. Then $0 < \phi^*(t) \le (i+1)\pi$ for $t \in [a,b]$. We claim that

$$\phi^*(t) \not\equiv k\pi + \frac{\pi}{2} \quad \text{on } (a,b], \tag{10.46}$$

for any $0 \le k \le i$. If not, for any $a_1 \in (a,b]$ and $\epsilon > 0$, there exists $s^* > 0$ such that for $s \ge s^*$

$$\phi(a_1,s) \in (k\pi + \pi/2 - \epsilon, k\pi + \pi/2),$$

which follows that

$$\phi(t,s) \in (k\pi + \pi/2 - \epsilon, k\pi + \pi/2) \quad \text{for } t \in [a_1,b].$$

This implies that

$$0 < \phi(b,s) - \phi(a_1,s) < \epsilon. \tag{10.47}$$

However, from (10.44) we see that for s^* sufficiently large,

$$\phi'(t,s) \ge \frac{1}{2}(f_* - q^*)w_* \quad \text{for } t \in [a_1,b].$$

This contradicts (10.47) and hence verifies (10.46).

It is easy to see that $\phi'(t,s) \to \phi^*(t)$ uniformly on $[a_1,b]$ as $s \to \infty$. Thus $\phi^*(t)$ is continuous on $[a_1,b]$. From (10.46), we can find nontrivial closed interval $[c,d] \subset [a_1,b]$ such that $\cos^2\phi^*(t) \ge k > 0$ for $t \in [c,d]$. Then from (10.44)

$$\phi'(t,s) \ge \left[\frac{1}{p(t)} + sh(t)\right]k \to \infty \quad \text{uniformly for } t \in [c,d] \text{ as } s \to \infty.$$

Therefore,

$$\begin{aligned}\phi(b,s) \ge \phi(d,s) \ge \phi(c,s) &+ \int_c^d \left[\frac{1}{p(t)} + sh(t)\right]k\,ds\\ &\ge \int_c^d \left[\frac{1}{p(t)} + sh(t)\right]k\,ds \to \infty \quad \text{as} \quad s \to \infty.\end{aligned}$$

This contradicts (10.45) and hence completes the proof. □

10.2 Nodal Solutions of Nonlocal BVP's — I

In this section, we study the BVP consisting of the equation

$$-y'' = \sum_{i=1}^{m} w_i(t) f_i(y), \quad t \in (a, b), \tag{10.48}$$

and the BC

$$\begin{aligned} &\cos\alpha \; y(a) - \sin\alpha \; y'(a) = 0, \quad \alpha \in [0, \pi), \\ &y(b) - \textstyle\int_a^b y(s) d\xi(s) = 0, \end{aligned} \tag{10.49}$$

where $m \geq 1$ is an integer, $a, b \in \mathbb{R}$ with $a < b$ and the integral in BC (10.49) is the Riemann-Stieltjes integral with respect to $\xi(s)$ with $\xi(s)$ a function of bounded variation. In the case where $\xi(s) = s$, the Riemann-Stieltjes integral in the second condition of (10.49) reduces to the Riemann integral. In the case that $\xi(s) = \sum_{j=1}^{d} k_j \chi(s - \eta_j)$, where $d \geq 1$, $k_j \in \mathbb{R}$, $j = 1, \ldots, m$, $\{\eta_j\}_{j=1}^{d}$ is a strictly increasing sequence of distinct points in (a, b), and $\chi(s)$ is the characteristic function on $[0, \infty)$, i.e.,

$$\chi(s) = \begin{cases} 1, & s \geq 0, \\ 0, & s < 0, \end{cases}$$

the second equation in (10.49) reduces to the multi-point BC

$$y(b) - \sum_{j=1}^{d} k_j y(\eta_j) = 0. \tag{10.50}$$

We assume throughout, and without further mention, that the following conditions hold:

(K1) For each $i \in \{1, \ldots, m\}$, $w_i \in C^{(1)}[a, b]$ and $w_i(t) > 0$;

(K2) for each $i \in \{1, \ldots, m\}$, $f_i \in C(\mathbb{R})$, f_i is locally Lipschitz on $(-\infty, 0) \cup (0, \infty)$, and $f_i(y) > 0$ and $f_i(-y) = -f_i(y)$ for all $y > 0$;

(K3) for each $i \in \{1, \ldots, m\}$, there exist extended real numbers $(f_i)_0, (f_i)_\infty \in [0, \infty]$ such that

$$(f_i)_0 = \lim_{y \to 0} f_i(y)/y \quad \text{and} \quad (f_i)_\infty = \lim_{|y| \to \infty} f_i(y)/y.$$

Remark 10.4. The oddness assumptions for f_i in (K2) is only for convenience. Our results can be extended to the case when $y f_i(y) > 0$ for $y \neq 0$ and $i \in \{1, \ldots, m\}$, without the oddness assumption.

Our results for BVP (10.48), (10.49) are established using the eigenvalues, $\{\lambda_n\}_{n=0}^{\infty}$, of the linear SLP consisting of the equation

$$-y'' = \lambda \sum_{i=1}^{m} w_i(t)y, \ t \in (a,b), \tag{10.51}$$

and the two-point BC

$$\begin{aligned} \cos\alpha \ y(a) - \sin\alpha \ y'(a) = 0, \quad \alpha \in [0,\pi), \\ y'(b) = 0. \end{aligned} \tag{10.52}$$

It is well known that any eigenfunction associated with λ_n has n simple zeros in (a,b), see [Zettl (2005), Theorem 4.3.2].

Note that the function $\xi(s)$ given in BC (10.49) is of bounded variation on $[a,b]$. Thus, there are two nondecreasing functions $\xi_1(s)$ and $\xi_2(s)$ such that

$$\xi(s) = \xi_1(s) - \xi_2(s), \quad s \in [a,b]. \tag{10.53}$$

We study the solutions of BVP (10.48), (10.49) in the following function class T_k^γ, which is a variation of the definition in [Rynne (2007)].

Definition 10.1. A solution y of BVP (10.48), (10.49) is said to belong to the class $\mathcal{T}_k^\gamma$ for $k \in \mathbb{N}_0 := \{0,1,2,\dots\}$ and $\gamma \in \{+,-\}$ if

(i) y and y' have only simple zeros in $[a,b]$,
(ii) y' has exactly $k+1$ zeros in (a,b) if $\alpha \in [0,\pi/2)$ and has exactly k zeros in (a,b) if $\alpha \in [\pi/2,\pi)$,
(iii) there is exactly one zero of y strictly between any two consecutive zeros of y',
(iv) $\gamma y(t) > 0$ in a right-neighborhood of a.

Remark 10.5. It is easy to see that for $y \in \mathcal{T}_k^\gamma$ with $k \in \mathbb{N}_0$ and $\gamma \in \{+,-\}$, y may have k or $k+1$ zeros in (a,b).

10.2.1 *Main results*

We will use the notation $h_\pm(t) := \max\{0, \pm h(t)\}$ for any function h. For each $i \in \{1,\dots,m\}$, let $F_i(y) = \int_0^y f_i(\xi)\,d\xi$ for $y \in \mathbb{R}$ and denote

$$H(t,y) := \sum_{i=1}^{m} w_i(t)F_i(y) \quad \text{and} \quad k_0 = \int_a^b l(t)dt, \tag{10.54}$$

where

$$l(t) := \max\left\{ \frac{(w_1')_-(t)}{w_1(t)}, \dots, \frac{(w_m')_-(t)}{w_m(t)} \right\}.$$

By (K2), each F_i is strictly increasing on $[0, \infty)$. Thus, for any fixed $t \in [a, b]$, $H(t, y)$ is strictly increasing in y on $[0, \infty)$, and hence, is invertible in y on $[0, \infty)$. We denote by $H_+^{-1}(t, y)$ its inverse. Similarly, $H(t, y)$ has an inverse $H_-^{-1}(t, y)$ in y on $(-\infty, 0]$.

Note that the assumption (K2) implies that F_i is even. Therefore, for $t \in [a, b]$,

$$H_-^{-1}(t, y) = -H_+^{-1}(t, y), \quad y \in [0, \infty). \tag{10.55}$$

It is useful to note that for the special case with $m = 1$, we have $H(t, y) = w(t)F(y)$. Then, for $t \in [a, b]$,

$$H_\pm^{-1}(t, y) = F_\pm^{-1}\left(\frac{y}{w(t)}\right), \quad y \in [0, \infty). \tag{10.56}$$

We now present our main results on the existence and nonexistence of nodal solutions of BVP (10.48), (10.49) with the proofs given in a later section. In the sequel, we let λ_n be the nth eigenvalue of the SLP (10.51), (10.52) for $n \in \mathbb{N}_0$. The first theorem concerns the existence of certain types of nodal solutions.

Theorem 10.8. *Assume for some $n \in \mathbb{N}_0$ and all $t \in [a, b]$, either*

$$\begin{cases} \displaystyle\sum_{i=1}^m w_i(t)(f_i)_0 < \lambda_n \sum_{i=1}^m w_i(t), \\ \displaystyle\lambda_{n+1} \sum_{i=1}^m w_i(t) < \sum_{i=1}^m w_i(t)(f_i)_\infty, \end{cases} \tag{10.57}$$

or

$$\begin{cases} \displaystyle\sum_{i=1}^m w_i(t)(f_i)_\infty < \lambda_n \sum_{i=1}^m w_i(t), \\ \displaystyle\lambda_{n+1} \sum_{i=1}^m w_i(t) < \sum_{i=1}^m w_i(t)(f_i)_0. \end{cases} \tag{10.58}$$

Suppose further that, for any $c > 0$,

$$\int_a^b H_+^{-1}\left(s, ce^{k_0}\right) d\left(\xi_1(s) + \xi_2(s)\right) < H_+^{-1}(b, c), \tag{10.59}$$

where $\xi_1(s)$ and $\xi_2(s)$ are given by (10.53). Then BVP (10.48), (10.49) has a solution $y_n^\gamma \in \mathcal{T}_n^\gamma$ for $\gamma \in \{+, -\}$.

As consequences of Theorem 10.8, we have the following corollaries. The first one is for the special case of Eq. (10.48) with $m = 1$.

Corollary 10.3. *Consider equation*

$$-y'' = w(t)f(y). \tag{10.60}$$

Assume, for some $n \in \mathbb{N}_0$,

$$\text{either } f_0 < \lambda_n \text{ and } \lambda_{n+1} < f_\infty \quad \text{or} \quad f_\infty < \lambda_n \text{ and } \lambda_{n+1} < f_0. \tag{10.61}$$

If, for any $c > 0$,

$$\int_a^b F_+^{-1}\left(\frac{ce^{k_0}}{w(s)}\right) d(\xi_1(s) + \xi_2(s)) < F_+^{-1}\left(\frac{c}{w(b)}\right), \tag{10.62}$$

then BVP (10.60), (10.49) *has a solution* $y_n^\gamma \in \mathcal{T}_n^\gamma$ *for* $\gamma \in \{+, -\}$.

In particular, when $w(t) \equiv w_0 > 0$, *if* (10.61) *holds and*

$$\int_a^b d(\xi_1(s) + \xi_2(s)) < 1, \tag{10.63}$$

then BVP (10.60), (10.49) *has a solution* $y_n^\gamma \in \mathcal{T}_n^\gamma$ *for* $\gamma \in \{+, -\}$.

Remark 10.6. Let the second condition of BC (10.49) be replaced by the multi-point BC (10.50), then we may choose $\xi(s) = \xi_1(s) - \xi_2(s)$ with

$$\xi_1(s) = \sum_{j=1}^{d} (k_j)_+ \chi(s - \eta_j) \quad \text{and} \quad \xi_2(s) = \sum_{j=1}^{m} (k_j)_- \chi(s - \eta_j),$$

where $(k_j)_\pm = \max\{\pm k_j, 0\}$. Hence, $\xi_1(s) + \xi_2(s) = \sum_{j=1}^d |k_j| \chi(s - \eta_j)$. Then, it is easy to see that (10.62) reduces to

$$\sum_{j=1}^{d} |k_j| F_+^{-1}\left(\frac{ce^{k_0}}{w(\eta_j)}\right) < F_+^{-1}\left(\frac{c}{w(b)}\right). \tag{10.64}$$

In particular, when $f(y) = |y|^{r-1}y$ for $r > 0$, then (10.64) reduces to

$$\sum_{j=1}^{d} |k_j| \left(\frac{w(b)e^{k_0}}{w(\eta_j)}\right)^{1/(r+1)} < 1;$$

and when $w(t) \equiv w_0 > 0$ and $m = 1$, then (10.64) reduces to

$$\sum_{j=1}^{d} |k_j| < 1.$$

Therefore, it is easy to see that for the case when $m = 1$, Theorem 10.8 covers the main results in [Kong *et al.* (2010)] for BVPs with multi-point BCs.

Corollary 10.4. *Assume that* (10.59) *holds and either* (10.57) *or* (10.58) *holds with* $n = 0$. *Then*

(i) *BVP* (10.48), (10.49) *has positive and negative solutions in* $\mathcal{T}_0^\gamma$ *for* $\gamma \in \{+,-\}$ *if* $\xi(s)$ *is increasing on* $[a,b]$;
(ii) *BVP* (10.48), (10.49) *has solutions in* $\mathcal{T}_0^\gamma$ *for* $\gamma \in \{+,-\}$ *with exactly one zero in* (a,b) *if* $\xi(s)$ *is decreasing on* $[a,b]$ *and such that* $\int_a^{b-\epsilon} d\xi(s) < 0$ *for some small* $\epsilon > 0$.

Corollary 10.5. *Let* $\alpha \in [0,\pi/2)$. *Assume* (10.59) *holds, and either*

$$\sum_{i=1}^{m}(f_i)_0 = 0 \quad \text{and} \quad \sum_{i=1}^{m}(f_i)_\infty = \infty, \tag{10.65}$$

or

$$\sum_{i=1}^{m}(f_i)_\infty = 0 \quad \text{and} \quad \sum_{i=1}^{m}(f_i)_0 = \infty. \tag{10.66}$$

Then for all $n \geq 0$ *and* $\gamma \in \{+,-\}$, *BVP* (10.48), (10.49) *has a solution* $y_n^\gamma \in \mathcal{T}_n^\gamma$.

Here we observe that (10.65) and (10.66) mean that

$(f_i)_0 = 0$ for all $i \in \{1,2,\ldots,m\}$, and $(f_j)_\infty = \infty$ for some $j \in \{1,2,\ldots,m\}$

and

$(f_i)_\infty = 0$ for all $i \in \{1,2,\ldots,m\}$, and $(f_j)_0 = \infty$ for some $j \in \{1,2,\ldots,m\}$,

respectively.

The next theorem is about the nonexistence of certain types of nodal solutions.

Theorem 10.9. (i) *Assume, for some* $n \in \mathbb{N}_0$,

$$\sum_{i=1}^{m} w_i(t) f_i(y)/y < \lambda_n \sum_{i=1}^{m} w_i(t)$$

for all $t \in [a,b]$ *and* $y \neq 0$. *Then BVP* (10.48), (10.49) *has no solution in* $\mathcal{T}_j^\gamma$ *for all* $j \geq n$ *and* $\gamma \in \{+,-\}$.
(ii) *Assume, for some* $n \in \mathbb{N}_0$,

$$\sum_{i=1}^{m} w_i(t) f_i(y)/y > \lambda_{n+1} \sum_{i=1}^{m} w_i(t)$$

for all $t \in [a,b]$ *and* $y \neq 0$. *Then BVP* (10.48), (10.49) *has no solution in* $\mathcal{T}_j^\gamma$ *for all* $j \leq n$ *and* $\gamma \in \{+,-\}$.

10.2.2 *Proofs of the main results*

To prove Theorem 10.8, we need some preliminaries. Lemmas 10.3, 10.4, and 10.5 below have been proven in [Kong (2007)] for the case when $m = 1$ and can be extended to the general case with essentially the same proofs. The first lemma is a generalization of [Kong (2007), Propositions 3.1 and 3.2 and Corollary 3.1].

Lemma 10.3. *Any initial value problem associated with Eq.* (10.48) *has a unique solution which exists on the whole interval* $[a, b]$. *Consequently, the solution depends continuously on the initial condition.*

As a direct result from Lemma 10.3, we have the following corollary.

Corollary 10.6. *For any nontrivial solution* y *of Eq.* (10.48), y *and* y' *have only simple zeros in* $[a, b]$.

Proof. Assume y has a non-simple zero $t_0 \in [a, b]$, i.e., $y(t_0) = y'(t_0) = 0$. Then, from Lemma 10.3, $y(t) \equiv 0$ on $[a, b]$. Assume y' has a non-simple zero $t_0 \in [a, b]$, i.e., $y'(t_0) = y''(t_0) = 0$. From Eq. (10.48) and noting that $w_i(t) > 0$ for $i = 1, \ldots, m$, we have $f_i(y(t_0)) = 0$, $i = 1, \ldots, m$. Hence, $y(t_0) = 0$ and as above, $y(t) \equiv 0$ on $[a, b]$. In either case, we have reached a contradiction. □

For $\gamma \in \{+, -\}$, let $y(t, \rho)$ be the solution of Eq. (10.48) satisfying

$$y(a) = \gamma\rho \sin\alpha \quad \text{and} \quad y'(a) = \gamma\rho\cos\alpha, \tag{10.67}$$

where $\rho > 0$ is a parameter. Let $\theta(t, \rho)$ be the Prüfer angle of $y(t, \rho)$, i.e., $\theta(\cdot, \rho)$ is a continuous function on $[a, b]$ such that

$$\tan\theta(t, \rho) = y(t, \rho)/y'(t, \rho) \quad \text{and} \quad \theta(a, \rho) = \alpha.$$

By Lemma 10.3, $\theta(t, \rho)$ is continuous in ρ on $(0, \infty)$ for any $t \in [a, b]$. The following results are generalizations of [Kong (2007), Lemmas 4.1, 4.2, 4.4, and 4.5].

Lemma 10.4. (i) *Assume, for some* $n \in \mathbb{N}_0$,

$$\sum_{i=1}^{m} w_i(t)(f_i)_0 < \lambda_n \sum_{i=1}^{m} w_i(t)$$

for all $t \in [a, b]$. *Then there exists* $\rho_* > 0$ *such that* $\theta(b, \rho) < n\pi + \pi/2$ *for all* $\rho \in (0, \rho_*)$.

(ii) *Assume, for some* $n \in \mathbb{N}_0$,

$$\lambda_n \sum_{i=1}^{m} w_i(t) < \sum_{i=1}^{m} w_i(t)(f_i)_\infty$$

for all $t \in [a, b]$. *Then there exists* $\rho^* > 0$ *such that* $\theta(b, \rho) > n\pi + \pi/2$ *for all* $\rho \in (\rho^*, \infty)$.

Lemma 10.5. (i) *Assume, for some* $n \in \mathbb{N}_0$,

$$\sum_{i=1}^{m} w_i(t)(f_i)_\infty < \lambda_n \sum_{i=1}^{m} w_i(t)$$

for all $t \in [a, b]$. *Then there exists* $\rho^* > 0$ *such that* $\theta(b, \rho) < n\pi + \pi/2$ *for all* $\rho \in (\rho^*, \infty)$.
(ii) *Assume, for some* $n \in \mathbb{N}_0$,

$$\lambda_n \sum_{i=1}^{m} w_i(t) < \sum_{i=1}^{m} w_i(t)(f_i)_0$$

for all $t \in [a, b]$. *Then there exists* $\rho_* > 0$ *such that* $\theta(b, \rho) > n\pi + \pi/2$ *for all* $\rho \in (0, \rho_*)$.

Proof of Theorem 10.8. We first prove it under the assumption that (10.57) holds. Without loss of generality, we assume $\gamma = +$. The case with $\gamma = -$ can be proved in the same way. Let $y(t, \rho)$ be the solution of Eq. (10.48) satisfying (10.67) with $\gamma = +$, and $\theta(t, \rho)$ its Prüfer angle. By Lemma 10.4, there exist $0 < \rho_* < \rho^* < \infty$ such that

$$\theta(b, \rho) < n\pi + \pi/2 \text{ for all } \rho \in (0, \rho_*)$$

and

$$\theta(b, \rho) > (n + 1)\pi + \pi/2 \text{ for all } \rho \in (\rho^*, \infty).$$

By the continuity of $\theta(t, \rho)$ in ρ, there exist $\rho_* \le \rho_n < \rho_{n+1} \le \rho^*$ such that

$$\theta(b, \rho_n) = n\pi + \pi/2 \quad \text{and} \quad \theta(b, \rho_{n+1}) = (n + 1)\pi + \pi/2, \tag{10.68}$$

and

$$n\pi + \pi/2 < \theta(b, \rho) < (n + 1)\pi + \pi/2 \text{ for } \rho_n < \rho < \rho_{n+1}. \tag{10.69}$$

Then, for all $t \in [a, b]$ and all $\rho > 0$, we define an energy function $E(t, \rho)$ for $y(t, \rho)$ by

$$E(t, \rho) = \frac{1}{2}\big(y'(t, \rho)\big)^2 + H\big(t, y(t, \rho)\big), \tag{10.70}$$

where $H(t,y)$ is defined in (10.54). By (iH1) and (iH2), $F_i(y) \geq 0$ on $\mathbb{R}$ for each $i \in \{1,\ldots,m\}$. Then, $E(t,\rho) \geq 0$ on $[a,b]$. By (10.48) and the definition of $l(t)$, we have

$$\begin{aligned} E'(t,\rho) &= \sum_{i=1}^{m} w_i'(t)F_i(y(t,\rho)) \geq \sum_{i=1}^{m} \frac{w_i'(t)}{w_i(t)} w_i(t)F_i(y(t,\rho)) \\ &\geq -l(t)\Big(\frac{1}{2}\big(y'(t,\rho)\big)^2 + H\big(t,y(t,\rho)\big)\Big) = -l(t)E(t,\rho). \end{aligned}$$

Thus, $E'(t,\rho) + l(t)E(t,\rho) \geq 0$ for all $t \in [a,b]$ and $\rho > 0$. By solving this inequality, we obtain

$$E(s,\rho) \leq E(b,\rho)e^{\int_s^b l(\tau)d\tau} \leq E(b,\rho)e^{k_0}, \quad s \in [a,b]. \tag{10.71}$$

We observe from (10.70) that, for $\rho = \rho_n$ and $\rho = \rho_{n+1}$,

$$E(s,\rho) \geq H\big(s,y(s,\rho)\big) \text{ and } E(b,\rho) = H\big(b,y(b,\rho)\big). \tag{10.72}$$

Recall that, for fixed t, $H_+^{-1}(t,y)$ is increasing in y on $[0,\infty)$. Thus, from (10.72) and (10.55), we see that for $\rho = \rho_n$ and $\rho = \rho_{n+1}$ and $s \in [a,b]$,

$$|y(s,\rho)| \leq H_+^{-1}(s,E(s,\rho)) \text{ and } |y(b,\rho)| = H_+^{-1}(b,E(b,\rho)). \tag{10.73}$$

Define

$$\Gamma(\rho) = y(b,\rho) - \int_a^b y(s,\rho)d\xi(s). \tag{10.74}$$

Let $n = 2k$ with $k \in \mathbb{N}_0$. Since $y(b,\rho_{2k}) > 0$ and $y(b,\rho_{2k+1}) < 0$, then using a similar argument as in [Kong *et al.* (2010)] (see (2.11) and (2.12) there), from (10.71), (10.73), and (10.59), we can see that

$$\begin{aligned} \Gamma(\rho_{2k}) &= y(b,\rho_{2k}) - \int_a^b y(s,\rho_{2k})d\xi(s) \\ &\geq y(b,\rho_{2k}) - \int_a^b |y(s,\rho_{2k})|d(\xi_1(s)+\xi_2(s)) > 0 \end{aligned} \tag{10.75}$$

and

$$\begin{aligned} \Gamma(\rho_{2k+1}) &= y(b,\rho_{2k+1}) - \int_a^b y(s,\rho_{2k+1})d\xi(s) \\ &\leq -|y(b,\rho_{2k+1})| + \int_a^b |y(s,\rho_{2k+1})|d(\xi_1(s)+\xi_2(s)) < 0. \end{aligned} \tag{10.76}$$

By the continuity of $\Gamma(\rho)$, there exists $\bar\rho \in (\rho_{2k},\rho_{2k+1})$ such that $\Gamma(\bar\rho) = 0$. Similarly, for $n = 2k+1$ with $k \in \mathbb{N}_0$, there exists $\bar\rho \in (\rho_{2k+1},\rho_{2k+2})$ such that $\Gamma(\bar\rho) = 0$. In both cases, from (10.69),

$$n\pi + \pi/2 < \theta(b,\bar\rho) < (n+1)\pi + \pi/2.$$

Note that for $t \in (a,b)$ with $y(t) \neq 0$, $\theta(t)$ satisfies the equation

$$\theta'(t,\rho) = \cos^2\theta(t,\rho) + \sum_{i=1}^{m} w_i(t)\frac{f_i(y(t,\rho))}{y(t,\rho)}\sin^2\theta(t,\rho). \tag{10.77}$$

By (K1) and (K2), $\theta(\cdot,\rho)$ is strictly increasing on $[a,b]$. We note that $y(t) = 0$ if and only if $\theta(t,\rho) = 0 \pmod{\pi}$ and $y'(t) = 0$ if and only if $\theta(t,\rho) = \pi/2 \pmod{\pi}$. Thus, y' has exactly $n+1$ zeros in (a,b) if $\alpha \in [0,\pi/2)$ and n zeros in (a,b) if $\alpha \in [\pi/2,\pi)$, and y has exactly one zero strictly between any two consecutive zeros of y'. Initial condition (10.67) implies that $y(t,\bar{\rho}) > 0$ in a right-neighborhood of a. Therefore, $y(t,\bar{\rho}) \in \mathcal{T}_n^+$.

The proof when (10.58) holds is essentially the same as above except that the discussion is based on Lemma 10.5 instead of Lemma 10.4. □

Proof of Corollary 10.3. Since $m = 1$, by (10.56), $H_\pm^{-1}(t,ce^{k_0}) = F_\pm^{-1}(ce^{k_0}/w(t))$. In this case, conditions (10.57)–(10.59) reduce to (10.61)–(10.62). Then the conclusion follows from Theorem 10.8.

In addition, when $w(t) \equiv w_0 > 0$, from the definition of $l(t)$, $l(t) \equiv 0$. Hence, $k_0 = \int_a^b l(t)dt = 0$. By (10.56), $H_\pm^{-1}(t,ce^{k_0}) = F_\pm^{-1}(c/w_0)$. In this case, (10.62) reduces to (10.63). Then the conclusion follows from Theorem 10.8. □

Proof of Corollary 10.4. Without loss of generality, let $\gamma = +$. The case for $\gamma = -$ can be proved in the same way. Let $y(t,\rho)$ be the solution of Eq. (10.48) satisfying (10.67) with $\gamma = +$, and $\theta(t,\rho)$ its Prüfer angle. Let ρ_0 and ρ_1 be given in (10.68) with $n = 0$. Then for the function $\Gamma(\rho)$ defined by (10.74), from (10.75) and (10.76), we have $\Gamma(\rho_0) > 0$ and $\Gamma(\rho_1) < 0$.

(i) Assume $\xi(s)$ is increasing on $[a,b]$. Since $\xi(s) = \xi_1(s) - \xi_2(s)$, we may take $\xi(s) = \xi_1(s)$. From (10.68) and (10.69) with $n = 0$ and by the continuity of $\theta(t,\rho)$ in ρ, there exists $\tilde{\rho} \in (\rho_0,\rho_1)$ such that $\theta(b,\tilde{\rho}) = \pi$ and $\pi/2 < \theta(b,\rho) < \pi$ for $\rho \in (\rho_0,\tilde{\rho})$. Note that $\theta(s,\tilde{\rho}) < \pi$ for $s \in [a,b)$. Thus, $y(b,\tilde{\rho}) = 0$ and $y(s,\tilde{\rho}) > 0$ for $s \in [a,b)$. By the Mean Value Theorem for Riemann-Stieltjes integrals and the continuity of the solution y, there exists $t_0 \in [a,b]$ such that

$$\Gamma(\tilde{\rho}) = -\int_a^b y(s,\tilde{\rho})d\xi_1(s) = -y(t_0,\tilde{\rho})\int_a^b d\xi_1(s) \leq 0.$$

Therefore, there exists $\bar{\rho} \in (\rho_0,\tilde{\rho}]$ such that $\Gamma(\bar{\rho}) = 0$. This means that $y(t,\bar{\rho})$ is a solution of BVP (10.48), (10.49). Note that $\theta(b,\bar{\rho}) \in (\pi/2,\pi)$, we have $y(t,\bar{\rho}) \in \mathcal{T}_0^+$ and is a positive solution.

(ii) Assume $\xi(s)$ is decreasing on $[a, b]$ such that $\int_a^{b-\epsilon} d\xi(s) < 0$ for some small $\epsilon > 0$. Since $\xi(s)$ is decreasing, we may take $\xi(s) = -\xi_2(s)$. As in (i), there exists $\tilde{\rho} \in (\rho_0, \rho_1)$ such that $\theta(b, \tilde{\rho}) = \pi$ and $3\pi/2 > \theta(b, \rho) > \pi$ for $\rho \in (\tilde{\rho}, \rho_1)$. We note that $\theta(s, \tilde{\rho}) < \pi$ for $s \in [a, b)$. Thus, $y(b, \tilde{\rho}) = 0$ and $y(s, \tilde{\rho}) > 0$ for $s \in [a, b)$. Thus, for some $t_0 \in [a, b - \epsilon]$,

$$\begin{aligned}\Gamma(\tilde{\rho}) &= -\int_a^b y(s, \tilde{\rho}) d\xi_2(s) \\ &\geq -\int_a^{b-\epsilon} y(s, \tilde{\rho}) d\xi_2(s) = -y(t_0, \tilde{\rho}) \int_a^{b-\epsilon} d\xi_2(s) > 0.\end{aligned}$$

Therefore, there exists $\bar{\rho} \in (\tilde{\rho}, \rho_1)$ such that $\Gamma(\bar{\rho}) = 0$. This means that $y(t, \bar{\rho})$ is a solution of BVP (10.48), (10.49). Note that $\theta(b, \bar{\rho}) \in (\pi, 3\pi/2)$, we have $y(t, \bar{\rho}) \in \mathcal{T}_0^+$ and has exactly one zero in (a, b). $\square$

Proof of Corollary 10.5. It is easy to see that $\lambda_0 = 0$ is the first eigenvalue of the BVP consisting of Eq. (10.51) and the BC

$$y'(a) = y'(b) = 0,$$

i.e., the BC (10.52) with $\alpha = \pi/2$. In fact, $y_0(t) \equiv 1$ is an associated eigenfunction. From [Kong and Zettl (1996), Theorem 4.2], we see that λ_0, as a function of α, is strictly decreasing. This shows that $\lambda_0 > 0$ for $\alpha \in [0, \pi/2)$. Then, by Theorem 10.8, BVP (10.48), (10.49) has a solution $y_n^\gamma \in \mathcal{T}_n^\gamma$ for all $n \geq 0$ and $\gamma \in \{+, -\}$. $\square$

Finally, the proof of Theorem 10.9 is essentially the same as that of [Kong *et al.* (2010), Theorem 2.2] and hence is omitted.

10.3 Nodal Solutions of Nonlocal BVP's — II

In this section, we are concerned with the nonlocal BVP consisting of the equation

$$-(p(t)y')' + q(t)y = w(t)f(y), \quad t \in (a, b), \tag{10.78}$$

and the BC

$$\begin{aligned}&\cos\alpha\ y(a) - \sin\alpha\ (py')(a) = 0, \quad \alpha \in [0, \pi), \\ &(py')(b) - \textstyle\int_a^b (py')(s) d\xi(s) = 0,\end{aligned} \tag{10.79}$$

where $a, b \in \mathbb{R}$ with $a < b$ and the integral in BC (10.79) is the Riemann-Stieltjes integral with respect to $\xi(s)$ with $\xi(s)$ a function of bounded variation. In the case that $\xi(s) = s$, the Riemann-Stieltjes integral in the second condition of (10.79) reduces to the Riemann integral. In the case that $\xi(s) = \sum_{i=1}^{m} k_i\chi(s-x_i)$, where $m \geq 1$, $k_i \in \mathbb{R}$, $i = 1, \ldots, m$, $\{x_i\}_{i=1}^{m}$ is a sequence of distinct points in $[a, b]$, and $\chi(s)$ is the characteristic function on $[0, \infty)$, i.e.,

$$\chi(s) = \begin{cases} 1, & s \geq 0, \\ 0, & s < 0, \end{cases}$$

the second equation in (10.79) reduces to the multi-point BC

$$(py')(b) - \sum_{i=1}^{m} k_i(py')(x_i) = 0. \tag{10.80}$$

We assume throughout, and without further mention, that the following conditions hold:

(L1) $p, q, w \in C^1[a, b]$ such that $p(t) > 0$, $w(t) > 0$, and $q'(t) + q^* \leq l(q^* - q(t))$ on $[a, b]$ with

$$q^* := \max_{t\in[a,b]} \{q(t), 0\} \quad \text{and} \quad l := \max_{t\in[a,b]} \left\{ \left(\frac{p'(t) + q^*}{p(t)}\right)_+, \frac{w'_-(t)}{w(t)} \right\},$$

where $h'_-(t) := \max\{0, -h'(t)\}$ and $h_+(t) := \max\{0, h(t)\}$;

(L2) $f \in C(\mathbb{R})$ such that $yf(y) > 0$ for $y \neq 0$, and f is locally Lipschitz on $(-\infty, 0) \cup (0, \infty)$;

(L3) there exist extended real numbers $f_0, f_\infty \in [0, \infty]$ such that

$$f_0 = \lim_{y\to 0} f(y)/y \quad \text{and} \quad f_\infty = \lim_{|y|\to\infty} f(y)/y.$$

Remark 10.7. For $p, w \in C^{(1)}[a, b]$, the following are examples of the function classes for q satisfying (L1):

(i) $q \in C^{(1)}[a, b]$ such that $q'(t) \leq -q^*$ on $[a, b]$. It is easy to see that any non-positive, non-increasing function q belongs to this class. In particular, any non-positive constants belong to this class.

(ii) $q \in C^{(1)}[a, b]$ such that $q'(t) \leq -lq(t)$ on $[a, b]$ with $l \geq 1$. For $c \geq 0$, it is easy to see that

$$q_1(t) = ce^{-kt} \text{ for } t \in [0, 1] \text{ with } k \geq l \geq 1 \text{ and}$$
$$q_2(t) = -ce^{-kt} \text{ for } t \in [0, 1] \text{ with } 0 \leq k \leq l \text{ and } l \geq 1$$

belong to this class.

We obtain results on the existence and nonexistence of nodal solutions of BVP (10.78), (10.79) by relating it to the eigenvalues of an associated linear SLP with a two-point separated BC. The shooting method and a generalized energy function play key roles in the proofs. We also discuss the changes of the existence of different types of nodal solutions when some parameters in the problem change, more precisely, when the interval $[a, b]$ shrinks, when the functions w, p, and q increase in certain directions, and when the boundary condition angle α changes. Note that our results are for the general BVP (10.78), (10.79) with variable w, p, and q, a separated BC at the left endpoint prescribed by an arbitrary α, and a BC given by a Riemann-Stieltjes integral with respect to $\xi(s)$.

10.3.1 *Existence and nonexistence of nodal solutions*

We study the nodal solutions of BVP (10.78), (10.79) in the following classes.

Definition 10.2. A solution y of BVP (10.78), (10.79) is said to belong to class $\mathcal{S}_n^\gamma$ for $n \in \mathbb{N}_0 := \{0, 1, 2, \dots\}$ and $\gamma \in \{+, -\}$ if
(i) y has exactly n zeros in (a, b),
(ii) $\gamma y(t) > 0$ in a right-neighborhood of a.

Our results on the existence and nonexistence of nodal solutions of BVP (10.78), (10.79) are established utilizing the eigenvalues of the linear SLP consisting of the equation

$$-(p(t)y')' + q(t)y = \lambda w(t)y, \quad t \in (a, b), \tag{10.81}$$

and the two-point BC

$$\begin{aligned} &\cos\alpha\ y(a) - \sin\alpha\ (py')(a) = 0, \quad \alpha \in [0, \pi), \\ &y(b) = 0. \end{aligned} \tag{10.82}$$

It is well known that SLP (10.81), (10.82) has an infinite number of eigenvalues $\{\lambda_n\}_{n=0}^{\infty}$ satisfying

$$-\infty < \lambda_0 < \lambda_1 < \cdots < \lambda_n < \cdots, \quad \text{and } \lambda_n \to \infty \text{ and } n \to \infty,$$

and any eigenfunction associated with λ_n has n simple zeros in (a, b), see [Zettl (2005), Theorem 4.3.2].

Note that the function $\xi(s)$ given in BC (10.79) is of bounded variation on $[a, b]$. Thus, there are two nondecreasing functions $\xi_1(s)$ and $\xi_2(s)$ such that

$$\xi(s) = \xi_1(s) - \xi_2(s), \quad s \in [a, b]. \tag{10.83}$$

In the following we assume (10.83) holds. We now present our main results with the proofs given later in this section after several technical lemmas are derived. The first theorem is about the existence of certain types of nodal solutions.

Theorem 10.10. *Assume either (i) $f_0 < \lambda_n$ and $\lambda_{n+1} < f_\infty$, or (ii) $f_\infty < \lambda_n$ and $\lambda_{n+1} < f_0$, for some $n \in \mathbb{N}_0$. Suppose*

$$1 - \int_a^b \sqrt{\frac{p(s)}{p(b)}}\, e^{l(b-a)/2}\, d(\xi_1(s) + \xi_2(s)) > 0. \tag{10.84}$$

Then BVP (10.78), (10.79) *has two solutions $y_{n,\gamma} \in \mathcal{S}_{n+1}^{\gamma}$ for $\gamma \in \{+,-\}$.*

Remark 10.8. (a) Note that for the multi-point case, i.e., BVP (10.78), (10.79) with the second condition in (10.79) replaced by (10.80), we have that $\xi(s) = \sum_{i=1}^{m} k_i \chi(s - x_i)$. Thus $\xi(s) = \xi_1(s) - \xi_2(s)$ with

$$\xi_1(s) = \sum_{i=1}^{m} (k_i)_+ \chi(s - x_i) \quad \text{and} \quad \xi_2(s) = \sum_{i=1}^{m} (k_i)_- \chi(s - x_i),$$

where $(k_i)_\pm = \max\{\pm k_i, 0\}$. Hence $\xi_1(s) + \xi_2(s) = \sum_{i=1}^{m} |k_i| \chi(s - x_i)$. It is easy to see that condition (10.84) then becomes

$$1 - \sum_{i=1}^{m} |k_i| \sqrt{\frac{p(x_i)}{p(b)}}\, e^{l(b-a)/2} > 0.$$

(b) When $\xi_i \in C^{(1)}[a,b]$ for $i = 1, 2$, condition (10.84) becomes

$$1 - \int_a^b \sqrt{\frac{p(s)}{p(b)}}\, e^{l(b-a)/2}\, (\xi_1'(s) + \xi_2'(s))ds > 0.$$

In particular, if $p(t) \equiv 1$, $q(t) \equiv 0$, $w(t) > 0$ is increasing, and $\xi(t) = t$, then it is reduced to $b - a < 1$.

As a consequence of Theorem 10.10, we have the following corollary on the existence of an infinite number of different types of nodal solutions for a special case of BVP (10.78), (10.79).

Corollary 10.7. *Consider the special case that $p(t) \equiv 1$ and $q(t) \equiv 0$ on $[a, b]$. Assume* (10.84) *holds and*

$$\textit{either } f_0 = 0 \textit{ and } f_\infty = \infty, \quad \textit{or} \quad f_\infty = 0 \textit{ and } f_0 = \infty.$$

Then there exists $\alpha^* \in (\pi/2, \pi)$ such that

(i) *if* $\alpha \in [0, \alpha^*)$, *then BVP* (10.78), (10.79) *has a solution* $y_n^\gamma \in \mathcal{S}_{n+1}^\gamma$ *for each* $n \geq 0$ *and* $\gamma \in \{+, -\}$;

(ii) *if* $\alpha \in [\alpha^*, \pi)$, *then BVP* (10.78), (10.79) *has a solution* $y_n^\gamma \in \mathcal{S}_{n+1}^\gamma$ *for each* $n \geq 1$ *and* $\gamma \in \{+, -\}$.

Remark 10.9. The number α^* in the above theorem can be explicitly computed using the fundamental solutions of (10.81) see [Cao *et al.* (2007), Theorem 2.2] for details.

The next theorem is about the nonexistence of certain types of nodal solutions.

Theorem 10.11. (i) *Assume* $f(y)/y \leq \lambda_n$ *for some* $n \in \mathbb{N}_0$ *and all* $y \neq 0$. *Then BVP* (10.78), (10.79) *has no solution in* $\mathcal{S}_i^\gamma$ *for all* $i \geq n + 1$ *and* $\gamma \in \{+, -\}$.

(ii) *Assume* $f(y)/y \geq \lambda_n$ *for some* $n \in \mathbb{N}_0$ *and all* $y \neq 0$. *Then BVP* (10.78), (10.79) *has no solution in* $\mathcal{S}_i^\gamma$ *for all* $i \leq n$ *and* $\gamma \in \{+, -\}$.

To prove Theorem 10.10, we need some preliminaries. The lemmas below are on the initial value problems (IVPs) associated with Eq. (10.78) and are simple generalizations of [Kong (2007), Corollary 3.1, Lemmas 4.1, 4.2, 4.4, and 4.5] originally for the case where $p(t) \equiv 1$ with essentially the same proofs. The first one is on the global existence of solutions of IVPs associated with Eq. (10.78).

Lemma 10.6. *Any initial value problem associated with Eq.* (10.78) *has a unique solution which exists on the whole interval* $[a, b]$. *Consequently, the solution depends continuously on the initial condition.*

For $\gamma \in \{+, -\}$, let $y(t, \rho)$ be the solution of Eq. (10.78) satisfying

$$y(a) = \gamma\rho \sin\alpha \quad \text{and} \quad (py')(a) = \gamma\rho\cos\alpha, \tag{10.85}$$

where $\rho > 0$ is a parameter. Let $\theta(t, \rho)$ be the Prüfer angle of $y(t, \rho)$, i.e., $\theta(\cdot, \rho)$ is a continuous function on $[a, b]$ such that

$$\tan\theta(t, \rho) = y(t, \rho)/(py')(t, \rho) \quad \text{and} \quad \theta(a, \rho) = \alpha.$$

By Lemma 10.6, $\theta(t, \rho)$ is continuous in ρ on $(0, \infty)$ for any $t \in [a, b]$.

The next two lemmas provide some estimates for the Prüfer angle.

Lemma 10.7. (i) *Assume* $f_0 < \lambda_n$ *for some* $n \in \mathbb{N}_0$. *Then there exists* $\rho_* > 0$ *such that* $\theta(b, \rho) < (n + 1)\pi$ *for all* $\rho \in (0, \rho_*)$.

(ii) *Assume* $\lambda_n < f_\infty$ *for some* $n \in \mathbb{N}_0$. *Then there exists* $\rho^* > 0$ *such that* $\theta(b, \rho) > (n + 1)\pi$ *for all* $\rho \in (\rho^*, \infty)$.

Lemma 10.8. (i) *Assume $f_\infty < \lambda_n$ for some $n \in \mathbb{N}_0$. Then there exists $\rho^* > 0$ such that $\theta(b,\rho) < (n+1)\pi$ for all $\rho \in (\rho^*, \infty)$.*

(ii) *Assume $\lambda_n < f_0$ for some $n \in \mathbb{N}_0$. Then there exists $\rho_* > 0$ such that $\theta(b,\rho) > (n+1)\pi$ for all $\rho \in (0, \rho_*)$.*

Proof of Theorem 10.10. We first prove it for the case where $f_0 < \lambda_n$ and $\lambda_{n+1} < f_\infty$. Without loss of generality we assume $\gamma = +$. The case with $\gamma = -$ can be proved in the same way. Let $y(t,\rho)$ be the solution of Eq. (10.78) satisfying (10.85) with $\gamma = +$ and $\theta(t,\rho)$ its Prüfer angle. By Lemma 10.7, there exist $0 < \rho_* < \rho^* < \infty$ such that

$$\theta(b,\rho) < (n+1)\pi \quad \text{for all } \rho \in (0, \rho_*)$$

and

$$\theta(b,\rho) > (n+2)\pi \quad \text{for all } \rho \in (\rho^*, \infty).$$

By the continuity of $\theta(t,\rho)$ in ρ, there exist $\rho_* \leq \rho_{n+1} < \rho_{n+2} \leq \rho^*$ such that

$$\theta(b,\rho_{n+1}) = (n+1)\pi \quad \text{and} \quad \theta(b,\rho_{n+2}) = (n+2)\pi, \tag{10.86}$$

and

$$(n+1)\pi < \theta(b,\rho) < (n+2)\pi \quad \text{for} \quad \rho_{n+1} < \rho < \rho_{n+2}. \tag{10.87}$$

Then, for all $t \in [a,b]$ and all $\rho > 0$, we define an energy function $E(t,\rho)$ for $y(t,\rho)$ by

$$E(t,\rho) = \frac{1}{2p(t)}\big(p(t)y'(t,\rho)\big)^2 + \frac{1}{2}\big(q^* - q(t)\big)y^2(t,\rho) + w(t)F\big(y(t,\rho)\big). \tag{10.88}$$

where $F(y) = \int_0^y f(s)ds$. By (L1) and (L2), $F(y) \geq 0$ on $\mathbb{R}$ yielding $E(t,\rho) \geq 0$ on $[a,b]$. For ease of notation, in the following, we use $p = p(t)$, $q = q(t)$, $w = w(t)$, $y = y(t,\rho)$, $E = E(t,\rho)$. Then, by (10.78) and (L1), we find that

$$\begin{aligned}
E' &= -\frac{p'}{2p^2}(py')^2 - \frac{1}{2}q'y^2 + q^*yy' + w'F(y) \\
&\geq -\frac{p'}{2p^2}(py')^2 - \frac{1}{2}q'y^2 - \frac{q^*}{2}\left(y^2 + y'^2\right) + w'F(y) \\
&= -\frac{(p' + q^*)}{2p^2}(py')^2 - \frac{1}{2}(q' + q^*)y^2 + w'F(y) \\
&\geq -\left(\frac{p' + q^*}{p}\right)_+ \left(\frac{1}{2p}(py')^2\right) - l\left(\frac{1}{2}(q^* - q)y^2\right) - \frac{w'_-}{w}wF(y) \\
&\geq -l\frac{(py')^2}{2p} - l\left(\frac{1}{2}(q^* - q)y^2\right) - lwF(y) \\
&= -lE(t,\rho).
\end{aligned}$$

Thus, $E'(t,\rho)+lE(t,\rho)\geq 0$ for all $t\in[a,b]$ and $\rho>0$. By solving this inequality, we obtain that

$$E(s,\rho)\leq E(b,\rho)e^{l(b-s)},\quad s\in[a,b]. \tag{10.89}$$

We observe that for $\rho=\rho_{n+1}$ and $\rho=\rho_{n+2}$

$$E(s,\rho)\geq \frac{1}{2p(s)}\left[p(s)y'(s,\rho)\right]^2,\quad s\in[a,b],$$

and

$$E(b,\rho)=\frac{1}{2p(b)}\left[p(b)y'(b,\rho)\right]^2.$$

Thus, for $\rho=\rho_{n+1}$, $\rho=\rho_{n+2}$, and $s\in[a,b]$,

$$|(py')(s,\rho)|\leq\sqrt{2p(s)E(s,\rho)}\ \text{ and }\ |(py')(b,\rho)|=\sqrt{2p(s)E(b,\rho)}. \tag{10.90}$$

Define

$$\Gamma(\rho)=(py')(b,\rho)-\int_a^b (py')(s,\rho)d\xi(s). \tag{10.91}$$

Assume $n=2k-1$ with $k\in\mathbb{N}_0$. Since $(py')(b,\rho_{2k})>0$ and $(py')(b,\rho_{2k+1})<0$, by (10.83), (10.89), (10.90), and (7.29) we have

$$\begin{aligned}\Gamma(\rho_{2k})&=(py')(b,\rho_{2k})-\int_a^b(py')(s,\rho_{2k})d\xi(s)\\&\geq\left|(py')(b,\rho)\right|-\int_a^b|(py')(s,\rho_{2k})|d\big(\xi_1(s)+\xi_2(s)\big)\\&\geq\sqrt{2p(b)E(b,\rho_{2k})}-\int_a^b\sqrt{2p(s)E(s,\rho_{2k})}d\big(\xi_1(s)+\xi_2(s)\big)\\&\geq\sqrt{2p(b)E(b,\rho_{2k})}-\int_a^b\sqrt{2p(s)E(b,\rho_{2k})e^{l(b-a)}}d\big(\xi_1(s)+\xi_2(s)\big)\\&=\sqrt{2p(b)E(b,\rho_{2k})}\left(1-\int_a^b\sqrt{p(s)/p(b)}e^{l(b-a)/2}d\big(\xi_1(s)+\xi_2(s)\big)\right)\\&>0\end{aligned}$$

and

$$\Gamma(\rho_{2k+1}) = (py')(b, \rho_{2k+1}) - \int_a^b (py')(s, \rho_{2k+1}) d\xi(s)$$

$$\begin{aligned}
&\le -\left|(py')(b, \rho_{2k+1})\right| + \int_a^b (|(py')(s, \rho_{2k+1})| d(\xi_1(s) + \xi_2(s)) \\
&\le -\sqrt{2p(b)E(b, \rho_{2k+1})} + \int_a^b \sqrt{2p(s)E(s, \rho_{2k+1})}\, d(\xi_1(s) + \xi_2(s)) \\
&\le -\sqrt{2p(b)E(b, \rho_{2k+1})} + \int_a^b \sqrt{2p(s)E(b, \rho_{2k+1})e^{l(b-a)}}\, d(\xi_1(s) + \xi_2(s)) \\
&= -\sqrt{2p(b)E(b, \rho_{2k+1})}\left(1 - \int_a^b \sqrt{p(s)/p(b)}\, e^{l(b-a)/2}\, d(\xi_1(s) + \xi_2(s))\right) \\
&< 0.
\end{aligned}$$

By the continuity of $\Gamma(\rho)$, there exists $\bar{\rho} \in (\rho_{2k}, \rho_{2k+1})$ such that $\Gamma(\bar{\rho}) = 0$. Similarly, for $n = 2k$ with $k \in \mathbb{N}_0$, there exists $\bar{\rho} \in (\rho_{2k+1}, \rho_{2k+2})$ such that $\Gamma(\bar{\rho}) = 0$. In both cases, from (10.87)

$$(n+1)\pi < \theta(b, \bar{\rho}) < (n+2)\pi.$$

Since

$$\theta'(t, \rho) = \frac{1}{p(t)} \cos^2 \theta(t, \rho) + w(t) \frac{f(y(t, \rho))y(t, \rho)}{r^2(t, \rho)} - q(t) \sin^2 \theta(t, \rho) \quad (10.92)$$

for $t \in [a, b]$, where $r = (y^2 + py')^{1/2}$, we have that $\theta(\cdot, \rho)$ is strictly increasing at the points t where $\theta(t, \rho) = 0 \pmod{\pi}$. We note that $y(t) = 0$ if and only if $\theta(t, \rho) = 0 \pmod{\pi}$. Thus, y has exactly $n + 1$ zeros in (a, b). Initial condition (10.85) implies that $y(t, \bar{\rho}) > 0$ in a right-neighborhood of a. Therefore, $y(t, \bar{\rho}) \in \mathcal{S}_{n+1}^+$.

The proof for the case where $f_\infty < \lambda_n$ and $\lambda_{n+1} < f_0$ is essentially the same as above except that the discussion is based on Lemma 10.8 instead of Lemma 10.7. □

Proof of Corollary 10.7. Consider the SLP consisting of Eq. (10.81) with $p(t) \equiv 1$, $q(t) \equiv 0$, and the BC

$$\begin{aligned}
\cos\alpha\; y(a) - \sin\alpha\; y'(a) = 0, &\quad \alpha \in [0, \pi), \\
\cos\beta\; y(b) - \sin\beta\; y'(b) = 0, &\quad \beta \in (0, \pi].
\end{aligned}$$

Denote by $\lambda_n(\alpha, \beta)$ the nth eigenvalue of this problem for $n \in \mathbb{N}_0$. It is easy to see that $\lambda_0(\pi/2, \pi/2) = 0$. In fact, $y_0(t) \equiv 1$ is an associated eigenfunction. From [Kong and Zettl (1996), Theorem 4.2] and [Kong *et al.* (1999), Lemma 3.32], we see that $\lambda_0(\alpha, \beta)$ is a continuous function of (α, β) on $[0, \pi) \times (0, \pi]$, and is strictly decreasing in α and strictly increasing in β. Furthermore, for any $\beta \in (0, \pi]$,

$$\lim_{\alpha \to \pi-} \lambda_0(\alpha, \beta) = -\infty \quad \text{and} \quad \lim_{\alpha \to \pi-} \lambda_{n+1}(\alpha, \beta) = \lambda_n(0, \beta) \text{ for } n \in \mathbb{N}_0.$$

This shows that $\lambda_0(\pi/2, \pi) > 0$, and hence there exists $\alpha^* \in (\pi/2, \pi)$ such that $\lambda_0(\alpha, \pi) > 0$ for $\alpha \in [0, \alpha^*)$, and $\lambda_0(\alpha, \pi) \leq 0$ and $\lambda_1(\alpha, \pi) > 0$ for $\alpha \in [\alpha^*, \pi)$. Note that $\beta = \pi$ if and only if $y(b) = 0$. Then the conclusion follows from Theorem 10.10. □

Proof of Theorem 10.11. (i) Assume to the contrary that BVP (10.78), (10.79) has a solution $y \in \mathcal{S}_i^\gamma$ for some $i \geq n+1$ and $\gamma \in \{+, -\}$. Let $\tilde{w}(t) = w(t) f(y(t))/y(t)$. Then $\tilde{w}(t)$ is continuous on $[a, b]$ by the continuous extension since $f_0 < \infty$. Let $\theta(t)$ be the Prüfer angle of $y(t)$ with $\theta(a) = \alpha$. Then $\theta(t)$ satisfies Eq. (10.92) and $\theta(b) > i\pi$. Note, from the assumption that $\tilde{w}(t) \leq \lambda_n w(t) \leq \lambda_{i-1} w(t)$ on $[a, b]$, we have that for $t \in [a, b]$

$$\begin{aligned}\theta'(t) &= \frac{1}{p(t)} \cos^2 \theta(t) + [\tilde{w}(t) - q(t)] \sin^2 \theta(t, \rho) \\ &\leq \frac{1}{p(t)} \cos^2 \theta(t, \rho) + [\lambda_{i-1} w(t) - q(t)] \sin^2 \theta(t, \rho).\end{aligned}$$

Let $u(t)$ be an eigenfunction of SLP (10.81), (10.82) associated with the eigenvalue λ_{i-1} and $\phi(t)$ its Prüfer angle with $\phi(a) = \alpha$. Then

$$\phi'(t) = \frac{1}{p(t)} \cos^2 \phi(t) + [\lambda_{i-1} w(t) - q(t)] \sin^2 \phi(t)$$

and $\phi(b) = i\pi$. By the theory of differential inequalities, we find that $\theta(b) \leq \phi(b) = i\pi$. We have reached a contradiction.

(ii) It is similar to (i) and hence omitted. □

10.3.2 *Dependence of nodal solutions on the problem*

We investigate the changes of the existence of different types of nodal solutions of BVP (10.78), (10.79) as the problem changes. Our work is based on the following lemma for the dependence of the nth eigenvalue of SLP (10.81), (10.82) on the problem which can be excerpted from [Kong *et al.* (2008), Theorems 2.2 and 2.3], [Kong and Zettl (1996), Theorem 4.2], and [Kong *et al.* (1999), Lemma 3.32].

Lemma 10.9. *For any $n \in \mathbb{N}_0$, we have the following conclusions.*

(a) *Consider the nth eigenvalue of SLP (10.81), (10.82) as a function of b for $b \in (a, \infty)$, denoted by $\lambda_n(b)$. Then $\lambda_n(b) \to \infty$ as $b \to a^+$.*

(b) *Consider the nth eigenvalue of SLP* (10.81), (10.82) *as a function of w for $w \in C^{(1)}[a, b]$, denoted by $\lambda_n(w)$. Then $\lambda_n(w)$ is decreasing as long as it is positive, i.e., for $w_1, w_2 \in C^{(1)}[a, b]$ such that $w_1(t) \le w_2(t)$ for $t \in [a, b]$, we have $\lambda_n(w_1) \ge \lambda_n(w_2)$ as long as $\min\{\lambda_n(w_1), \lambda_n(w_2)\} \ge 0$.*

(c) *Consider the nth eigenvalue of SLP* (10.81), (10.82) *as a function of q for $q \in C^{(1)}[a, b]$, denoted by $\lambda_n(q)$. Then $\lambda_n(q)$ is increasing, i.e. for $q_1, q_2 \in C^{(1)}[a, b]$ such that $q_1(t) \le q_2(t)$ for $t \in [a, b]$, we have $\lambda_n(q_1) \le \lambda_n(q_2)$.*

(d) *Consider the nth eigenvalue of SLP* (10.81), (10.82) *as a function of $1/p$ for $1/p \in C^{(1)}[a, b]$, denoted by $\lambda_n(1/p)$. Then $\lambda_n(1/p)$ is decreasing, i.e. for $1/p_1, 1/p_2 \in C^{(1)}[a, b]$ such that $1/p_1(t) \le 1/p_2(t)$ for $t \in [a, b]$, we have $\lambda_n(1/p_1) \ge \lambda_n(1/p_2)$.*

(e) *Consider the nth eigenvalue of SLP* (10.81), (10.82) *as a function of the boundary condition angle α, denoted by $\lambda_n(\alpha)$. Then $\lambda_n(\alpha)$ is a continuous and decreasing function on $[0, \pi)$. Furthermore,*

$$\lim_{\alpha \to \pi^-} \lambda_0(\alpha) = -\infty \quad \text{and} \quad \lim_{\alpha \to \pi^-} \lambda_{n+1}(\alpha) = \lambda_n(0) \ \text{ for } \ n \ge 1.$$

The first result is about the changes as the interval $[a, b]$ shrinks, more precisely, as $b \to a^+$. We discuss both the cases when one of f_0 and f_∞ is infinite and when both of them are finite.

Theorem 10.12. *Let Eq.* (10.78) *and BC* (10.79) *be fixed and let* (10.84) *hold.*

(i) *Assume either $f_0 < \infty$ and $f_\infty = \infty$, or $f_\infty < \infty$ and $f_0 = \infty$. Then for any $n \in \mathbb{N}_0$, there exists $b_n > a$ such that for any $b \in (a, b_n)$ and for any $i \ge n$, BVP (10.78), (10.79) has a solution $y_i^\gamma \in \mathcal{S}_{i+1}^\gamma$ for $\gamma \in \{+, -\}$.*

(ii) *Assume $f_0 < \infty$ and $f_\infty < \infty$. Then for any $n \in \mathbb{N}_0$, there exists $b_n > a$ such that for any $b \in (a, b_n)$ and for any $i \ge n + 1$, BVP (10.78), (10.79) has no solutions in $\mathcal{S}_i^\gamma$ for $\gamma \in \{+, -\}$.*

Proof. (i) Without loss of generality assume $f_0 < \infty$ and $f_\infty = \infty$. Let $\lambda_n(b)$ be defined as in Lemma 10.9(a). By Lemma 10.9(a), for any $n \in \mathbb{N}_0$, there exists $b_n > a$ such that for any $b \in (a, b_n)$, we have $f_0 < \lambda_n(b) < f_\infty$ and hence $f_0 < \lambda_i(b) < f_\infty$ for all $i \ge n$. Then the conclusion follows from Theorem 10.10.

(ii) By Lemma 10.9(a), for any $n \in \mathbb{N}$, there exists $b_n > a$ such that for any $b \in (a, b_n)$, we have that $\lambda_n(b) > f^* := \sup\{f(y)/y : y \in (0, \infty)\}$. Then the conclusion follows from Theorem 10.11(i). □

We then present a result on the nonexistence of certain types of nodal solutions of BVP (10.78), (10.79) as the function w increases in a given direction. More precisely, let $s \geq 0$ and $h \in C^1[a, b]$ such that $h(t) > 0$ on $[a, b]$, and consider the equation

$$-(p(t)y')' + q(t)y = [w(t) + sh(t)]f(y). \tag{10.93}$$

Theorem 10.13. *Let the interval* $[a, b]$ *and BC* (10.79) *be fixed and let* (10.84) *hold. Assume* $f(y)/y \geq f_* > 0$ *for all* $y \neq 0$. *Then for any* $n \in \mathbb{N}_0$, *there exists* $s_n \geq 0$ *such that for any* $s > s_n$ *and for any* $i \leq n$, *BVP (10.93), (10.79) has no solution in* $\mathcal{S}_i^\gamma$ *for* $\gamma \in \{+, -\}$.

Proof. For $s \geq 0$ and $i \in \mathbb{N}_0$, we denote by $\lambda_i(s)$ the ith eigenvalue of the SLP consisting of the equation

$$-(p(t)y')' + q(t)y = \lambda\big[w(t) + sh(t)\big]y$$

and BC (10.82). Let $h_* = \min\{h(t)/w(t) : t \in [a, b]\}$, and denote by $\mu_i(s)$ the ith eigenvalue of the SLP consisting of the equation

$$-(p(t)y')' + q(t)y = \mu(1 + sh_*)w(t)y$$

and BC (10.82). Since

$$w(t) + sh(t) \geq (1 + sh_*)w(t) \quad \text{for } s \geq 0,$$

by Lemma 10.9(b),

$$\lambda_i(s) \leq \mu_i(s) \ \text{ for all } s \geq 0 \text{ and } i \geq 0, \text{ whenever } \lambda_i(s) \geq 0. \tag{10.94}$$

Note that for $i \geq 0$, $\mu_i(s)(1 + sh_*) = \mu_i(0)$, we have

$$\mu_i(s) = \frac{\mu_i(0)}{1 + sh_*} \to 0 \ \text{ as } s \to \infty.$$

This together with (10.94) implies that $\lambda_i(s) < f_*$ as $s \to \infty$. Then, for any $n \in \mathbb{N}_0$, there exists $s_n \geq 0$ such that $\lambda_n(s) < f_*$ for $s > s_n$. Therefore, the conclusion follows from Theorem 10.11(ii). □

The next result is on the nonexistence and existence of certain types of nodal solutions of BVP (10.78), (10.79) as the function q changes in a given direction. More precisely, let $s \in \mathbb{R}$ and $h \in C^1[a, b]$ such that $h(t) > 0$ on $[a, b]$, and consider the equation

$$-(p(t)y')' + \big[q(t) + sh(t)\big]y = w(t)f(y). \tag{10.95}$$

Theorem 10.14. *Let the interval* $[a, b]$ *and BC* (10.79) *be fixed and let* (10.84) *hold.*

(i) *For any $n \in \mathbb{N}_0$, there exists $s_n \leq 0$ such that for any $s < s_n$ and for any $i \leq n$, BVP* (10.95), (10.79) *has no solutions in $\mathcal{S}_i^\gamma$ for $\gamma \in \{+,-\}$.*
(ii) *Assume either $f_0 < \infty$ and $f_\infty = \infty$, or $f_\infty < \infty$ and $f_0 = \infty$. Then for any $n \in \mathbb{N}_0$, there exists $s_n \geq 0$ such that for any $s > s_n$ and for any $i \geq n$, BVP* (10.95), (10.79) *has two solutions $y_{i,\gamma} \in \mathcal{S}_{i+1}^\gamma$ for $\gamma \in \{+,-\}$.*
(iii) *Assume $f_0 < \infty$ and $f_\infty < \infty$. Then for any $n \in \mathbb{N}_0$, there exists $s_* \geq 0$ such that for any $s > s_*$, BVP* (10.95), (10.79) *has no solution in $\mathcal{S}_i^\gamma$ for all $i \geq n+1$ and $\gamma \in \{+,-\}$.*

Proof. For $s \in \mathbb{R}$ and $i \in \mathbb{N}_0$, we denote by $\lambda_i(s)$ the ith eigenvalue of the SLP consisting of the equation

$$-(p(t)y')' + \big[q(t) + sh(t)\big]y = \lambda w(t)y$$

and BC (10.82). Let $h_* = \min\{\frac{h(t)}{w(t)} : t \in [a,b]\}$, and denote by $\mu_i(s)$ the ith eigenvalue of the SLP consisting of the equation

$$-(p(t)y')' + [q(t) + sh_*(t)w(t)]y = \mu w(t)y \tag{10.96}$$

and BC (10.82).

(i) Since for $s \leq 0$,

$$q(t) + sh(t) \leq q(t) + sh_* w(t),$$

by Lemma 10.9(c), $\lambda_i(s) \leq \mu_i(s)$ for all $s \leq 0$ and $i \geq 0$. Note that Eq. (10.96) yields

$$-(p(t)y')' + q(t)y = (\mu - sh_*)w(t)y.$$

Thus, for $s \leq 0$ and $i \geq 0$, $\mu_i(0) = \mu_i(s) - sh_*$, which implies that

$$\mu_i(s) = \mu_i(0) + sh_* \to -\infty \quad \text{as } s \to -\infty,$$

and hence $\lambda_i(s) \to -\infty$ as $s \to -\infty$ for all $i \geq 0$. Then, for any $n \in \mathbb{N}_0$ there exists $s_n \leq 0$ such that $\lambda_n < 0$ for all $s < s_n$. Therefore, the conclusion follows from Theorem 10.11(ii).

(ii) Without loss of generality, assume $f_0 < \infty$ and $f_\infty = \infty$. Similar to the argument in (i), we have $\lambda_i(s) \to \infty$ as $s \to \infty$ for all $i \geq 0$. Then for any $n \in \mathbb{N}_0$ there exists $s_n \geq 0$ such that for any $s > s_n$ we have $f_0 < \lambda_n(s) < f_\infty$ and hence $f_0 < \lambda_i(s) < f_\infty$ for $i \geq n$. Therefore, the conclusion follows from Theorem 10.10.

(iii) As we can see from Part (ii), for any $n \in \mathbb{N}_0$, there exists $s_* \geq 0$ such that for all $s > s_*$ we have $\lambda_n(s) > f^* := \sup\{f(y)/y : y \in (0,\infty)\}$. Thus, the conclusion follows from Theorem 10.11(i). □

Similar to Theorem 10.13, we show a result on the nonexistence of certain types of nodal solutions of BVP (10.78), (10.79) as the function $1/p(t)$ increases in a certain direction. More precisely, let $s \geq 0$ and $h \in C[a, b]$ such that $h(t) > 0$ on $[a, b]$, and consider the equation

$$-\left(\frac{1}{1/p(t) + sh(t)} y'\right)' + q(t)y = w(t)f(y). \tag{10.97}$$

Theorem 10.15. *Let the interval $[a, b]$ and BC (10.79) be fixed and let (10.84) hold. Define $\hat{q} := \max\{q(t)/w(t) : t \in [a, b]\}$ and assume $f(y)/y \geq f_* > \hat{q}$ for all $y \neq 0$. Then for any $n \in \mathbb{N}_0$, there exists $s_n \geq 0$ such that for any $s > s_n$, BVP (10.97), (10.79) has no solution in $\mathcal{S}_i^\gamma$ for all $i \leq n$ and $\gamma \in \{+, -\}$.*

Proof. For $s \geq 0$ and $i \in \mathbb{N}_0$, we denote by $\lambda_i(s)$ the ith eigenvalue of the SLP consisting of the equation

$$-\left(\frac{1}{1/p(t) + sh(t)} y'\right)' + q(t)y = \lambda w(t)y$$

and BC (10.82) with an eigenfunction $u_i(t, s)$. Let $\theta_i(t, s)$ be the Prüfer angle of $u_i(t, s)$ satisfying $\theta_i(a, s) = \alpha$. Then

$$\theta_i'(t, s) = \left[\frac{1}{p(t)} + sh(t)\right] \cos^2 \theta_i(t, s) + [\lambda_i w(t) - q(t)] \sin^2 \theta_i(t, s). \tag{10.98}$$

By Lemma 10.9(d), $\lambda_i(s)$ is decreasing and hence

$$\lim_{s \to \infty} \lambda_i(s) = \lambda_i^* \in [-\infty, \infty).$$

We show that $\lambda_i^* < f_*$ and then the conclusion follows from Theorem 10.11(ii). Assume the contrary, i.e., $\lambda_i^* \geq f_*$. Let $w_* = \min\{w(t) : t \in [a, b]\}$. By (10.98),

$$\begin{aligned}
\theta_i'(t, s) &\geq \left[\frac{1}{p(t)} + sh(t)\right] \cos^2 \theta_i(t, s) + [\lambda_i^* w(t) - q(t)] \sin^2 \theta_i(t, s) \\
&= \left[\frac{1}{p(t)} + sh(t)\right] \cos^2 \theta_i(t, s) + [\lambda_i^* - q(t)/w(t)] w(t) \sin^2 \theta_i(t, s) \\
&\geq \left[\frac{1}{p(t)} + sh(t)\right] \cos^2 \theta_i(t, s) + [f_* - \hat{q}] w_* \sin^2 \theta_i(t, s).
\end{aligned}$$

Let $\phi(t, s)$ be the solution of the equation

$$\phi'(t, s) = \left[\frac{1}{p(t)} + sh(t)\right] \cos^2 \phi(t, s) + [f_* - \hat{q}] w_* \sin^2 \phi(t, s) \tag{10.99}$$

satisfying $\phi(a,s) = \alpha$. By the theory of differential inequalities, we have $\phi(t,s) \le \theta_i(t,s)$. In particular,

$$\phi(b,s) \le \theta_i(b,s) = (i+1)\pi. \tag{10.100}$$

We observe from (10.99) that $\phi(t,s)$ is strictly increasing in t and s, and $0 < \phi(t,s) \le (i+1)\pi$ for $t \in [a,b]$ and $s \ge 0$. Let $\phi^*(t) = \lim_{s\to\infty} \phi(t,s)$. Then $0 < \phi^*(t) \le (i+1)\pi$ for $t \in [a,b]$. We claim that

$$\phi^*(t) \not\equiv k\pi + \frac{\pi}{2} \quad \text{on } (a,b] \tag{10.101}$$

for any $0 \le k \le i$. If not, for any $a_1 \in (a,b]$ and $\epsilon > 0$, there exists $s^* > 0$ such that for $s \ge s^*$,

$$\phi(a_1,s) \in (k\pi + \pi/2 - \epsilon, k\pi + \pi/2),$$

which yields that

$$\phi(t,s) \in (k\pi + \pi/2 - \epsilon, k\pi + \pi/2) \quad \text{for} \quad t \in [a_1,b].$$

This implies that

$$0 < \phi(b,s) - \phi(a_1,s) < \epsilon. \tag{10.102}$$

However, from (10.99), we see that for s sufficiently large,

$$\phi'(t,s) \ge \frac{1}{2}(f_* - \hat{q})w_* \quad \text{for} \quad t \in [a_1,b].$$

This contradicts (10.102) and hence verifies (10.101).

It is easy to see that $\phi(t,s) \to \phi^*(t)$ uniformly on $[a_1,b]$ as $s \to \infty$. Thus, $\phi^*(t)$ is continuous on $[a_1,b]$. From (10.101), we can find a nontrivial closed interval $[c,d] \subset [a,b]$ such that $\cos^2\phi^*(t) \ge \nu > 0$ for $t \in [c,d]$. Then from (10.99),

$$\phi'(t,s) \ge \left[\frac{1}{p(t)} + sh(t)\right]\nu \to \infty \quad \text{uniformly for} \quad t \in [c,d] \quad \text{as} \quad s \to \infty.$$

Therefore,

$$\begin{aligned}\phi(b,s) \ge \phi(d,s) &\ge \phi(c,s) + \int_c^d \left[\frac{1}{p(t)} + sh(t)\right]\nu ds \\ &\ge \int_c^d \left[\frac{1}{p(t)} + sh(t)\right]\nu ds \to \infty \quad \text{as} \quad s \to \infty.\end{aligned}$$

This contradicts (10.100) and hence completes the proof. □

The last result is on the existence of certain types of nodal solutions of BVP (10.78), (10.79) as the boundary condition angle α changes.

Theorem 10.16. *Let Eq.* (10.78) *and the interval* $[a, b]$ *be fixed and let* (10.84) *hold. Assume either* $f_0 = 0$ *and* $f_\infty = \infty$, *or* $f_\infty = 0$ *and* $f_0 = \infty$. *For* $n \in \mathbb{N}_0$ *denote* $\lambda_n(\alpha)$ *the nth eigenvalue of the SLP* (10.81), (10.82). *Suppose* k *is the first nonnegative integer such that* $\lambda_k(\alpha^*) > 0$ *for some* $\alpha^* \in (0, \pi)$. *Then*

(i) *for* $\alpha \in [0, \alpha^*)$, *BVP* (10.78), (10.79) *has a solution* $y_n^\gamma \in \mathcal{S}_{n+1}^\gamma$ *for all* $n \geq k$ *and* $\gamma \in \{+, -\}$;
(ii) *for* $\alpha \in [\alpha^*, \pi)$, *BVP* (10.78), (10.79) *has a solution* $y_n^\gamma \in \mathcal{S}_{n+1}^\gamma$ *for all* $n \geq k + 1$ *and* $\gamma \in \{+, -\}$.

Proof. By assumption, $\lambda_k(\alpha^*) > 0$. Then Lemma 10.9(e) shows that $\lambda_k(\alpha) > 0$ for $\alpha \in [0, \alpha^*]$; and for $\alpha \in (\alpha^*, \pi)$, $\lambda_k(\alpha) < 0$ and $\lambda_{k+1}(\alpha) > 0$. Therefore, the conclusion follows from Theorem 10.10. □

Remark 10.10. Theorems 10.12–10.16 show that we can "create" or "eliminate" certain types of nodal solutions by changing the interval $[a, b]$, the coefficient functions q, p, w, and the boundary condition angle α. Since the eigenvalues of SLP (10.81), (10.82) can be easily computed using computer software such as that in [Bailey *et al.* (2006)], we are able to construct specific BVPs (10.78), (10.79) which have or do not have nodal solutions in $\mathcal{S}_n^\gamma$ for a prescribed $n \in \mathbb{N}_0$.

Bibliography

Ambrosetti A. and Hess P. (1980). Positive solutions of asymptotically linear elliptic eigenvalue problems, *J. Math. Anal. Appl.* **73**, pp. 411–422.

Bailey P., Everitt N. and Zettl A. (2006). SLEIGN2 package for Sturm-Liouville problems, *http://www.math.niu.edu/SL2/*.

Cao X., Kong Q., Wu H. and Zettl A. (2007). Geometric aspects of Sturm–Liouville problems III. Level surfaces of the nth eigenvalue, *J. Comput. Appl. Math.* **208**, pp. 176–193.

Chamberlain J., Kong L. and Kong, Q. (2011). Nodal solutions of boundary value problems with boundary conditions involving Riemann-Stieltjes integrals, *Nonlinear Anal.* **74**, pp. 2380–2387.

Chamberlain J., Kong L. and Kong, Q. (2009). Nodal solutions of nonlocal boundary value problems, *Math. Model. Anal.* **14**, pp. 435–450.

Kong Q. (2007). Existence and nonexistence of solutions of second-order nonlinear boundary value problems, *Nonlinear Anal.* **66**, pp. 2635–2651.

Kong L. and Kong Q. (2009). Nodal solutions of second order nonlinear boundary value problems, *Math. Proc. Camb. Phil. Soc.* **146**, pp. 747–763.

Kong L., Kong Q. and Wong J. S. W. (2010). Nodal solutions of multi-point boundary value problems, *Nonlinear Anal.* **72**, pp. 382–389.

Kong Q., Wu H. and Zettl A. (2008). Limits of Sturm-Liouville eigenvalues when the interval shrinks to an end point, *Proc. Roy. Soc. Edinburgh Sect. A*, **138**, pp. 323–338.

Kong Q., Wu H. and Zettl A. (1999). Dependence of the nth Sturm-Liouville eigenvalue on the problem, *J. Differential Equations* **156**, pp. 328–354.

Kong Q. and Zettl A. (1996). Eigenvalue of regular Sturm-Liouville problems, *J. Differential Equations* **131**, pp. 1–19.

Ma R. (2007). Nodal solutions of second-order boundary value problems with superlinear or sublinear nonlinearities, *Nonlinear Anal.* **66**, pp. 950–961.

Ma R. and Thompson B. (2004). Nodal solutions for nonlinear eigenvalue problems, *Nonlinear Anal.* **59**, pp. 707–718.

Naito Y. and Tanaka S. (2004). On the existence of multiple solutions of the boundary value problem for nonlinear second-order differential equations, *Nonlinear Anal.* **56**, pp. 919–935.

Rabinowitz P. H. (1971). Some global results for nonlinear eigenvalue problems, *J. Funct. Anal.* **7**, pp. 487–513.

Rynne B. P. (2007). Spectral properties and nodal solutions for second-order, m-point, boundary value problems, *Nonlinear Anal.* **67**, pp. 3318–3327.

Weidmann J. (1987). *Spectral Theory of Ordinary Differential Equations*, Lecture Notes in Mathematics, Vol. 1258 (Springer).

Zettl A. (2005). *Sturm-Liouville Theory*, in Mathematical Surveys and Monographs, Vol. 121 (American Mathematical Society, Providence).

Index

www.ingramcontent.com/pod-product-compliance
Lightning Source LLC
LaVergne TN
LVHW052034200926
848481LV00005B/1670